Praise for *The Measure of God*

"An insightful and colorful narrative catching the pivotal themes of the world's best-known lecture series. A delightful mix of intellectual analysis and stimulating after-dinner conversation."
  —Holmes Rolston III, University Distinguished Professor, Colorado State University, author of *Genes, Genesis and God,* Gifford Lecturer 1997-98

"Larry Witham gives a beautifully clear and readable portrayal of various forms of natural theology—the attempt to base religious beliefs on reason, science, and human experience—as he traces the history of the Gifford Lectures in their responses to successive challenges from materialism, science, technology, and religious pluralism."
  —Ian Barbour, Winifred and Atherton Bean Professor Emeritus of Science, Technology and Society, Carleton College, author of *When Science Meets Religion,* Gifford Lecturer 1989 and 1990

"Larry Witham's excellent overview of the history of the Gifford Lectures gives the reader a fascinating snapshot of how the Gifford Project itself has come to represent and embody the past one hundred years of Western intellectual culture. In this sense the Gifford Lectures portray a complex century of thought in which natural science has encountered religion in full force. But Witham's book gives us even more than that: what unfolds before our eyes is the intellectual and social history of ideas that has infused the science and religion struggle to this day. This book carefully tracks this remarkable interdisciplinary journey through the lives of the real life international scholars who have risen to Lord Gifford's challenge over time, and who are now bound together by this shared experience."
  —Dr. J. Wentzel van Huyssteen, James I. McCord Professor of Theology and Science, Princeton Theological Seminary, author of *The Shaping of Rationality,* Gifford Lecturer 2004

"Scholars are well aware of the prestige accorded the Gifford Lectures, which have informed and formed Western thought. Witham examines the who, when, and where of these ground-breaking and enduring contributions in a skillful narrative that accurately and informatively traces an intellectual heritage that engages the humanities, social sciences, and natural sciences. The author captures ideas in an inviting prose that sacrifices neither complexity nor depth. If you want to understand these most influential intellectual forebears, this book is a must-read."
  —Chester Gillis, chair of the Department of Theology Georgetown University, author of *Roman Catholicism in America*

"For over a century the Gifford Lectures have remained the most prestigious forum for discussion of natural theology and the philosophy of religion. They have brought some of the most important philosophers and theologians to the roundtable of debate on these issues and also on the central question of the relationship between religion and science. *The Measure of God* succeeds to present clearly and in an engaging manner the history of these lectures, bringing to life the many crucial issues discussed over the decades by Gifford Lectures, issues that are so pertinent for all human beings searching for deeper understanding of the meaning of existence."

—Dr. Seyyed Hossein Nasr, University Professor of Islamic Studies, George
Washington University, author of *Islam,* Gifford Lecturer 1980-81

"The sequence of Gifford lectures affords a moving image of the fluctuating interaction between science and religion. Larry Witham offers a fluent and accessible guide to this fascinating intellectual history."

—Sir John Polkinghorne, former President of Queens' College, Cambridge University, author of *Science and the Trinity,* Gifford Lecturer 1993-94

"Spanning more than a century, the Gifford Lectures have contained some of the most innovative and influential scholarship in defining the scope, character and limitations of discourse about God. The exciting news that this "museum of intellectual conflict," as it has been called, is to be a living museum, with the text republished on the Web, makes Larry Witham's accessible guide to their contours and contents all the more welcome."

—John Hedley Brooke, Andreas Idreos Professor of Science and Religion, Oxford
University, author of *Science and Religion,* Gifford Lecturer 1995

"Since their inception in the late 19th century, the Gifford Lectures have attained international significance both through the subject matter of successive series and the calibre of those who have delivered them. In his historical overview, Larry Witham recounts the changes in personal style, theme, and context that have characterised the lectures in the four ancient universities of Scotland. Lord Gifford intended his lectures on natural theology to reach a wide public audience; he has surely been vindicated by this vivid and engaging account."

—David Fergusson, Professor of Divinity, University of Edinburgh, author of
The Cosmos and the Creator

"Using the fabled Gifford Lectures as a door that opens upon the vista of over a century of debate on religion and science, Witham entertains and educates the reader. Anyone interested in the past, present, and (possible) future of this vital ongoing dialogue should read his book."

—Jean Bethke Elshtain, Laura Spelman Rockefeller Professor of Social and
Political Ethics, University of Chicago Divinity School, author of *Democracy
on Trial,* Gifford Lecturer 2005-06

"The Gifford Lectures are a testimony to the best efforts of many of our wisest modern scientists to reconcile their understanding of the natural with their sense of the supernatural. In *The Measure of God,* Witham shows us what we can still learn from their efforts."

—Edward J. Larson, Herman E. Talmadge Chair of Law, University of Georgia,
Pulitzer Prize winner and author of *Evolution: The Remarkable History of a
Scientific Theory*

The
MEASURE
of GOD

Our Century-Long Struggle
to Reconcile Science & Religion

Larry Witham

HarperSanFrancisco
A Division of HarperCollins*Publishers*

HarperCollins books may be purchased for educational, business, or sales
promotional use. For information please write: Special Markets Department, HarperCollins Publishers, 10 East 53rd Street, New York, NY
10022.

HarperCollins Web site: http://www.harpercollins.com

HarperCollins®, ■®, and HarperSanFrancisco™ are trademarks of
HarperCollins Publishers.

FIRST EDITION

Library of Congress Cataloging-in-Publication Data

Witham, Larry
   The measure of God : our century-long struggle to reconcile
science & religion / Larry Witham.— 1st ed.
      p.   cm.
   Includes bibliographical references (p.) and index.
   ISBN 10: 0-06-059191-9 (cloth)
   ISBN 13: 978-0-06-059191-5 (cloth)
   1. Religion and science. I. Title.
   BL240.3.W585 2005
201'.65'0904—dc22   2005046077

05 06 07 08 09   ❖/RRD(H)   10 9 8 7 6 5 4 3 2 1

CONTENTS

ACKNOWLEDGMENTS

I N A BOOK SUCH AS THIS, the first thing to acknowledge is
what had to be left out, so enormous is the legacy of the Gifford
Lectures over a century. Of the 220 people who gave the lectures, only a
fraction appear in these pages—the most famous, perhaps, and those who
illustrate broader trends. Admirers of a particular Gifford lecturer may
find that person inexplicably absent, and it may be small consolation that
all the speakers are listed in the appendix. Writing for an American audience, I have also highlighted the American greats, although, by far, more
characters in this story are European and, for that matter, nearly all men.
Both the omissions and the biases are regrettable, but I hope not in vain,
if so short a book can convey the spiritual and scientific life of an entire
century.

I first heard of the "Gifford Lectures" in the mid-1970s when I
bought a used version of William James's *Varieties of Religious Experience* in
graduate school. In the mid-1990s, when I took an active interest in science and religion topics, I realized that the Giffords were perhaps the best
repository on earth for this discussion, more than a hundred years old,

and filled with famous names. For a writer, it was a remarkable story, if it could be told simply enough.

In completing that job, I am first indebted to the people at the four Scottish universities who hosted my visits in 2003. At the University of Edinburgh, Paul McGuire, Stewart "Jay" Brown, David Fergusson, and university archivist Arnott Wilson were generous to a fault. Professor Fergusson offered valuable corrections and suggestions on the early manuscript. At Glasgow University, my thanks go to Eileen Reynolds and the staff at the archives. The St. Andrews University special library staff were particularly helpful on short notice. Cheryl Croydon was my host at Aberdeen, and Professors Gordon Graham and John Webster were gracious in providing interviews. The staff at the Edinburgh Public Library were extremely helpful, and in Glasgow Michael K. Abram, whose shipping firm is at 17 Sandyford Place, took me in on a dark, rainy evening for a tour of the crime scene in the Sandyford murder mystery. Also abroad, the crew of the Karl Barth Archives in Basel, Switzerland, were kind in sending excerpts from Barth's letters on the Giffords.

Stateside, other Gifford enthusiasts offered their guidance as well. Professor John Snarey at Emory University, who uncovered the facts of Williams James's talks in Edinburgh, shared that treasure, and Hendrika Vande Kemp, a scholar of psychology, passed on her list and thoughts on the Giffords. John Clendenning, the biographer of Josiah Royce, provided guidance on that topic and Werner Seubert of Virginia translated German texts for me. Alvin Plantinga and Nicholas Wolterstorff, who both gave the Giffords, provided helpful interviews on the closing topics of this book.

Fortunately, efforts to tackle the entire Gifford corpus have been made before I got to them. Although an American doctoral dissertation of 1941 analyzed four decades of Giffords on the topic of "theism and reason," the most significant scholarship came in 1966 with the British dissertation of Bernard E. Jones, a Methodist scholar who surveyed the Giffords over eighty years. Separately, Jones published the first anthology of Gifford material. Also of note is John Macquarrie's *Twentieth Century Religious Thought* (1963), which cites many Gifford speakers. The book of record, however, came in 1986, when physicist and Benedictine priest

Stanley Jaki, also a Gifford lecturer, wrote *Lord Gifford and His Lectures* to mark their centenary. That volume, although a personal interpretation, remains the standard source (updated in 1995) with its readable overview and reprinting of Gifford's will, his brother's recollections, excerpts from Lord Gifford's own public talks, and a list of the Gifford lecturers.

Finally, I owe a great deal to Eric Brandt, senior editor at HarperSan-Francisco, who believed in this audacious book and then improved it greatly with attentive reading and good advice. Cheers also to my agent, Giles Anderson, who makes the author's trade friendly and fun. And for my wife, Kazui, the gratitude is beyond words.

INTRODUCTION

SCIENCE *and* RELIGION

A Century in Four Acts

M ODERN SCIENCE came of age at the cusp of the twentieth century. It was a period marked by discovery of radio waves and X rays, the first skyscraper, automobiles, cinema, and vaccines. The quantum theory of the atom and the deciphering of genetics both sprang from the Western mind. In a word, the close of the Victorian age unleashed a great wave of scientific challenges to the religious beliefs of the Christian world.

To navigate this tidal wave of doubt, some Christian thinkers argued that science could not possibly undermine religious faith, for they were two different worlds. Others took a different tack, saying that the critical tools of science—facts and reason—could even advance the knowledge of God. In the English-speaking world, many public figures espoused such efforts to reconcile science and religion, but one solitary figure stands out. Before his death in 1887, the Scottish judge Adam Gifford endowed the Gifford Lectures, which encouraged a lively and perpetual debate on science and "all questions about man's conception of God or the Infinite."[1]

Since that bequest was made, Scotland's four historic universities have hosted a Who's Who of those whom the Gifford will designated as "able reverent men, true thinkers, sincere lovers of and earnest inquirers after truth." They could be believers, "skeptics, agnostics or freethinkers," but their fraternal task was to take up the God question "as a strictly natural science." Given that mandate, the Gifford project has become a window on a century in which natural science encountered biblical religion with full force. That is the story of this book, a century-long story of science and religion told in four acts through the Gifford Lectures.

When Gifford conceived of his project, he was one of the archetypal characters of his British milieu: a Scottish Presbyterian, a freethinker, and a scientific optimist. He emerged in a time in which the term *natural theology* stood for the attempt to use nature and reason to find God. In his will, Gifford referred to natural theology, but he also used his own phrase of "science . . . of Infinite Being." He even cited chemistry and astronomy as fields worthy of imitation by God-seekers.

Despite this sense of progress, however, the Gifford project was launched in "one of the most dangerous and fretful periods of human intellectual history, the period that gave rise to our own age," as one writer described the time.[2] Everywhere, a mood of doubt was in the air, even if its meaning was not entirely understood. "By 1900 schoolboys decided not to have faith because Science, whatever that was, disproved Religion, whatever that was," said a later Gifford lecturer.[3] The sciences would challenge virtually every cherished belief society held about God, humanity, morality, and history.

Yet the lecturers came, their motivations as mixed as those of any group. They came by horse-drawn carriages and coal-powered trains, by rickety automobiles, steamships, and finally jet airplanes. Although most came from Britain, others crossed the English Channel or the Atlantic. In one case, the trek began with a riverboat in West Africa. Over the century, lecturers addressed Old World audiences, as if speaking ex cathedra, competed with warplanes overhead, and battled the television age. The number of lectures each gave shrank from a series of twenty down to ten and finally six, and in some cases turned into panel forums or roundtables at which the public fired off questions.

In all these ways the Gifford Lectures embody a century of change. At first glance they represent just a corner of a library, a set of volumes that are usually darkly bound and dusty from disuse. As a mere quantity, the Giffords amount to 220 speakers who showed up for some 207 lecture series since 1888. More than 200 books have flowed from about 150 of these lecturers (some as two volumes) and perhaps 2,000 news reports have appeared, mostly before the advent of television. Yet the Giffords are palpable in other ways. They contain the ideas—spoken, reported, and published—that have been the lifeblood of a particular class of men and women, mature thinkers who wedded their lives to science, philosophy, or theology. But they were no less human, for their efforts to conceive, produce, and finally deliver the lectures reveal a remarkable drama of mortal hopes, fears, victories, defeats, vanities, and frailties.

Early on, when the lectures were hardly known, one appointee happily took the very large cash award and commented, "The honor is not great but the honorarium is colossal."[4] That would change, of course. If the value of the payment decreased, the honor became something like the Nobel Prize in philosophy or theology. Eight people who were Nobel laureates delivered the lectures. Yet the challenge of keeping faith with Gifford's vision was always difficult. "Lord Gifford's wish," said one Oxford don, "is by its very grandeur impossible of fulfillment."[5]

When, in the 1920s, the invitation went to the first Roman Catholic candidate, he enthused at how it was "the finest Lectureship on these great subjects in the world."[6] Sadly he never arrived, because of a nervous breakdown. A religious philosopher in the 1980s, however, felt an almost cathartic "relief!" at getting the invitation: "What if I had never been invited to do the Giffords, when some who have less to say have given them?"[7] Those who had said less may have preferred to dance around Gifford's apparent desire for resounding proofs about God and other great mysteries. "Lord Gifford preferred lecturers who prove that witches are powerless," said a Cambridge historian, who explained that historians cannot supply such proofs. Historians can only say why men once viewed witches as powerful and why that belief changed over time.[8]

The audiences came in all sizes, with a variety of dispositions, and for all kinds of reasons, as three American examples suggest. When the

American psychologist William James finished his series in 1902, the packed hall sang, "He's a jolly good fellow." Carl Sagan, eight decades later, drew overflow crowds to a visually stunning audio-visual series at Glasgow. Being the famed host of the recent *Cosmos* television series surely helped. In Aberdeen in the 1980s, the American philosopher of God Alvin Plantinga spoke at the peak of the North Sea oil boom. An oil worker and his wife apparently relished the dinner-hour lectures. "They came, they said, because they liked to hear an American accent."[9]

Local success did not always predict the fate of the published lectures. Luckily for James, he not only had enthusiastic audiences, but his published *Varieties of Religious Experience* made the Giffords famous on both sides of the Atlantic. Nearly one hundred years later, when the Modern Library, an imprint of Random House, polled the reading public on the best nonfiction books of the twentieth century, *Varieties* came in second. The 1939 Gifford Lectures of American theologian Reinhold Niebuhr ranked eighteenth. Neither James nor Niebuhr would have done the work if not for the Scottish invitation.

When Alfred North Whitehead lectured on his new process philosophy in Edinburgh in the 1920s, the abstractness soon scattered the crowd. His disconsolate wife was among the six or so loyalists who stuck it out in the echoing hall. When published, however, Whitehead's *Process and Reality* stirred a revolution in ideas. The lectures "turned out to be a happy decision for philosophy, if not for the Edinburgh public."[10] Etienne Gilson's lectures in Aberdeen, published as *The Spirit of Mediaeval Philosophy,* may have spurred popular study of Thomas Aquinas even more than a papal decree. His sheer act of speaking on Catholic thought, moreover, may have been enough to persuade Swiss theologian Karl Barth to present a diametrically opposite stance in the Giffords.

Certainly not every lecture in the Gifford legacy was noteworthy. Many were over the heads of mere mortals. Gifford himself could be quite abstruse, speaking in his will of "the Infinite, the All, the First and Only Cause, the One and Sole Substance, the Sole Being, the Sole Reality, and the Sole Existence." But even though there "may well be much chaff among the grain," as one historian said earlier in the century, the

Giffords were a showcase of some of "the most outstanding achievements of speculation" in British, and perhaps Western, philosophy.[11]

All told, the sheer number and diversity of thinkers who have taken to the Scottish university rostrums have made the lectures an unparalleled exhibition of modern thinking about God. Few would contest a claim that the legacy produced "the most notable group of scholarly thinkers ever assembled by the fiat of a single man."[12] Although this book presumes no law of history at work, it will try to make sense of the Gifford legacy by presenting a story in four acts, separate dramas made up of biography, ideas, and events that span a century—or about 117 years—of Western religious thinking in the face of science. The first act opens with the story of Adam Gifford and the era in which the Gifford Lectures started. In this drama, the great philosophical systems that included God clashed with scientific materialism, and as materialism seemed to prevail, the religious side of the West experienced an "end to philosophy," at least for a time.

Then comes act two, the story of the materialist sciences, which consumes five chapters of this book. Natural science arrived dramatically on the scene with the twentieth century, and the first generation or two of Gifford Lectures make up the fascinating story of how religious thought met this new challenge. For this story, science has been divided into five fields: anthropology, psychology, physics, sociology, and historical criticism. All of them adopted the mode of a scientific discipline, a process of observing nature, collecting facts, comparing data, testing hypotheses, and developing theories. When these fields applied their art on the human specimen, they sought to replace traditional spiritual, moral, and supernatural explanations with material ones. This was as much a part of that "dangerous and fretful" period as were any of the new technologies, from dynamite to the machine gun.

Anthropology, for example, sought to explain the material origins of human belief and behavior. Psychology tried to penetrate the physical process of the mind. Sociology, in turn, claimed to be deciphering the "laws" or "social forces" that manipulated individuals and groups. Physics had always been the science par excellence because of its measurable certainty, but around 1900 its material certitudes evaporated in the realm of quantum particles. Also at this time, historians gained professional status by adopting scientific methods, and although critical study of the Bible was not new, Albert Schweitzer's 1906 work, *The Quest of the Historical Jesus,* marked a new era in analyzing sacred scriptures.

Act three tells the story, set mostly between the First and Second World Wars, of a great rebellion against science and reason in the West, a movement that could be called *subjectivism* for its revolt against the vaunted "objectivity" of science. It was a plea for divine revelation and personal experience, and from the vantage of the Giffords, it included a denunciation of natural theology's attempt, like Icarus flying too near the sun, to reach the Almighty. In this view, God is entirely separate from nature, a "wholly other" known only by the divine fiat of revelation. Some Gifford lecturers argued this point, despite Lord Gifford's request that speakers explain God without "any supposed special exceptional or so-called miraculous revelation." No one represented this revolt better than Barth, who declared, "God is known by God and by God alone." When Barth gave the Giffords in 1937, he came as natural theology's "avowed opponent."[13]

Act four, which is consummated in the final three chapters, covers the Gifford Lectures after the two world wars. In this period, shattered utopias, rapid advances in technology, and exposure to diverse religious beliefs and world cultures characterized the West. Yet if the last four decades of the twentieth century were a hot house for both nihilism and religious fundamentalism, they were also a happy time for natural theology. The rational search for God was resurrected, and as theologians tried to close the gap between God and nature, the idea of revelation took on a broader meaning. Still, the days of any one dominant belief system in the West were over. As a result, the Giffords came into the twenty-first century reflecting the times, with their great diversity of ideas, but also a kind of revived interest in religious thought, in both its more traditional and its innovative forms.

Quite naturally, the story of the Giffords builds a bridge between two major actors in its legacy, Scotland and the United States. Natural theology had been called the "the sick man of Europe," but it remained healthy in America, one reason why more Americans were giving the Gifford Lectures after the Second World War. Just eight North American residents had the honor before the war; more than five times that many (forty-one) have received the invitation since the war's end. The American interest in religion, moreover, has links to the Scottish past, and to conclude, this book will consider the impact of philosopher Thomas Reid's common sense "Scottish Philosophy" on the American founding and religious outlook. Like many thinkers of the eighteenth-century Enlightenment, Reid argued for "self-evident" truth. In the twenty-first century, a time when all the wraps are off and the God question is wide open, the notion of "self-evident" truth is worth revisiting, especially to see what it implies about the future of natural theology.

Gifford had insisted that the lectures be for "the whole community," the ordinary man, and believed the "subject should be studied and known by all." Although that popularization did not happen, the themes of many lecturers inescapably leavened popular culture. C. S. Lewis, one of the most popular Christian writers of the twentieth century, said that reading a 1914 Gifford Lecture on theism helped him abandon atheism.[14] One also thinks of the bumper sticker distributed by a small group of academic enthusiasts, "One More Family for Whitehead." And when the British novelist Iris Murdoch published her 1982 Giffords, *Metaphysics as a Guide to Morals*, the perennial discussion about Plato, art, morals, and God enjoyed a rare popular eloquence. Most significant of all is how the Giffords became a mainstay of higher education. "One can hardly conceive that there is a university in any land which does not have on the shelves of its library at least some volumes of Gifford lectures," one expert on the legacy said in the 1960s.[15]

The vision of Lord Gifford came in a simpler time. He was a son of "metaphysical" Scotland (compared to "utilitarian" England or "rationalist" France). The lectures began as, and largely remain, a British philosophical institution, and one might also say male and Protestant. The first Jewish thinker spoke in 1914, the first Catholic in 1930, the first woman in 1972, and the first Muslim in 1980. Although Gifford had hoped that knowledge about God would increasingly harmonize as time passed, the lectures became a "magnificent array of fundamental and unresolved disagreements, a kind of museum of intellectual conflict," says Alasdair MacIntyre, who gave the Giffords in 1987.[16]

And yet such cacophony has not rent the lectures asunder. To this day, the Gifford legacy bridges two unfamiliar worlds, the end of the Victorian age and the modern space age. As mundane, yearly events in Scotland— indeed, as long-winded talks given by mere mortals—the lectures themselves dissolve away and expose our more general, heroic, and modern desire to reconcile God and science. They represent, perhaps, our optimism about knowing the truth, and even about taking the measure of God.

ONE

THE SANDYFORD
MYSTERY

How the Gifford Lectures Began

I N THE SUMMER OF 1862, when the United States was embroiled in its second year of the Civil War, the Scottish city of
Glasgow was enthralled by its own bloody episode. On July 4, the maid at
the prosperous Fleming household at Sandyford Place was brutally murdered.

The Sandyford Mystery had "everything requisite to a great criminal
drama," a crime writer said, especially its two unlikely suspects, a debauched
grandfather and a sailor's wife. Old James Fleming lived with the corpse
for three days while his family summered at their coastal villa up the
Clyde, the river that had made Glasgow a hub of trade and shipbuilding.
When investigators eventually arrived, they found that Jessie McLachlan,
a twenty-eight-year-old mother who was a friend of the maid and a
former employee at the house, had left three bloody footprints behind.

By modern standards, this was not exactly the high-stakes criminal
drama of an O. J. Simpson murder trial. But for 1860s Scotland it came

close enough. There was wealth, class, politics, and, as could happen only in Scotland, a bit of metaphysical philosophy mixed in as well. The Flemings' spacious flat in Sandyford, scene of the murder, was just one token of Glasgow's growing affluence. Old Fleming, who at eighty-seven was a hunched and balding figure with sideburns and a hooked nose, was a rustic who had turned to textile manufacture. Now his son headed an accounting firm and had joined the nouveau riche. From Glasgow to Edinburgh, not a few clerks and lawyers had become rich by managing the new commercial vitality—lawyers such as Adam Gifford, who would be the chief prosecutor of Jessie McLachlan.

But if commercial times set the mood in the 1860s, the legalities of Scottish law would turn the trial into an extraordinary controversy. Old Fleming's son had "intimate" ties to the chief investigator. While that must have smiled on the fate of the elder Fleming, who collected rents for his son and attended church twice the Sunday after the murder, he also had Scottish law on his side. Scots Law is perhaps most famous for its third courtroom verdict—besides "guilty" or "not guilty" a jury may declare "not proven" and set a suspect free. Yet Scottish law also had a unique emphasis in its prosecution of murders: although all parties to a murder shared equal guilt, a suspect turned prosecution witness was guaranteed total immunity. In the end, Old Fleming was made legally white as snow: once a prime suspect with blood spattered on his nightshirt, the old man became the Crown's chief witness in the murder and robbery trial of the hapless Mrs. McLachlan.

In those ballad-like days of "the trial of Jessie McLachlan," the law, prosperity, and much else about Scotland could still be traced to the single most important political event for the nation: the Union of 1707. In that year Scotland abolished its parliament, and while keeping its legal system and established Presbyterian church, the nation merged with a single British parliamentary system. Even though it was a difficult political bargain, the historical windfall was great, for now Scotland dramatically expanded trade with England. By the time of the McLachlan trial, Scotland was called "the workshop of the empire," a world of looms, ships, bridges, and train systems, financed in turn by tobacco, coal, steel, textiles, and banking.

Before the Union of 1707, Scotland had exported its native philoso-phy here and there, but after the political union, Scottish ideas became something of a world commodity. Over at the University of Glasgow, the moral philosophy professor Adam Smith began his musings on this proj-ect. To compensate for Scotland's loss of political independence, Smith believed, the nation should dominate the world with its ideas. The nation had already begun to export its inventions, tackle foreign explorations, and show the muscle of sheer commercial tenacity. In addition to that, an event called the Scottish Enlightenment would blossom between 1740 and 1790, and now Scotland had its ideas to export as well, a joint venture of tough-minded clergy, libertine literati, and stern philosophers.

Although the ships left mostly from Glasgow, Edinburgh played its role, and it soon became known as the "Athens of the North," both for its rocky cliffs with neoclassical architecture and for its learning. From this Scottish period came the skepticism of David Hume of Edinburgh and the Christian realism of Thomas Reid of Aberdeen. Adam Smith offered a historical mode of thinking, known for its four stages of civilization, rising up from the primitive hunter-gatherer to, of course, the Scottish professor. Having watched the shipping trade in Glasgow as he taught moral philosophy, Smith proffered the idea of joining free markets with moral sympathies. The idea was published as *The Wealth of Nations* in 1776, the year of the American Revolution. More than a century later, Winston Churchill claimed of Scotland that no "small nation" besides Greece had contributed so much to human progress.

In September of 1862, however, even the academic rumblings of a "second" Enlightenment paled next to agitation over the McLachlan murder trial.[1] It was "the topic not only of the hour, but of the year," said crime writer William Roughead. Newspaper circulations had grown fivefold, and while the *Glasgow Herald* sided with Fleming as "the old in-nocent," the others railed at the court. "It was manifestly impossible that the accused could have a fair trial in Glasgow," they editorialized.

Down at Jail Square, where the Glasgow Green sat alongside the River Clyde, legal teams had to fight their way through crowds to enter the court. Dozens of witnesses testified; floorboards with a bloody foot-print, ripped up as evidence, were displayed before a rapt audience of fashionable ladies, reporters, and even city officials. McLachlan sat stoical in her white straw bonnet with ribbons and veil, her hands tucked under a black woolen shawl. As advocate-depute for the Crown, Adam Gifford wore black robes and a wig. On the final day, he disclosed his own "deep feeling of anxiety and responsibility" as he dispensed the final measure of Scots Law. "It will be my duty to ask a verdict against the prisoner," he told the jury, exhorting it to narrow its mind to a single question, "Is the prisoner guilty or is she not guilty? Not, had she confederates?" While Old Fleming was indeed under "the gravest possible suspicion," a crime by multiple parties was not at issue. "If guilt be brought home to one, it will not be enough to say, 'Somebody else had a share in it.'"

The jury took fifteen minutes to find McLachlan guilty. Although the verdict was dramatic enough, next came "one of the greatest sensations in Scottish legal history." Lord Deas, a judge known as Lord Death for his willingness to hang, allowed McLachlan a final statement. She stood in the dock, pulled back her veil, and said firmly, "I desire to have it read, my Lord." For the next forty minutes, her lawyer told her story "amid the breathless attention of the court." She wiped her eyes at mention of the victim. She shook her head at the name of Fleming.

On the Friday night of the murder, she had visited her maid-friend Jess McPherson. She found Jess in the downstairs kitchen with a drunken Fleming. His whiskey jug had run dry. He asked McLachlan to replenish it at a local pub. When she returned, Jess was lying in her bedroom moaning. She had resisted the drunkard's advances, and now she had cuts about her face and there was a "large quantity of blood on the floor."

McLachlan begged to go for a doctor, but Fleming made her swear secrecy on a Bible. In return "he would make her comfortable all her life." She again began to leave, but hearing a noise in the kitchen rushed back to see "the old man striking" the maid with a "meat chopper." Fleming cornered her: "Only if you tell you know about her death you will be taken in for it as well as me." Dawn had arrived, and before she left, Fleming gave her a few pounds of hush money and silverware to pawn. It was to look like a robbery.

Lord Deas was unmoved, his mind having not "a shadow of suspicion" about Old Fleming. He put on his black cap and sentenced

McLachlan to hang, directing that until then she live on bread and water and asking that God have mercy on her soul. "Mercy!" she said. "Ay, He'll hae mercy, for I'm innocent."

But even in metaphysical Scotland, even under Scots Law, and with all the evidence paraded, who could really know the truth? What could truly be known, and what would always remain a mystery? Nobody knows Gifford's thoughts at the time. He was well aware of the approximation to truth that the law tried to achieve. But what is the ultimate truth? Truth about life? Truth about God? If Gifford began his own intellectual and spiritual journey of no return at that time, it was surely cloaked behind the self-confidence, robes, and wig of the young lawyer.

Public opinion could not be hidden, however. McLachlan's eleventh-hour account in the courtroom hit the presses and galvanized the nation. Lawyers called the testimony so credible as to "render its fabrication incredible." According to Roughead, "The demand was urgent that its truth or falsity must be determined before she was put to death." Fifty thousand Glaswegians signed a petition. Old Fleming was so harassed that he fled to the villa. His family would move away from the city.

When McLachlan's sympathizers took their pleas to London, it rocked the House of Commons with two furious debates. Whitehall opened a "secret inquiry." Finally, new evidence put the verdict in limbo "until further significance of Her Majesty's pleasure." The queen's pleasure was to commute the death sentence. In the fall of 1862, Mrs. McLachlan walked through the gloomy gates of Her Majesty's General Prison at Perth for a "life" imprisonment, which meant fifteen years in those days. She was released in 1877.

A potato famine and a slump in shipping had forced her son and her husband, once a second mate on the steamship *Pladda*, to join a mass migration to the United States. With the thirty pounds the Crown gave her, she joined them. On New Year's Day in 1899, at age sixty-five, she died at Port Huron, Michigan. Soon after, in 1901, the queen who had given McLachlan back her life would also die. The Victorian age came to a close and a new century began.

Adam Gifford had preceded them both to the grave by more than a decade. But his name would endure, and not only in Victorian crime anthologies. In his youth, Gifford had been called the "philosopher" by his siblings, but this was an interest that could emerge only after his long law career was curtailed, opening the way for a far-future time when "Gifford" would become almost a brand name for philosophizing in the West.

Gifford had begun his career as a lawyer around the Edinburgh court-house. Clean-shaven, he wore thick sideburns and his hair longer than usual, with an "abnormal superabundance of long, streaming, dark brown locks." He had built a reputation for lengthy and dramatic rhetorical flourishes to win his cases. An image was conjured by his foes, or the envious. "The wags of the Parliament House laughed at it, according to their merry, wicked, way, and characterized these out-bursts of eloquence as 'Gifford letting down his back hair.' "[2] A photograph of Gifford, probably when he was a judge, shows him in tuxedo-like dress holding spectacles. He has the strong smooth face of an American Indian chief, the nose long, cheekbones high, the mouth thin, and his jaw large.

In a time of landed gentry, Gifford had acquired considerable wealth as the son of a leather-goods manufacturer and as a lawyer who settled commercial disputes. "He knew the value of money. He knew how to make it," said an acquaintance. The money for the bequest, he went on, was earned "not in the big lotteries of commerce, but in the exercise of the most laborious profession of the law, in which the famous and successful practitioner often pockets more than he truly works for, but in which no practitioners can ever be successful without a great deal of enormous hard work."[3]

Despite his average social station and decidedly liberal views, Gifford was tapped by the government as an advocate-depute in 1861, the year before the McLachlan case. Soon after that trial, the pundits had vigorously dissected the performances by Lord Deas, the defense, and Gifford, whose final argument for the jury to convict Mrs. McLachlan went on for two and a half hours. Still, he had apparently lacked his usual zeal. In his "white-washing [of] old Fleming," Roughead recounts, Gifford had been "surprisingly half-hearted."

A year after the trial, Gifford was happily married, but the bliss would not last long. Their only child, Herbert, was born in 1864 and then his wife, Maggie, died of a sudden illness in 1868. As Gifford's brother re-counted, he had a large family around him but had made no close friends. A turn of mind came to Gifford, now forty-eight, and as he focused more intensely on his work, his "enjoyments were found in intellectual pur-suits."

Although Gifford also had a few health worries, such as a numbness in his limbs, when he was asked to be a judge at the national Court of Session in 1870, it was hard for him to refuse. What he did refuse, his brother said, was "to act as a criminal judge."[4] Advocates around Parlia-ment House today doubt whether a judge could have escaped such duties. But if he did, it was for this reason, his brother said: "I believe pity for the poor criminals, and a deep conviction that the wrongdoers had generally been deeply wronged, made him most unwilling to be their judge." Gifford much preferred the cut-and-dried sorting out of financial claims in the relentless commercial life of Scotland. He was sought out by litigators, his brother claimed, because of his "fearless application of equity in guiding his decisions. Merchants prefer common sense law."

One other happiness for the family came in 1871, when the Gifford household, including his aged mother, moved into a grand estate called Granton House, which stood seventy feet above the Firth of Forth in north Edinburgh with a majestic view beyond the backyard garden. The antiquarian home had been built by the Lord Hope family, which had held sway in city politics and the courts for much of the nineteenth cen-tury. The year after they set up house at Granton, Gifford began to suffer paralysis and before long had to drag his right leg to walk. Even so, he took fresh-air strolls around Granton, and one resident recalled his strug-gling along, "holding his head bravely aloft and looking imperturbably before him."[5]

This was not the man of only a few memories past. As a student and lawyer, Gifford had been an avid ice skater down at Duddingston Loch, where he had joined the Skating Club. One can see him moving across the ice, his long hair streaming and his breaths of steam, men and women in mufflers, and the early sunset of an Edinburgh winter (though he often skated in the early morning before studies or work). Whatever the hour, the image of an Edinburgh skater is not hard to imagine. That feeling of movement and freedom is captured in Scotland's most famous painting, *Reverend Walker Skating on Duddingston Loch,* executed by Sir Henry Raeburn in 1795. The clergyman in black, his arms crossed, glides effortless through the world on one skate, his other leg slightly raised. The image proved, it was said, that even a nation with strict Presbyterian clergy and harsh winters could enjoy life.

As Gifford walked about Granton, he may have been thinking of those times of freedom, for ice skating was like a liberation of the body and the soul. Every such dream ended in 1881, however, during a routine day at Parliament House. After the court session closed and the Lord judges stood, Gifford summoned help, his legs powerless and his right hand soon to numb as well. Even his eloquence, according to news reports, was perhaps forever gone: the seizure "seriously affected his speech."[6] Lord Gifford never walked again, and having resigned the court, his brother said, "his mind became more and more absorbed in Philosophy and Theology or rather Philosophic Theology." Or as Gifford himself put it, he had been "for many years deeply and firmly convinced that the true knowledge of God . . . and the true foundation of all ethics and morals . . . is the means of man's highest well being, and the security of his upward progress."

Even as the years went by, Gifford left behind no particular thoughts on the McLachlan trial, or on crime and punishment in general. Encomiums on God, ethics, and progress were far more his style. They set the stage for a final bold action: he willed nearly half his fortune—80,000 pounds of an estate worth 190,000—to fund a permanent lecture series at Scotland's four historic universities. At $5 million by today's standards, the endowment was so great and its topic so striking—"the knowledge of God"—that it was posted in full in the *Times* of London and the *Scotsman,* preached on from a London pulpit, and both decried and praised in print. Calvinist clergy warned of pantheism or heresy. Modernist Christian thinkers praised philanthropy. And materialists in science denounced such a waste of money. In faraway New England, "my father read aloud from a Boston newspaper that part of Lord Gifford's will which founded these four lectureships," William James recalled.[7]

Adam Gifford had been a man of affairs and a man of his time, and part of that story was about coming of age in the Presbyterian tradition of Scotland. For generations after the Reformation turned the nation into a Protestant domain, the churches still operated under a patronage system in which landowners appointed the clergy, in effect controlling not only church life but the pulpits as well.

Presbyterian history in Scotland was replete with large and small rebellions against the patronage system of the Presbyterian Church of Scotland. One such withdrawal created a small "succession" church in which the Gifford family held membership. Adam Gifford's father was a church elder and Sunday school teacher and his uncle a minister for a time. This desire for independence matched well that of the merchant class in which the Giffords thrived, and the affiliation was certainly mainstream enough, for Gifford's father had been a member of the Edinburgh town council. As a young man, Adam Gifford helped at poorer Sunday schools. He was not so different from other middle-class Victorian youth, reared in no-nonsense Protestant households, where churchgoing, commerce, and sobriety were the moral plumb line, and the Bible was life's textbook.

The issue of patronage finally came to a dramatic showdown in 1843 for all the Presbyterian churches in Scotland, an event so large that it has become known as "the Disruption." When the British Parliament refused to reform the patronage system, half of the national Presbyterian Church's membership and more than a third of its clergy broke away to form an independent, evangelical Free Church. The Free Church, which shared all the same sentiments as the smaller "succession" groups, became a major denomination, gathering its considerable resources and building a whole new system of churches, clergy homes, and parish schools.

The Free Church movement also complained that the patronage clergy had become too liberal, so the Free Church cause rallied around the strict Calvinist beliefs of the Westminster Confession, a Protestant creed written in 1647. The creed reflected the teachings of John Calvin, the French Reformation figure who emphasized the doctrines of a sovereign God, human sinfulness, biblical law on earth, and divine predestination of the saved and the lost. With particular strength in the lowlands around Edinburgh and Glasgow, the Free Church not only took over foreign missions but built its own system of colleges, the most famous being New College, whose two gothic towers have looked down from an Edinburgh hillside since 1850.

In those years surrounding the McLachlan trial, a time in which the Free Church and "succession" churches talked of merger, Adam Gifford's busy life made his affiliation ambiguous at best. But he was surely an eyewitness to these contentious events of church and state. He would have been a keen spectator as the Free Church itself, with its emphasis on learning, began its own liberalization. One such turning point came in 1873, three years after Gifford became a judge, when the theologian Robert Rainy became the principal of New College, just a block down the hill from the courts. In his inaugural lecture, Rainy—remembered today by the stately Rainy Hall on the campus—argued that evangelicals could accept, if properly understood, the Darwinian theory of evolution. Before long, Free Church biblical scholars adopted the methods of "historical criticism," which analyzed how the ancient world had influenced the writing of the Old and New Testaments. Indeed, well into its founding century, New College taught natural theology and even had a natural theology museum.

In addition to this upheaval in Scotland's national church, two other features of that period must have made a deep impression on the mind of Adam Gifford—the new philosophical clubs and lectures, and the growing importance of science, exemplified in one of Scotland's greatest liberal arts inventions, the *Encyclopaedia Britannica,* first published in Edinburgh in 1768.

The young Gifford had been exposed to all of this living in the shadow of Scotland's two largest cities, Edinburgh and Glasgow. Both cities had contributed to the Scottish Enlightenment, but they were different in one important respect. While Edinburgh was more artistic and literary, Glasgow was practical and commercial. If Edinburgh styled itself as the most freethinking Scottish town, Glasgow was a stronghold of the

Calvinistic attitudes of Presbyterianism and the Free Church. Indeed, in 1862, when old James Fleming took the witness stand in the trial of Jessie McLachlan, the Glasgow newspapers said he looked like nothing so much as a sober Free Church elder, dressed as he was in a severe black suit.

Adam Gifford was born and reared, of course, in Edinburgh, where cramped city streets, speculative philosophy, and the drinking of a little rum had always mixed together amicably. In the mid-1700s, for example, Edinburgh had rivaled even Paris and London for having a "close knit community of scholars and thinkers, who were willing to take up new ideas," according to one historian.[8] The city, with its seven hills, squeezed in more citizens by erecting multistoried shops and tenements, and in such close quarters the taverns became the great centers of intellectual ferment. Taverns were home to the Tuesday Club, Poker Club, Oyster Club, Mirror Club, and Rankerian Club. Most exclusive was the Select Society, home to the editorial board for the *Edinburgh Review*. By Gifford's era, groups with names such as the New Speculative Society had grown in number and popularity.

In 1848, when Gifford was a twenty-eight-year-old clerk and law student in Edinburgh, and just five years after the Disruption of the Scottish church, he sauntered down to one such city club called the Philosophical Institution, a fashionable neoclassical row house in the newer part of the city, appropriately called New Town. There, for four evenings, he listened to a series of lectures by the American philosopher Ralph Waldo Emerson. Although Emerson's "appearance was far from impressive," Gifford recalled in old age, "I listened to him with a youthful and overflowing enthusiasm." It was not what the local clergy would have advised him to do, he seemed to say, but looking back, "I rejoice to think that my early admiration was not misplaced."[9]

Emerson lectured on topics such as "Genius of the Age" and "Eloquence." He was to America what Thomas Carlyle was to his native Scotland: both had caught the romantic spirit of the age, typified in German literature and philosophy, and both emphasized the individual, or hero, in history. Ordained as a Unitarian minister, Emerson himself had been nurtured in fairly traditional American Calvinism, but he had extensively diluted it—especially its claims of original sin, the trinitarian God, and predestination—with an admixture of German philosophy, Hinduism, self-reliance, free thinking, poetry, and the noblesse oblige of the educated classes in New England. The Calvinist clergy protested his visit to Glasgow, and in Edinburgh an opposition pamphlet warned: *Emerson's Orations to the Modern Athenians,* or *Pantheism.*[10]

By contrast, Adam Gifford seemed to find not only the uproar, but the very times in which he lived exhilarating. The sermon in church was one thing, but the traveling lecturer and, indeed, the philosophical clubs were quite another. Scotland had pioneered the college format of a master professor lecturing before a large classroom, and a classroom open to anyone who paid the enrollment fee. To be modern, in some ways, was to lecture about modern ideas. As a judge in the 1870s, Gifford was summoned to the speaking circuit as well. "He most willingly became a lecturer at many of the Literary and Philosophical Institutions of Edinburgh, and other towns of Scotland," said his brother. From courtroom experience, Gifford "knew the power of impassioned declamation over a jury or over a public meeting." During the McLachlan trial, he had spoken of how a murderous deed in Glasgow had "made the ears of the public to tingle," and perhaps he now believed that more uplifting kinds of mysteries could do the same.

The lectures, of course, could also be about the new sciences and their impact on every other sector of life. Many of Scotland's leading religious thinkers, such as Robert Rainy at New College, were acutely aware of how technology, urban life, and theories in natural science such as biological evolution would be changing the Christian worldview forever more. To the biblical knowledge of God, they believed, must be added knowledge of the created world. Such was the view, of course, of the *Encyclopaedia Britannica.*

In the years in which Gifford became a judge and Rainy the principal of New College, the *Britannica* was about to make a great shift in emphasis—lengthening its articles and giving pride of place to the scientific point of view in its ninth edition. Encyclopedic knowledge, it was believed, was essential to all progress, whether social, physical, or moral. In the ninth edition, which held sway from 1875 to 1911, the entry on "Theology" seemed more concerned with "truly scientific knowledge in Theology" than with tradition. The entry on "Bible" also drew attention. In Gifford's last years, the entry would foment a string of disputes that showed just how difficult it was for traditional Christian belief to accommodate the new historical "criticism," or analysis, of how the ancient world had influenced the writing of sacred texts such as the Bible.

After the new *Britannica* came out, at least four biblical scholars in the Free Church's three colleges—Aberdeen, Edinburgh, and Glasgow—faced complaints or heresy charges for using historical criticism on the Old or New Testament.[11] The most celebrated case was that of William Robertson Smith, professor of Hebrew language and the Old Testament at the Free Church college in Aberdeen. As a seminary professor, Smith held that the Old Testament was divine revelation. But as a critical scholar he said that its first five books had several authors, not just Moses, and that the entire work reflected ancient Middle Eastern culture. In 1875, when Smith wrote the "Bible" entry for the ninth edition of the *Britannica,* a five-year church investigation ensued. The Free Church General Assembly cleared him, but his next *Britannica* article, "Hebrew Language and Literature," led to his ouster in 1881. He left Scotland for Cambridge and, in 1887, became the chief editor of the *Britannica.*

What Gifford thought of the case is unknown. But in close-knit Scotland, how could the drama *not* mingle with his bequest. Many of the contributors to the ninth edition would be early Gifford lecturers. There was also the case of Alexander Balmain Bruce, a New Testament scholar who taught at the Free Church college in Glasgow. Shortly after Gifford's death, Bruce also faced heresy complaints for using scientific analysis on texts. He was easily the most conservative of the critical scholars, however, and no action was taken. Years earlier, in 1871, Bruce had published the popular book *The Training of the Twelve,* still in print today. It told the story of the apostles, their characters and missions, and how Jesus instructed them. When an apostle searched the scriptures and asked questions to find the very "knowledge of God," he was the proverbial "earnest inquirer after truth."[12] Lord Gifford's will similarly defined future lecturers as "earnest inquirers after truth." In 1896, Bruce was also the first Bible scholar to give the Giffords.

Unlike the Bible scholars of his era, Gifford did not struggle with the historical problems of the Bible texts but enjoyed them for their spiritual insights and poetry. He loved all kinds of poetry, and said that the more supernatural, and least historical, Gospel of John was his favorite of the four Gospels. Yet he did oppose the materialistic philosophies of his time. Some of them were virulently anti-Christian, whereas others simply undercut spiritual values by presenting a loveless, mechanistic universe produced by the mindless evolution of matter.

In old age, as if an apostle looking for his own complete knowledge of God, Gifford became deeply attracted to the rationalist philosophies of continental Europe, especially the pantheistic science of the Jewish thinker Baruch Spinoza. By his brother's account, Gifford "studied and admired" Spinoza's crisp, clear, almost mathematical approach—with one proposition after another—to writing about God. In one of his public lectures, Gifford emphasized the greatness of Spinoza's idea of God as the One Substance. One proof that such a substance exists, Gifford argued, is the "reasonable and intelligent soul" of man himself. All things "are but forms of the Infinite, the shadow of the Substance that is One forever," the judge said.[13] Privately, he used Spinoza as a foil to explain his own beliefs. "Spinoza holds that every thing is God," Gifford told his brother. "I hold that God is everything; if I were to assume a name descriptive of my belief, I should be called a Theopanist."

Although some of Gifford's philosophical interests may have been known around Edinburgh already, his will further reminded people of how open he was to all kinds of constructive spiritual ideas, an openness that dogmatists in neither religion nor science would support. When, in March 1887, the full text of Gifford's will was published in both the *Times* and the *Scotsman,* the lack of any Christian terminology in the text made clear that he was not funding a particularly orthodox project. For this reason, according to one account, the criticisms came fast and furious, "anointing him after their fashion with such epithets as vanity, heresy, atheism, pantheism."[14]

Gifford's personal quest to reconcile the old theology and the new philosophy had moved him to the side of tolerance and free thinking. With an unshakable conviction that God existed, he did not worry about what philosophy or science might turn up. Like the encyclopedists, he may have believed that accumulated knowledge would lead reasonable people inexorably to a belief in God and godly ethics. His plea for tolerance had been evident much earlier when he lectured at Glasgow City Hall, making clear to his churchgoing audience that the long-gone Emerson had believed merely in an Eternal Cause, not a personal God. Still, "Is one who occupies this position to be blamed?" Gifford asked. "I cannot condemn that wise and humble skepticism [and so] meantime let us rejoice in every high and holy aspiration."[15]

The rejoicing was worth it, after all, because the drudgery of life and finally death were the fate of every human being, including Gifford. By his diligent studies, Gifford had ended up a lawyer and judge, and by his labors a wealthy estate holder in Edinburgh, taken to work each day in a coach with gray horses. But he knew the tedium of life as much as anyone. "His heart never was entirely with his profession," his brother said. The law could be technical, abrasive, and petty. As Gifford complained one day, "All last night and all this day I have had to investigate and make up my mind on a wretched, paltry dispute between two unreasonable men about a trifle." The old judge remarked that he had loved money far too much in life. Given that lament, Gifford must have felt a great relief when he decided to give his wealth away. He did so by funding a perpetual lectureship, dedicated to what he believed was the ultimate topic: knowledge of God by way of science and "natural theology."

The idea of an endowed lecture, already an English tradition, is traced most prominently to Robert Boyle, "the father of modern chemistry," who in 1691 left part of his wealth to pay for regular public talks in London. Shortly before he passed from this world, Boyle lamented the everyday doubt and disbelief on the street, now filled with grubby commerce and materialism. The natural theologian who gave the first Boyle Lecture spoke on "Matter and Motion Cannot Think," subtitled "A Confutation of Atheism."

In time, the number of lectureships would grow. At his death in 1751, the Reverend John Bampton, Anglican canon of Salisbury, endowed the Bampton Lectures at Oxford University to "confirm and establish the Christian faith, and to confute all heretics and schismatics."[16] Robert

Hibbert, a wealthy merchant and ardent Unitarian, established his lectures to vent a more liberal view.

Often, a series of published talks became intellectual benchmarks. In
the 1850s, a series of lectures in Dublin by the Oxford don turned
Catholic convert John Henry Newman became a historic book in itself,
The Idea of a University. Thomas H. Huxley's argument for a moral war
against the "gladiatorial theory of existence" in Darwinism, an appeal
made in the Romanes Lectures, has been quoted ever since.[17] And there
were enough Scottish cases, lectures endowed in the names of beloved
clergy or famed politicians, for Gifford to get the idea.

Undergirding that idea was a hope to reach the common man. The
Gifford Lectures were to be class levelers, a cornucopia palatable to a
Jessie McLachlan as well as a Lord Deas. Perhaps most consequential in
this respect was the mandate that the lectures be published. For as Gifford
once said, "There on the printed page stands the man at his very best.
His imperfections, his frailties, his defects, are not printed."[18] And so it
might be said of Gifford's will. He signed it in 1885, wheelchair-bound
but stationed on the back lawn of his old mansion, looking over the Firth
of Forth. He had wealth, but perhaps by now lonely wealth. His last close
associates were an upper-class doctor and a working-class cab driver. They
witnessed his will, and they were doubtless the kind of people Gifford
wanted his bequest to edify.

For the last six years of his life, Gifford had been bound to that
wheelchair. Propped up by pillows, he was surrounded by philosophy,
theology, and poetry books. His ability to read Greek, Latin, German, and
French put him in an upper echelon of a remarkably literate nation. Yet
like Everyman, he took the measure of his soul. Sleep did not always
come easy. In those hours he cursed his paralyzed limbs, evoking biblical
allusions to the "vile" and "corruptible" body, indeed a tabernacle that
was finally "the body of this death." Yet he never complained, his brother
said. "He wished to be done with his frail, useless body forever."

Like a metaphysical treatise, Gifford's will was a flourish of John's
Gospel and Spinozistic philosophy, showing piety of the heart and praise
of the rational intellect. Bestowing his body's "enduring blocks of materials" to the earth, he next gave his "soul to God, in Whom and with

Whom it always was, to be in Him and with Him forever in closer and more conscious union." When he died on January 20, 1887, the family "kissed his cold brow." They said farewell, hoping he had found what was on his lips to the end: "To be happier or wiser, that just means to have more of God."

What the judge left behind would test the wisdom, and even felicity, of many future savants. His reference to natural theology was conventional enough. But he urged an approach to God akin to chemistry and astronomy, two exacting sciences. As most theologians knew, natural theology was no longer the stalwart fellow it had once been. By the 1880s, it was like a man who, having the audacity to prove God by logic and nature, had been beaten bloody to the floor. In fact, his assailants were old friends, Western philosophy itself and the natural sciences.

What was more, Gifford said, God was to be illumined "without reference to or reliance upon any supposed special exceptional or so-called miraculous revelation." More than a few theologians shook their heads at this caveat.

Wisely, though, Gifford had locked few doors on how to proceed. No "constraint" was to be imposed on lecture topics. Speakers would suffer "no test of any kind, and shall not be required to take any oath, or to emit or subscribe to any declaration of belief, or to make any promise of any kind." Still, they were urged to give two series of ten public talks, no small feat for popular consumption. As time would show, the sages who traveled to Edinburgh, Glasgow, Aberdeen, and St. Andrews would take up the Gifford chair with great ingenuity. The ceiling could not be higher—to "draw conclusions concerning the ultimate nature of the universe"—or the benefits more benign: "nothing but good can result from free discussion."

For that matter, the stipend could hardly have been more generous. The academic senates, which were the university bodies charged with running the Gifford Lectures, began with offerings as high as 800 pounds, easily twice a professor's annual salary in those days. In a few decades, of course, the badge of honor was to give the Giffords, not to receive its increasingly modest cash award. But for now, the four universities looked wide-eyed at their new resources in the bank.[19] And as the academic senates at Edinburgh, Glasgow, Aberdeen, and St. Andrews knew well, they had a friendly rivalry to keep up—indeed a kind of historic competition that the Giffords would only enhance.

The four "ancient" universities of Scotland had been central players in the nation's very colorful past. The schools at St. Andrews, Aberdeen, and Glasgow had all been founded in the fifteenth century by papal bull. When the Protestant Reformation arrived, St. Andrews—the oldest, founded in 1413—saw its cathedral burned to the ground. Both St. Andrews and Aberdeen are north of Scotland's industrial and population belt, occupying coastland that ends in steep cliffs on the North Sea. The universities are known by their small, ancient quadrangles, but like an Oxford or Cambridge, they are not so much walled-in campuses as college towns that mingle medieval and Victorian structures with cobblestone streets, shops, and sundry homes.

The urban settings of Edinburgh and Glasgow tell a different story of university life. Edinburgh is the youngest of the four schools, organized in 1583 by the city council and approved by a Protestant king. The Edinburgh University quadrangle, which still covers a large city block, went up in the 1820s. For all the city's literary acclaim, however, Edinburgh was known beyond the eighteenth century as "Auld Reekie" for its awful smells, rising from sewage and a swampy ravine below Old Edinburgh's rocky heights. That ravine would be drained. By the 1880s, the decade of Gifford's passing and the start of the lectures, a new kind of city was on display.

To the east of Edinburgh, Glasgow had long been called Britain's "second city," so important was its industry and commerce. Industrial smog, in fact, had forced its university to relocate to the western heights of Gillmorehill. When the castle-like university with a double quadrangle was completed in 1870, it was one of Britain's largest building projects ever.

By their histories, academic faculties, and local geographies, each of the four schools had much to boast of, and the Gifford Lectures were only the latest opportunity to show the school flag. With the endowments in hand, each school had to find the best lecturer it could, and preferably launch its maiden event before any of its three rivals. By the end of 1887, when the executors began releasing the bequest's funds, Edinburgh and Glasgow were the schools to watch in the race to be first, and both can plausibly make such a claim today. Edinburgh was first to secure a Gifford Lecture appointment (January 1888), but Glasgow was first to have its lecturer take the rostrum (November 1888).

The very first words of a Gifford talk thus issued from the mouth of Friedrich Max Müller, a famed German scholar who, retired in Oxford, had received the Glasgow appointment. Yet for our story, the Edinburgh appointment of James Hutchison Stirling, a Scottish medical doctor turned philosopher, is best suited to open act one of the Gifford drama. A Glasgow native, Stirling now lived in Edinburgh, a great convenience as he prepared the first Gifford extravaganza for that city (though his lectures came a month or so after Müller's in Glasgow). "These lectures have given me perplexity, and make me thin," Stirling wrote on November 2, 1888.[20] The opening day was January 12, 1889. Stirling was perplexed over Lord Gifford's interest in natural theology, as Stirling felt that in many ways that age had passed. Now was the age, he seemed to believe, of the philosophy of God.

Stirling may have been the only lecturer in history to know Gifford close up, a personal acquaintance not over legal matters but probably over an exotic new import—the philosophy of Germany, which had significantly shaped the way that people philosophized about God in Stirling's era. For the first act of the Gifford legacy, the spotlight falls on Edinburgh and the dramatic clash between the new philosophy of God and scientific materialism. This clash had a century-long pedigree in Edinburgh already, beginning in the days of David Hume during the Scottish Enlightenment and ending in Gifford's time, when German philosophy was all the rage in Scotland. The opening years of the Giffords brought that drama to a climactic end. In ways, it was the end of philosophy, which speculated about God, and the dawning of natural science, which asked about God amid the material facts of the world.

TWO

THE END *of* PHILOSOPHY

The Rise and Fall of Idealism

T HE WORLD WAS ARRIVING at Edinburgh's door in the 1880s, or so it seemed. The engineering feat of the day, a railroad bridge across the Firth of Forth, the great waterway on the city's north, was about to be completed. The fancy trestle earned it the name "Scotland's Eiffel Tower." Travel into Edinburgh mushroomed, and by 1892 the construction sounds of iron and glass rose from Waverley Station, the main rail terminus. The station was being rebuilt in Victorian splendor, thirteen acres of glass roof, iron ribs, and ornate filigree.

In the 1890s, to arrive under Waverley Station's great glass dome and disembark onto Edinburgh's cobblestone streets was still to journey into Scotland's philosophical past, evoked most surely by the name of that radical skeptic David Hume. The station sat in the great ravine, a valley of rails and parks that divided hilly Old Town on the south from New Town, built on a gradual slope to the north. Up the hill in Old Town was Parliament House, home to the Scottish courts. Hume was born almost across the street. In his last years, he built a Georgian row house in New Town, large enough for a bachelor author, as he would say in his brogue. The address became known as St. David Street. Not far away, by the ravine, stands Old Calton Cemetery. Hume left 100 pounds to build his mausoleum there, and a columned rotunda was erected at the heart of Old Calton in 1777, a year after his death.

The Gifford Lectures began more than a century after Hume was placed in the mausoleum. But like his memory and his monuments, the arguments Hume generated over the existence of God were still around in the 1880s, as palpable as the steam of train engines arriving from London, Glasgow, and points north. Even in that late Victorian age, in 1889, by which time Christianity, empire, and progress were bywords, the "temper of the times" did not bode well for religion, said James Hutchison Stirling in the city's first Gifford lecture. "In our large towns in these days, in our capitals, in our villages, we are confronted by a vast mass of unbelief."[1] And he cited the eighteenth-century Enlightenment as the start of it all.

Hume was part of that Enlightenment, which set out to defeat the influence of medieval theology on the Western mind. In its devotion to promoting science, the Enlightenment eventually downgraded every kind of theistic philosophy. At the beginning of the Enlightenment, the term *natural philosophy* stood for the discipline of science, but in time all philosophy was held suspect by the new scientific milieu. As science claimed the entire domain of human affairs, the age of philosophy came to an end. The early Gifford Lectures revealed this clash, particularly between the two great systems of German idealism and Charles Darwin's materialist theory of evolution. Whereas the first acknowledged God's existence on a philosophical basis, the second deemed God irrelevant on a scientific basis. This was the very heart of the drama in act one of a century in which science and religion met head on.

The story of the final clash returns invariably to David Hume, the "Great Infidel" of his day but always Edinburgh's favorite son. As a young man, Hume visited France and came under the sway of the French skeptics. They had borrowed the scientific ideas of the Englishmen John Locke and Isaac Newton and become the leading thinkers of the Enlightenment. As Voltaire (François-Marie Arouet) was the famed "philosophe" in France, Hume became the same kind of literary figure for Scotland—although it took some doing. According to Hume, when he returned to Edinburgh in 1745, his nomination to the university faculty was defeated by the "bigotry of the clergy, and credulity of the mob." He left town again, but finally in 1751 his supporters won: he was hired as Keeper of the Advocates' Library at Parliament House, despite a "violent cry of Deism, atheism, and skepticism [that] was raised against me."[2] Two years as librarian allowed him to begin writing a six-volume history of England that, when completed in 1762, ensured his fame and fortune.

As a philosopher, Hume had sided with the materialists in their belief that knowledge came only by way of experience: sense data produced ideas. Beyond sense and ideas, Hume would argue, nothing else could be proved to exist, neither a universe nor even a human mind. What people in the marketplace, and scientists in the laboratory, lived by were only habits of thought, which offered an adequate restraint on animal passions. Under Hume's radical form of skepticism, even the rising of the sun was not guaranteed. When one billiard ball hit another, the "cause" and "effect" could not really be proved.

He first explained his radical skepticism at age twenty-six in *Treatise of Human Nature.* Years later, it would grow in fame and infamy along with his own reputation as a historian. Such utter skepticism, however, was so impractical to live by that even Hume would laugh at its pretense and play the common man. "Though I throw out my speculation to entertain the learned and metaphysical world, yet in other things I do not think so differently from the rest."[3] He too lived as if the sun would rise. He too would have looked at the evidence in the McLachlan trial for the *cause* of the death. Still, Hume's mental entertainments began a century of havoc. His brilliant doubts demanded a philosophical response.

That would come from a Presbyterian minister, Thomas Reid, professor of philosophy at the University of Aberdeen, 120 miles to the northeast of swampy Edinburgh. After reading Hume's *Treatise,* Reid began producing a series of counterarguments. It was the beginning of "common sense" philosophy, indeed what in history would be called the Scottish Philosophy. Reid set out to demonstrate that the mind knows matter by direct and immediate knowledge. This power of "common sense," while presumed to be given by God, could be discussed, according to Reid, by observing detailed human psychology.

When Hume received a copy of the Aberdeen clergyman's manuscript, he wished "the parsons would confine themselves to their old task of worrying one another, and leave philosophers to argue with moderation, temper, and good manners." After reading Reid, however, Hume wrote a friendly letter in 1763 saying the work was "deeply philosophical" and "so much entertainment." If Reid were "able to clear up these abstruse and important topics, instead of being mortified," Hume said, "I shall be so vain as to pretend to share of the praise." Reid replied that his little philosophical society at Aberdeen was likewise indebted to Hume "for its entertainment," and even as "good Christians" they would prefer his company over that of many others.[4]

More than just Hume and Reid had joined in the intellectual entertainments of that period. The British Enlightenment had begun with John Locke, who first argued that ideas were produced by purely material experiences, which were conveyed to a blank mind by the senses. Bishop William Berkeley countered that, saying that only ideas in God's mind were real and that the material world was an illusion. In the seventy years between Locke and Reid, a perplexing circle was established. As one historian aptly described the cycle, "Locke starts with minds, ideas, and matter. Berkeley disproves matter and retains minds and ideas. Hume denies the existence of minds and preserves only ideas. And Reid in turn denies ideas."[5]

To restore order, Reid returned to the assumption of Christian belief, that God had created a knowable world. Locke's "way of ideas" was wrong, he said, because ordinary people seemed to grasp immediate physical realities, and so did philosophers for that matter. In "all the history of philosophy we never read of any skeptic that ever stepped into fire or water because he did not believe his senses," he said.[6] Although in many ways history would forget the contribution of Reid, so simple was its premise, his insights into the idea of "self-evident" truth have never gone away, as will be seen at the end of this book's story.

Hume, of course, became a towering figure in the history of Western thought. Although his skepticism about everyday causes was open to question, he also left behind significant arguments against knowing the ultimate cause, namely the benign personal God of the biblical universe. Before he died, Hume directed his nephew to publish a little book titled *Dialogues Concerning Natural Religion,* which came out two years after Hume's demise. In this work, considered one of Hume's finest literary pieces, the Scottish thinker imitated a famous tract by the Roman skeptic Cicero, who had told the imaginary story of three Roman characters and their debate about the gods of antiquity.

Hume brought the three-man debate into his own time with new characters. His *Dialogues* gathered Cleanthes, Philo, and Demea one summer in Cleanthes's library to debate the chief proofs of God that Christianity had offered for centuries. These were, for example, God as the "final cause" of all causes in the world, or God known by the "argument from design"—that a designed universe needed a Designer. The *Dialogues* cast heavy doubt on all such arguments, a doubt articulated especially well by Philo, thought to be the voice of Hume. Yet for all the skepticism, a narrator at the end of the work did not totally rule out an ultimate cause: "*the cause or causes of order in the universe probably bear some remote analogy to human intelligence.*"[7]

Whether Hume was finally an atheist, or a deist like Voltaire, the collective impact of his philosophical masterpieces began a chain reaction in Western philosophy. In addition to prompting the work of Thomas Reid, Hume stirred a new approach to philosophy in continental Europe, beginning with the work of Immanuel Kant, a Prussian university professor who was one of the last Enlightenment figures of the age. Kant had read Hume in the 1770s, and his response is considered the beginning of modern German philosophy. "To Hume we owe the Philosophy of Kant, and, therefore, also, in general, the later philosophy of Germany," William Hamilton, the last Scottish common sense philosopher, argued in 1836.[8] In the decades that followed, German philosophy made its way into Britain, setting up the encounter between philosophy and science reflected in the early Gifford Lectures.

Fortunately for Scots pride, Immanuel Kant (d. 1804) was of distant Scottish extraction, and it may have been by way of Glasgow. The city was a major trading and shipbuilding port when the British Empire had expanded across the seas. The empire had chosen Königsberg, Prussia, as a hub to extract premium shipbuilding lumber from the Baltic forests. An English and Scottish settlement had arisen there, mixed with the local population, and about a hundred years later Kant was born. Although a city of commerce and military fortification, it was also a place where ideas crossed paths. Kant would never set foot outside the little province (which today is Kaliningrad, Russia), but from his reading desk he could take in the whole world.

As a student, librarian, and finally professor of metaphysics at the university, Kant had been steeped in the philosophical systems of the Continent, a rationalist world in which philosophers from René Descartes through Spinoza and Gottfried Leibniz argued, for example, that God and the soul were proved by "clear and distinct ideas" in the mind, and that so benign and rational a universe was surely "the best of all possible worlds." Then in 1775, at age fifty-one, Kant read a German translation of Hume. The arguments for skepticism troubled Kant, but they also awoke him from the "dogmatic slumbers" that had been foisted on him by the likes of Leibniz, with his picture-perfect world. In the next decade, Kant produced a response to Hume that reestablished the mind as an ultimate reality, but not in the extreme way proposed by the rationalists. His work began a movement called German idealism.

Not for the first time, Western philosophy and theology had wandered into the debate between idealism and realism. Ever since Plato said perfect "ideas" preceded their transient worldly copies, and Aristotle described "real" external objects in the world, idealism and realism have been sparring. Although there have been constant overlaps, idealism has stood for belief that ultimate reality is spiritual, or that human knowledge began with powers of the mind. Idealism has tended to be highly rationalist, even mathematical. Realism took external matter to be the truest reality and limited human knowledge to sense experience. It was favored by the sciences, but also by various kinds of Christian realists, including Thomas Reid himself. Kant was not a pure idealist, for he recognized real objects independent of the mind. Yet he proposed a single consciousness in the universe, from which the human mind drew its power to organize the physical data of the world. Kant's solution has been given the name *transcendental idealism*.

From his theoretical heights, however, Kant had little use for Reid's self-evident truths. Common sense was "indeed a great great gift of God," but truth must be shown by "well-considered and reasonable thoughts and words, not by appealing to it as an oracle when no rational justification for one's position can be advanced." Around himself Kant saw charlatans, philosophers who catered to the "opinion of the multitude" and gloried in the applause of the unthinking masses.[9] When it came to true philosophy, Kant seemed to say, there would always be the mob against the elite. He could not have described better the age-old battle: a clash of ordinary people with the intellectuals. The Gifford Lectures would raise the topic again and again, how to reconcile the ivory tower with the opinion of the common man. When the natural scientists came to the fore, as will be seen in act two of the Gifford legacy, they must have seemed to the average citizen so very down-to-earth in explaining how the world worked, especially when contrasted with the abstract speculations of the idealist philosophers.

As with all great philosophers, Kant's life is easier to grasp than are his written tomes. Reared in German pietism, and the son of a strap maker, Kant internalized his age: its battle between faith, rationalism, and the new sciences. The mature Kant wanted to know how metaphysics, that philosophic bridge between God and the world, could be truly scientific. For so weighty a question, he himself was slight, hollow-chested, and small. He "lived an abstract, mechanical, old-bachelor existence" on a quiet street, wrote Heinrich Heine, the German poet. Heine doubted whether Königsberg's cathedral clock ticked "in a less passionate and regular way" than Kant, who daily walked eight laps on a path of linden trees, nodding to the townspeople. "When he passed at the appointed hour, they gave him friendly greetings and set their watches." They couldn't have imagined he was brewing a "world destroying thought."[10]

Kant published his famous *Critique of Pure Reason* in 1771, agreeing with Hume that pure reason could not know God, the universe, or the soul. For Kant, these were "things in themselves," or ultimate realities. Contrary to Hume, however, he showed that the mind has innate ways to organize sense data, arranging it in time and space and twelve other qualities. As the Scottish idealist Andrew (Seth) Pringle-Pattison put it, "the fluid and indeterminate data of sense must be poured into certain mental molds."[11] This mental power was a permanent existence, Kant argued, a "transcendental ego," a single consciousness in the universe that was somehow less than God. Wags would call it Kant's "machine-shop" view of the mind.

Kant had probably hoped that pure reason could comprehend God, but he concluded that the quest was impossible. So he offered a compromise: the mind can know cause and effect in the sensory world of phenomena, but not the ultimate world of "noumena." The reason was simple, Kant said. When human rationality takes on ultimate questions—such as infinity and finitude, the One and the many, free will or not, and ultimate Cause or not—it ends in contradictions. Since the days of the Greek philosopher Socrates, the contradiction, or "dialectic," had been a conversational way to frustrate the mind, pushing it to a new level of thinking. Kant's four ultimate contradictions amounted to a "transcendental dialectic" that made rational knowledge about God, the cosmos, or soul impossible. When it came to God, Kant said, it was necessary to "deny *knowledge* in order to leave room for *faith*."[12]

Kant finally held that two things in the universe truly awed him, the starry heavens above and the moral law within. He saw in humanity the basic faculties to comprehend both realms, one by the discipline of science and the other by what Kant ultimately called *practical reason,* a moral sense of how to live, the nature of moral duty, and knowing right from wrong. Ultimately, the universe was a moral system for Kant. The moral impulse in life did not prove God's existence, but it certainly harmonized with the belief that a moral lawgiver such as God stood behind the cosmos, even though such a belief was finally taken on faith.

For so small a man, Kant's pure reason and practical reason cast giant shadows over Western thought, especially Protestant thinking. Few Gifford Lectures would stop at the altar of Thomas Reid. Nearly everyone had to genuflect before Hume and Kant. These two figures were seen to have thwarted any final hopes of natural theology to prove, or elaborate on, God's existence. Still, pure skepticism could be its own worst enemy: one could also be skeptical of skepticism, in other words. This was an argument often used in defense of God throughout the Gifford Lectures.

Although Kant did not hold out the moral sense as *proof* of God, others taking the Gifford rostrum would.

In a word, the philosophy of Hume and Kant did not destroy the philosophical quest to know God, but they surely would make it less naive than it had been in the past. Kant's unmistakable impact on Protestantism, especially as it flowed from Germany, was seen in this two-part formula: there was faith in God, but everything else in religion was ethics.

From his obscure northern home, Kant had cheered on the American Revolution as the most salient product of the liberty and autonomy that for him characterized the "age of Enlightenment." For a time at least, America had more fortifications against German idealism than did Great Britain. In America, not only did Scottish common sense philosophy and Protestant orthodoxy combine as "a breakwater to resist the assaults of skepticism," said James McCosh, an early president of Princeton University, they also helped Christians resist the "speculative theology which is coming like a fog from the German Ocean."[13]

For better or worse, the same was not true in England. Between the 1770s and the 1840s, German idealism had been both born and exhausted on its native soil. Yet the scientific revolutions in travel and printing helped the German fog skirt the barriers of geography and language and, on the eve of the Gifford Lectures, penetrate to the very heart of British philosophy. "The younger generation of our University men," said one Scottish philosopher in 1885, "are more strongly impressed with a German than with a native stamp."[14] With the advent of the German philosophy called idealism, the stage was set for a final clash between the philosophy of God and materialist science.

When Gifford had endowed the lectures, he had commented on neither the scope of their influence, whether national or international, nor the rank of the candidate to speak. Faced with making such a choice for its first lectureship, the senate at the University of Glasgow decided the honor was "so important and unique, it would be well to look beyond Scotland."[15] So it tapped Friedrich Max Müller, the renowned German scholar who had taught at Oxford. Scotland could no longer boast the philosophical likes of a Hume and a Reid, who lived in a time when a traveler to Old Town Edinburgh said he could "in a few minutes, take fifty men of genius by the hand." On the eve of the Giffords, when the Victorian age of railroads and empire had overshadowed such intellectual interests, one future lecturer spoke of this philosophical void: "The more speculative topics were avoided; and great figures were scarce."[16]

The Giffords presented another kind of enduring "difficulty" to the university committees, as the principal of St. Andrews would point out in 1905. The difficulty was "getting men of sufficient eminence to be appointed to such a highly paid lectureship who have not already said in one form or another all they have to say upon the subject." Regarding young and brash thinkers, "the lectureship is too exceptional to be placed in the hands of an untried man."[17]

Talent enough was found, however, and although it leaned toward older figures, including some who died on the eve of giving their lectures, the ages of Gifford lecturers at times would also drop to the forties. In the early days of the Giffords, one talent that had cachet was knowledge of German philosophy and learning, usually obtained by travel, study, and mastering the foreign tongue. That bridging was done in the opening Gifford Lectures at Glasgow by the German scholar Müller, who had learned English, but most significantly by Edinburgh's first lecturer, the Scotsman James Hutchison Stirling, who used so many German terms in his talks that a friend wrote, "Can't you somehow English these?"[18]

Stirling was a self-made man of sorts, a surgeon-turned-philosopher who had delved into German thinker George W. F. Hegel by studying in Germany. In his later life, he had been on a short list to take the chair of moral philosophy at the University of Glasgow, the chair once occupied by Adam Smith. But it was not to be, for Stirling, a Glasgow native, never quite mastered the insider's game and never obtained an academic rank. Still, at age sixty-eight he was cosmopolitan Edinburgh's first Gifford lecturer. A stocky man with a square face, Stirling had snow white sideburns that turned into a beard under his chin. He began his lectures on a Saturday morning, January 12, 1889, in the largest lecture hall of the university, which was not large enough. He promised to repeat the lectures in the evening for those who could not get inside, a situation that did not last too long into the series.

According to the *Scotsman* and others, Stirling's early talks were lively enough, and the humor dry enough, to solicit both "applause" and "laughter," despite his occasional use of German terms—and despite the strange new philosophy of which he had become Britain's expert, the Absolute Idealism of George W. F. Hegel, a philosophical successor to the diminutive Kant.

Stirling was born in 1820, when Hegel was still alive, and graduated in science and medicine from his hometown's Glasgow University. He set up a medical practice in Wales and read philosophy as a hobby. When he read a book review about Hegel, a name until then unknown to him, however, his life changed. By now Hegel was dead, taken down at age sixty-one by cholera, but his ideas lived on. Stirling "heard this Hegel talked of with awe" by German students he met. They hinted that Hegel was the "deepest of all philosophers, but as equally, also, the darkest." Yet the great moment was still to come, for when Stirling heard the claim that Hegel had "reconciled to philosophy Christianity itself," it spurred him to action. As Stirling recounted it, "*That struck!*"[19]

Relying on a small inheritance, Stirling gave up his medical practice to spend four years in France and one in Germany. In Heidelberg, he immersed himself in Hegel and German. By 1865 he was living in Edinburgh and had published *The Secret of Hegel*. In Edinburgh's polite society, his renown as an author brought Stirling to meet Gifford at dinner parties and elsewhere. His somewhat abstruse tome, in fact, launched the serious study of Hegelianism in Great Britain.

The "secret" in Stirling's title prompted much speculation. So difficult is Hegel in any language that the jest came down that Stirling had kept the secret to himself. In one view, the secret was that Hegel had saved Christianity by giving it a philosophical underpinning. "There is for Hegel nothing but God; and this God is a personal God, and no mere Pantheistic Substance that just passively undergoes a mutation of necessity," Stirling insisted. "It is just on this scheme that he is able to perceive that Christianity is, must be, and can only be, the Revealed Religion."[20]

Another view of the secret plunged more deeply into the difficult world of idealistic thought, arguing that if Kant were *properly* understood, the secret was this: Kant's ideas were the logical precursor to Hegel's ideas. As Stirling explained: Hegel "made *explicit* the *concrete* universal that was *implicit* in Kant."[21] The "concrete universal" was no easy doctrine. Hegel had been trained in Lutheran theology, and although he ended an astute man of letters, keeping company with bohemian artists and stuffy government officials in Berlin, his interest had sprung from the mysteries of the New Testament. He wanted to make philosophical sense of the Gospel of John's "in the beginning was the Word." The Word became flesh, and for Hegel the Absolute would do likewise in the "concrete universal." The Absolute was an Individual that revealed itself in individuals.

While Hegel was developing his philosophy, Napoleon's French army was marching into Prussia. Hegel felt that a new dawning of ideas in the world should parallel such momentous military events. He wanted to do away with the static idea of the classical world and replace it with a philosophy about reality marching forward, overcoming, and do this all under the power of the Absolute Spirit, a kind of synonym for God, depending on how Hegel was interpreted. In this concept of God, the chief characteristic was divine thinking. Like Kant, Hegel put the dialectic at the heart of his philosophy, but in this case he was talking about the dialectic of God's mind, where divine thought produced the progress in the universe. As one historian explained, "While Kant used the word 'dialectic' to refer to the propensity to fall into contradictions, Hegel used it to mean the propensity to transcend them."[22]

In the dialectical mind of the Absolute, it was as if God had an idea, then a counter idea, and to resolve the two, God then had a third idea that was better still. This was the creative process that made the universe emerge and progress. Another way Hegel spoke of it was that a subject, namely God, produced an object, namely something in the world, and then the subject and object developed by changing each other in a relationship. By this process, the Absolute Spirit made the world and revealed itself in human consciousness. The result was human civilization, with religion as the supreme way to know the Absolute Spirit. With his grand dialectic, Hegel had proposed Absolute Idealism, the flowering of German philosophy.

Stirling knew this would be hard for British tastes. As one translator said to him: "What! Would you introduce that d———d nonsense into this country?"[23] The British were already suspicious of the "German party" in philosophy, for in the 1860s its anti-Christian element was making waves as such left-wing Hegelians as Ludwig Feuerbach and David F. Strauss wrote influential works questioning God's existence and the accuracy of the Bible. Indeed, the German journalist and economic theorist Karl Marx took Hegel's dialectic and located it in matter itself, proffering the powerful ideology of dialectical materialism in history, the very basis of communist political thought.

Still, the Hegelian ideas that were friendly to religion caught on in the English-speaking world. Idealism was a formidable, and perhaps the only, philosophical system around to defend Christian theism against the rising tide of materialist philosophy. While the philosophy of idealism proposed a kind of evolutionary development, it was an evolution of God's mind, not the godless, willy-nilly, and accidental evolution of Darwin's scientific theory. It was no surprise, therefore, that idealist philosophers swamped the early choices for Gifford lecturers.

The chief materialists of the late Victorian era were the philosopher and engineer Herbert Spencer and the naturalist and science advocate Thomas Henry Huxley. Espousing agnosticism and adopting the idea of materialist evolution, Spencer had coined the phrase "survival of the fittest," which later was borrowed by Darwin. Huxley in turn gained fame as "Darwin's bulldog," defending the materialist theory of evolution and seeking to unseat Anglican clergy from the academic chairs at British universities. Against this backdrop, the new materialists of Britain founded a cutting-edge journal called *Mind*. Now the idealists had a target, and *Mind* became the convenient foe of seven young-turk Hegelians who came on the scene, not a few from Scotland. These young idealists took their stand in an 1883 volume, *Essays in Philosophical Criticism,* edited by their Scottish leaders Pringle-Pattison and Richard B. Haldane. All seven would give the Giffords under an idealist or Hegelian banner.[24]

In this respect, Stirling was of the old guard. His own Gifford Lectures were not so much an espousal of idealism as a survey of the philosophical attempt to prove God and the arguments against those attempts. Given the vastness of a topic such as natural theology, Stirling told his audience, speakers for years to come were likely to be repetitive, "all declaiming, year after year, on the same text."[25] Yet his January-to-April talks in 1889 and 1890 were Edinburgh's inaugural Giffords, so it was only natural that they became a historical survey of natural theology from the Greeks to the Victorians.

As it would turn out, down through the history of the Gifford lectureships the ancient Greeks themselves became the center of a dispute over whether the earliest Greek natural theology was monotheistic. It certainly was, argued philologist and historian Werner Jaeger in later Giffords. A typical early Hellenic philosopher, "overcome with awe and amazement at the sight of nature's sublimities, becomes a herald for the godhead," Jaeger said. The Greeks did not invent polytheism but instead spoke "of a single god whom they call Zeus, whose mind embraces all things in its knowledge, and who guides all things and is king of all."[26]

For Stirling's purposes, it was enough to show that the Greeks offered Western thought the idea of a supreme cause behind the material veil of nature. Although the doctrine of Zeus offered an anthropomorphic deity, Plato argued for a creator called the demiurge. Aristotle described the supreme cause as the prime mover. In Plato's scheme, a creator had to turn the transcendent Ideas, the true reality, into a transient material world, whereas Aristotle needed God to finish his metaphysical system. He needed a first mover behind all motion.

The Romans too had a natural theology, probably borrowed from the Stoics. Early Christianity would build its parallel edifice on texts in Psalms and Romans about seeing God in nature. When scholasticism began, Saint Anselm of Canterbury (d. 1109) proposed a logical proof of God—that if the mind could conceive of a perfect God, it would be a contradiction for him not to exist. Thomas Aquinas was not so sure about that. So he offered his "five ways," five aspects of finite being (the existence of things) that required a necessary infinite Being.

Renaissance humanists found evidence of divine benevolence in nature, and rationalists from Descartes to Spinoza argued along Anselmian lines. But with the British Enlightenment, for the first time scientists listed natural wonders—plants, animals, organs, and planets—as proof of design. This culminated in Stirling's own century. English theologian William Paley wrote *Natural Theology* in 1802, and a series of treatises commissioned by the Earl of Bridgewater in the 1830s gave the most sophisticated scientific treatment of design in nature yet known.

Stirling, in his first lecture at Edinburgh, said the point was this: it would be foolish to "just give Paley over again." Everyone had heard it already; surely it would be "met by most of us with a shudder." This was an age, he said, in which natural theology meant something different from Paley. Now it meant rational theology. And rational theology was "equally known as the Metaphysic of God."[27] To this extent, Stirling was letting his European idealism show. As he covered the history of arguments for and against God, he saved the ultimate clash between a philosophical God and materialist evolution for last. Before that culmination, Stirling devoted three entire lectures to Hume's criticisms of proofs for God and then two more to Kant, who had called natural theology so much "transcendental shine."[28] Finally, Stirling's last four lectures took on the "superlative" argument against design, the argument made by Charles Darwin.

Eight years earlier Darwin had been buried in Westminster Abbey. Born an Englishman, Darwin truly got his professional start at Edinburgh. In 1825 he had arrived there to study medicine for two years. Still aimless in life, he found the lectures in chemistry, geology, and zoology "incredibly dull." Each summer he was "turning an idle sporting man." When his father realized he "did not like the thought of being a physician," Darwin said, "he proposed that I should become a clergyman." After four years of divinity at Cambridge, that vocation "died a natural death."[29] Darwin then was hired as naturalist on a survey vessel, the HMS *Beagle*.

At Cambridge, Darwin was enthusiastic about Paley's *Natural Theology*. It gave him "as much delight as did Euclid." Upon his return on the *Beagle,* however, he wondered whether nature, by a random accumulation of variety and then a lawlike "natural selection," did the designing itself. "The old argument of design in nature, as given by Paley, which formerly seemed to me so conclusive, fails, now that the law of natural selection has been discovered," Darwin said. Darwin made the choice clear enough, Stirling told his audience in the university's Natural History Lecture Hall. "If natural selection is true, design is false." Darwin "doubts the existence of God; he denies design."[30]

In the era of a great clash between philosophy and materialism, Stirling had not tried to use the likes of Paley to discredit Darwin's idea, but instead the likes of Kant and Hegel, who argued philosophically for an unseen and persistent order behind nature. This was the very idea, however, that Darwin had set out to destroy with the modest observational techniques of a naturalist on some equatorial islands in the Pacific.

If Darwin had lived to debate Stirling, neither of them would have had the best facts in hand. Yet Stirling set a pattern that lasted across a century of Gifford Lectures: he did not doubt evolution; he only challenged what seemed its anti-God forms. So it would be for nearly every Gifford lecturer except the avowed atheists. Lord Gifford himself had given a talk on Darwin, although it has not survived. He was doubtless amenable to theistic versions of evolution. In fact, Edinburgh had been a cauldron of evolutionary theory since 1844, when the local publisher Robert Chambers put out his evolutionist *Vestiges of the Natural History of Creation,* which became a public sensation.

For his part, Stirling attacked Darwin's antidesign on two well-known vulnerabilities. Natural selection needed immense time to produce so many creatures, and it needed a source of variation to work on. "What the devil determines each particular variation?" Darwin himself had asked.[31] And what about the time needed? Over at Glasgow University, the great man of science was William Thomson, better known as Lord Kelvin after the Crown raised him to the peerage in 1892 in recognition of his scientific achievements. Lord Kelvin said the earth was too young for evolution by accumulation. By calculating the cooling rate of a molten solar system, he had pegged the earth's age at 100 to 500 million years. Darwin said Kelvin's theory, which stood for forty years, was "one of my sorest troubles."[32]

Kelvin was sixty-six, and still nine years away from retirement, when Stirling wrapped up his Gifford Lectures. As young men, Kelvin and Stirling had been in the same mathematics class at the University of Glasgow, and they both lived to see the facts of science change—in favor of Darwin. In 1903 Pierre Curie found radioactivity, a source of continuous heat. Judged by radioactive decay, the earth was exceedingly ancient, today put at 4.5 billion years old. The next mystery was the source of biological variation, and while English-speaking naturalists followed Huxley's lead by searching for evolutionary fossils, German scientists such as the Darwinian biologist August Weismann began to figure out the nature of heredity. Years earlier, the obscure Austrian monk Gregor Mendel had found out how heredity—the combination of traits from two parents in an off-spring—worked in peas, but his paper had been lost. By the time it surfaced, Weismann had similarly concluded in 1904 that the variation in offspring was created by "an increase or decrease of living particles."[33] Both Mendel and Weismann had pointed toward "mutation" in a genetic code.

Although ignorant of future science, Stirling nevertheless raised questions that still haunt biology—and would haunt future Gifford Lectures as well. One was about Darwin's nebulous theorizing, using "mental pictures" and assertions such as "I cannot doubt." Most of all, Stirling objected to Darwin's faith in "chance" and "accident." To question Darwin's novelty, he divulged a German study on Darwin's grandfather, Erasmus Darwin. Erasmus had already theorized everything the grandson now said—except for the idea of chance. Erasmus had assumed God's mind in nature, and in this respect he may have come under some of the influences of idealistic philosophy. Whatever the case, Stirling quite naturally believed that Erasmus was, "in his reverence for design, much nearer to the truth than Charles."[34]

Charles Darwin had arrived at his seemingly more scientific conclusions on the HMS *Beagle* adventure, a story on which Stirling ended his Gifford Lectures. "Everyone who has at all approached this subject has heard of the Galapagos," he said. On those desolate islands, Darwin was inclined to see that one species had changed into another, for they surely did not arrive separately. Stirling was skeptical. If he stayed loyal to the ideas of Kant and Hegel, natural types had a permanence that diversified only in appearance. "There is development of difference," Stirling agreed with Darwin. "But a new species, made by man, or made by nature, has it been ever *proved?*"[35]

The time that Stirling had dedicated to Darwin and his theory of evolution was an echo of the wider debate in Europe. Although many of the great naturalists of the day rejected Darwinian theory, it was only partly for its empirical problems—such as proof, testability, and ability to explain the integrated complexity of living systems. Perhaps an even greater obstacle was the tendency of European naturalists to be idealists—as was the French biologist Georges Cuvier (d. 1832), a contemporary of Hegel—who believed there was Mind behind nature and certain types were inherent in the natural world. In his Giffords, Stirling had fought that battle as good as most, but in another generation the Darwinians, with their rejection of Mind and types, would sweep the field. For a time they would declare the end of philosophy, and an effective challenge to that would be a good time in coming.

What Stirling did not address was the lively debate among idealists themselves. Was every version of idealism part of the "German fog," or could some of them be of genuine help to Christian belief? That conundrum was handed down to a next generation of idealists who gave the Gifford Lectures. Pringle-Pattison, an Edinburgh native who would rise to the chair of metaphysics at the university, was a point man for this younger vanguard. For thirty years at Edinburgh University, he proved his knack for sorting out idealisms from one another, beginning with the distinctions between German and British forms.

In Germany "every university professor was said to carry a scheme of the universe in his pocket," he said. "No doubt this system-mongering was carried to a pitch of absurdity." Yet he also looked at Scotland's Reformation legacy, wherein a literal Bible was the answer to everything. Even those who "simply fell back upon the language of religion," Pringle-Pattison argued, still needed a philosophical whole, and that was why he favored idealism.[36] This melding of metaphysics and Calvinism, in fact, made Scottish Presbyterianism a mixture of critical *and* orthodox thinking, and outsiders could not decide whether Scottish thinkers were conservatives or liberals, or both.

When it came to idealism, the system building was done mainly in a circuit between Oxford and Glasgow. At the Glasgow end, the family name that towered over Absolute Idealism was the Cairds—two brothers who shaped Western thought and were the only siblings to give Gifford Lectures. The older by fifteen years was John Caird, a prominent Presbyterian minister who, in 1873, began his long reign as principal of Glasgow

University. In an age in which dogma seemed to be dying, he espoused a religion of rationality. He "preached Hegelianism from the pulpit."[37]

In his widely read *Introduction to the Philosophy of Religion,* John Caird argued that reason, not the "subjective caprice" of feeling, marks true religion. "Feeling is necessary in religion, but it is by the content or intelligent basis of a religion, and not by intensity of feeling that its character and worth are to be determined."[38] Such rationalist claims and Hegelian flavor guaranteed that ecclesiastical suspicions would arise about the orthodoxy of his Christian beliefs. Fortunately, preaching at a university was safer than in a local Scottish parish, especially around Glasgow, where a strict Calvinism was still preferred by many of the local church elders and clergy.

If John Caird had imbibed Kant or Hegel in any depth, it was probably served up by Edward Caird, his brother. Edward had studied at Oxford, a newly stoked hotbed of Absolute Idealism and Plato studies. After Edward returned to Glasgow to be professor of moral theology, he built up the reputation of British idealism. Both John and Edward Caird had tenures of more than thirty years, making their academic influence considerable.

They were credited with giving Scottish theology a more liberal and human face and arming it with rational arguments against skepticism. The brothers supported progressive politics and may have even plotted to radicalize the university. With some success, they supported admitting women, hiring young professors, offering more subjects, and providing divinity instruction for non-Presbyterians. They were also eager to establish a "science of religion" in their theological circles. By science they meant rational, but also universal—a kind of Hegelian law of religious consciousness that worked its way up through various world religions, with their many gods, to achieve the higher pinnacle of monotheism.

Nearing death, John Caird began his Gifford Lectures, *The Fundamental Ideas of Christianity,* in 1892, but his illness delayed their completion until 1896. His younger brother, Edward, had an easier go at the project. He was tapped in his prime, and his 1891 lectures were the second Gifford series to be given at St. Andrews. A master of German idealism, Edward had written a two-volume work on Kant and a popular book on Hegel. Unlike Stirling, Caird used his Gifford Lectures for system building. He gave the Giffords twice, in fact. The first set, *The Evolution of Religion,* was published just as he reached his career apex by becoming master of Balliol College, Oxford, his alma mater.

In writing on Kant, Caird said the sage of Königsberg was "only the first stage, though of course a necessary stage, in the transition of philosophy to higher forms of Idealism."[39] Hegel brought in those higher forms, but Caird wanted to go still further; he wanted to explain the origin of religion in a way that exceeded even Hegel's tentative effort. Caird found religious development wherever human experience became rational. "Every rational being as such is a religious being," he said. As rationality welled up dialectically, "difference continually increases, not at the expense of unity, but in such a way that the unity is also deepened."[40] He surveyed this universal process in Greek, Hebrew, and Christian religion, showing how reason moved from external deities (Greek), to an internal one (Hebrew), to a universal concept in Christian thought.

Such Absolute Idealism could easily look like pantheism, and while Caird retained a vague notion of immortality, he rejected miracles and mysticism. His religious philosophy was clearly not orthodox, said one historian, but it was "at least highly compatible with Christianity." In an age in which orthodox Christian belief had few philosophical friends, this was the gift of idealism. Because of that, Caird had his share of students and acolytes, such as his successor at Glasgow, Henry Jones, a Methodist minister who took idealist philosophy to its optimistic peak. When someone suggested that Jones's early sermons denied Christ's divinity, he said: "I do not deny the divinity of any man."[41] Such was the potential of the Absolute in the world. In giving his own Gifford Lectures in 1919, Jones urged his audience to see the world as "friendly and helpful, [and] God as an inspiring, and empowering, and guiding presence."[42]

Jones would succeed Caird as professor of moral philosophy at Glasgow and become a knight. But Sir Henry was not the only acolyte of Caird.

In America, during the Christmas season of 1898, another admirer was preparing to board a ship to cross the Atlantic. The Harvard philosopher Josiah Royce, America's leading exponent of Absolute Idealism, was about to be the first American to deliver Gifford Lectures. On his way to Aberdeen, he stopped at Oxford in hopes of meeting his two intellectual heroes, Caird and Francis H. Bradley, the chief British exponent of a secular Absolute Idealism. But both of the British gurus of the absolute were gone for the winter holiday, so Royce moved on north.

On his journey to Aberdeen by train, he passed first through Edinburgh, then traversed the great rail bridges over the Firth of Forth and next the Firth of Tay, the second watery obstacle to land travel on the eastern coast. There, he encountered another bit of Scottish history. Halfway between Edinburgh and Aberdeen, the great Firth of Tay bridge had been Scotland's first celebrated engineering feat. Built by laborers from around the world, it was completed in 1878. The next year, however, the great Tay bridge became a "national calamity."[43] A storm plunged it and a train with eighty passengers into the icy waters. Royce crossed over the rebuilt giant, around which the legend of death and destruction, and the bridge engineer's suicide, still swirled.

The philosophy of idealism tried to make sense of such tragedies, "errors," or evils in a world still shy of the Absolute. The best human ideas and hopes were always bound for "practical defeat," said Royce. But in Absolute Idealism there was no blame to be cast, for what seemed like a fragmented world was actually a process of "the Many," both good and bad, working toward "the One" in the eternal Absolute. For some idealists, suffering was ultimately unreal. For others, people never suffered alone for their "sufferings are God's sufferings, not his external work," as Royce once said. These were not always armchair speculations. When the idealist Pringle-Pattison dedicated his published Giffords, it was to his son, "who gave his life willingly on the Somme," a river in France, in the First World War. Royce knew suffering as well. In anguish he committed his eldest son to a mental asylum for the rest of his life.[44]

Growing up in northern California, Royce seemed fated for either philosophy or religion. His parents, English immigrants who left New York State for California in the early 1800s, were evangelicals, and while the mother founded a school the father spread religious tracts as he hawked wares. Among his peers, Royce stood out for his small body, large head, and red hair. He often defended himself by precocious argument. His philosophical bent was influenced by the parents' Bible literalism, the western community of his childhood, and his own striving to achieve. After attending the university in Berkeley and studying in Germany, he went to Johns Hopkins University in Baltimore, where Hegel and science cracked heads. His doctorate in philosophy was one of the first awarded in the United States.

When the Gifford invitation arrived in 1898, Royce was chairman of the philosophy department at Harvard, and by then Anglo-American idealism had been around for more than thirty years. For some, the chief originality of Royce's work in idealism was to argue that, by the fact of "error," the Absolute was proved. Truth is defined, he said in his Giffords, "by its contrast with the error that it excludes." Similarly, the "internal meaning" of every human idea—and Royce argued that ideas are willful plans of action—is fulfilled in the "divine life" of the Absolute, although the idea might not be completely embodied on earth. Thus, human beings suffered a kind of "finite bondage," yet they still could experience a wholeness in relation to the infinite Absolute and by turning to others. As Royce said, "Our fellows furnish us the constantly needed supplement to our fragmentary meanings."[45]

Idealists were notorious for avoiding everyday examples. Royce was no different, although he wrote a history and a novel about California, trying to make concrete the drama of individualism in a community, or how society shapes a person. In his philosophical writings, he also spoke of how children began life knowing its wholeness, but how experience— encounter with the finite "appearances" of the world—led to adult alienation. Thinking adults needed a philosophy to integrate their world once again, Royce obviously believed. The adults who were party to the Firth of Tay bridge tragedy, for example, could take comfort in the immortality usually held by idealism, but also in the belief that the very ideas of the victims were not in vain. The victims had sought to fulfill the "internal meaning" of their ideas—in this case, to build a perfect bridge and to take a completely satisfying train trip—and even though these ideas failed in life, they found an eternal fulfillment in the Absolute.

When Royce arrived in Aberdeen to deliver his lectures in January of 1899, he set out to put his idealistic philosophy into a complete system. Titled *The World and the Individual,* the lectures were published in two volumes. In the first series, Royce wanted to establish why "constructive idealism" was the best philosophical system of them all. To do that, he surveyed four "conceptions of being" in history. The first two were mysticism and realism, which he found too abstract in their reduction of reality to either momentary experience (mysticism) or independent objects (realism). He called the third conception of being critical rationalism, which judged reality by "particular observations" only. For Royce, the fourth option of constructive idealism was superior. It viewed reality as a rational whole that was nevertheless dynamic, optimistic, and moral. It explained the imperfections of life, but held out the "eternal triumph of the good over the evil" in the Absolute. Idealism, he summarized, sought "the unity of finite and of infinite, of temporal dependence and of eternal significance, of the World and all its Individuals, of the One and the Many, of God and Man."[46]

Royce delivered his second set of Giffords in Aberdeen the next winter, January 1900, and emphasized the ethical implications of Absolute Idealism, which have probably had more staying power than the abstract metaphysics. As evidence of this, Royce is remembered as "America's philosopher of community," not as a dialectical theorist. He coined the term *beloved community,* a phrase employed decades later by the civil rights leader Martin Luther King, Jr., who wanted to give a Christian *and* philosophical aura to the civil rights struggle.[47] He expanded his ethical arguments in *The Philosophy of Loyalty* (1907) and *The Problem of Christianity* (1913). Loyalty was the highest ethic, he said, because it was "the willing and thoroughgoing devotion of a person to a cause," the cause being "something that unifies many human lives in one."[48] Loyalty was not conformity but a form of dedication and self-realization. His book on Christianity, meanwhile, warned of a proverbial lost generation in the early twentieth century, individuals alienated from the cares of all humanity, a recipe for lost religious faith, global wars, ethnic strife, and political hypocrisy.

For all the emphasis that Anglo-American idealists put on community, in Royce's day they began to split apart, creating the warring factions of *absolute* idealists and *personal* idealists. The personal idealists wanted to save the individual personality from being consumed by the voracious Absolute, philosophically speaking. While absolute idealism did not always need a God, much less a personal God, personal idealism made such a Being paramount in the philosophical scheme. The personal idealists had their rebel leader in the same Pringle-Pattison who had led the young-turk idealists in the 1883 *Essays* against the materialists.

Royce would also give up his loyalty to absolute idealism and its European proponents and lean toward personal idealism, which was more amenable to Christian theism *and* democracy in America, where the creed of one man, one vote vied mightily with the conformity often suggested by absolute idealism. Like all idealists, Royce had struggled for rational consistency. Reconciling individuals and the Absolute was always a troublesome task. Bradley, the dean of absolute idealists at Oxford, had simplified the problem by saying that "self-realization" came as an individual was absorbed into the Absolute. For the personal idealist, the Absolute was not a void that devoured people, even philosophically. In a supplement to his Giffords, Royce took exception to the way Bradley rejected "every more detailed effort to interpret our own life in its relations to the Absolute."[49] In the long run, American individualism would naturally dictate that Royce become a personal idealist, although he never gave up on the Absolute as a whole.

Despite this clash over whether to emphasize the individual or the whole, philosophical idealism was always about community, whether as a web of individuals, a whole society, or the functioning of a state. It was rooted in the Hegelian ideas of how subjects are changed by objects, and of how the whole precedes the individual. Every individual is part of and shaped by some social whole. Such ideas sent chills down libertarian spines, especially when metaphysics was used to justify the modern nation-state.

Before long, the influence of idealism in British culture stirred a fiery debate, triggered as much by politics and patriotism as anything. Idealism had been born in Hegel's homeland of Germany, a nation whose political structure was taking a different path from English liberalism. More precisely, at the turn of the century, Germany had crossed Britain by giving military aid to the Dutch in South Africa, where Holland and England clashed in the colonial Boer War (1899–1902). In such an age of nationalism, German ideas became anathema, especially to British liberals such as Leonard T. Hobhouse, who wrote for the *Manchester Guardian*. "The Rhine had flowed into the Thames," he warned in 1904. He blamed idealists from Thomas Carlyle to Stirling, from the Cairds to the Oxford dons, for importing the foreign ideas. "The stream of German idealism has been diffused over the academical world of Great Britain," he said.[50]

In response, Glasgow's famed moral philosopher Sir Henry Jones, the Caird acolyte, agreed with the influence but not the dire warnings. Idealist writers had indeed engrossed the public. "Almost more important than their writings is the fact that they have occupied philosophical chairs in almost every university in the kingdom," Jones said in a 1907 essay, "Idealism and Politics." Taken rightly, idealism had aspired to unite souls with God and individuals with community, and in attacking such ideas, Jones said, Hobhouse had "gravely wronged a great cause."[51] When Hegel said the World Spirit would manifest itself in a state, he did not forecast a Prussian military power going to war. But by 1914 that was how it looked. When Hobhouse published a second criticism of idealism, *The Metaphysical Theory of the State,* in 1918, it was a postmortem on the calamity of turning states, rulers, and armies into divine incarnations.[52]

British idealists had always backed government-driven reform. They sometimes joined the Labour party. Others supported a class system. Yet the common vision was about the order and altruism of the whole. In such a system, the antidote to conformity was authentic morality rising up from each person. If it did not arise, then education and "persuasion" were fair game. Idealism was clearly more collectivist in Europe than in America, where Ralph Waldo Emerson's idealism, given the name Transcendentalism in 1842, took individualism to extremes. Yet the community-minded spirit also sank roots in the United States. Out in the heartland in 1867, the St. Louis Hegelians launched the socially conscious *Journal of Speculative Philosophy.* They saw idealism as a unifying social philosophy. The nation was deeply divided by region and sect. So here was a balm for an age of doubt, when science confronted religious dogmas. The journal's founder became U.S. commissioner of education under President Benjamin Harrison in 1889, the dawning age of the Giffords.

In Germany idealism became a war zone of dialectics, systems, and vying artistic passions. Disciples of Hegel would eventually denounce him, and the failed 1848 social revolutions, in which kings and princes beat philosophers underground, soured the speculative agenda. Germany became a platform of natural science; the philosophers turned into scientific materialists, literary atheists, and moral pessimists. British idealism ran its course, splitting between idealists of the absolute and personal varieties, being scorched by the First World War, but remaining a chief alternative to crass scientific materialism. One historian said idealism nevertheless remained "the dominant British philosophy" until the 1940s, when the Second World War finally turned it to cinders.[53]

During the philosophical roller-coaster ride that began with David Hume and Thomas Reid and ended with Kantian and Hegelian dialectics, many of the schools declared that they had captured "the spirit of modern philosophy." Yet each in its own way also contributed to the demise of the philosophical quest, especially when it came to the question of God. Common sense softened skepticism but claimed no philosophical proof of God. Kant in turn ruled out reason's grasp of a Creator, an ultimate reality known only by a person's faith. Then the idealists delivered a final arrangement, like a gift-wrapped package, that absorbed God and all things into an absolute rational system. "In the main the work has been done," said the Absolute Idealist Bernard Bosanquet in his 1911 Giffords.[54] What was left to philosophize about?

As future Gifford Lectures would show, the quest for a system that accounted for God was not over, despite the receding of idealism. What took its place was the search for an empirical, or scientific, system with God at its heart. Although this surely smacked of philosophy and metaphysics as much as the Hegelians did, the proponents of an empirical metaphysics argued that they began with the scientific facts of the world, not philosophy. In a word, they were realists, not idealists.

Even though idealistic philosophy met its demise, traditional philosophical systems would always experience brief revivals in the Gifford Lectures, whether through reassessments of Plato and Thomas Aquinas or William Paley and Thomas Reid. Moreover, if any philosophical method endured into the twentieth century, it was use of the dialectic. The dialectic was ancient, explained by Socrates long ago but now dressed up by the German idealists as the conflict of thesis and antithesis, which produced a synthesis—but also another dialectical conflict ad infinitum.

Although the philosophical quest for God seemed to exhaust itself in the early Gifford era, the dialectic proved more resilient, as if a sleeping giant. When it emerged later in the twentieth century, dialectical thinking vastly influenced the God-talk of natural theologians and philosophers.

Natural science, compared to this apparent retirement of philosophy after idealism, was feeling that its work had just begun at the turn of the twentieth century. For some, science had arrived as the ultimate substitute for religion and philosophy. For others, and probably the vast majority of the citizenry in the West, the musings of scientists and philosophers did not really matter. The masses viewed the world for what it was materially and religiously. They preferred plain talk about plain things, an art of rhetoric honed not by philosophers but by preachers, businessmen, generals—and especially politicians.

No Gifford Lectures represented this stance better than those given by Arthur J. Balfour, the former British prime minister and native of Scotland. When the Conservative politician lectured at Glasgow in early 1914, he was escorted to Bute Hall in a procession of pomp and academic raiment due a philosopher king. Yet in the next moment Balfour declared his loyalty to "the plain man." While "the metaphysician wants to rethink the universe; the plain man does not," he said.[55] An ordinary world of things, morals, physical laws, beauty, and hope led plain men to believe in God.

Still, the sciences lay in waiting, and thus begins the second act of the Gifford legacy, when anthropology, psychology, physics, sociology, and history were brought to bear on beliefs about the existence of God. While the natural sciences might have seemed a refreshing alternative after so much idealist speculation, with its political afterglow, they too brought challenges to conventional religious belief. The objects of science could usually be grabbed on to, but the implications did not stop there. Science was looking for natural and universal laws behind physical objects, and as will be seen, when impersonal laws explain everything, the relevance of God can be put in question. The sciences rushed out of the Victorian era and into the twentieth century shoulder to shoulder, but for the Gifford story, the leader of the pack seemed to be anthropology, the "science of man."

THREE

THE "SAVAGE HIGH GODS"

Anthropology

TWO DECADES BEFORE the Gifford Lectures commenced, the scientific search for the origins of humankind's religious belief had begun in Victorian earnest. In 1870 the German scholar Friedrich Max Müller stood before a London audience to declare a new field of study, "the science of religion." A year later, the Englishman and field researcher Edward B. Tylor published his groundbreaking *Primitive Culture,* a book that marked the beginning of British anthropology, the "science of man."

Though Müller and Tylor held different theories about primitive man's first inkling of a god, they were well suited to open the Gifford Lectures, Müller at Glasgow and Tylor at Aberdeen. Their nascent field of anthropology had staked out two remarkable territories. The first was the mythological past, found in ancient literature and artifacts, a past preserved on papyrus or stone or in objects such as fetishes, amulets, dolls, masks, and talismans. The second territory was far more immediate, for it was the life of primitive tribes still on the earth, "savage" peoples encountered by British traders, officials, missionaries, and ethnologists—the field-workers of anthropology—who traveled the globe.

In the nineteenth century, it seemed that the world had disgorged its ancient past. In Europe, researchers unearthed Neanderthal and Cro-Magnon man, ancient cave paintings turned up in Spain, and news of primitive customs was brought back from Oceania, Africa, and the Americas. As ancient Babylon and Troy were excavated, and Paul Gauguin returned from Tahiti with paintings of romantic primitives, the British read headlines about the 1870 "rescue" of the evangelical missionary doctor David Livingston in Africa. The Scottish doctor's letters home, in fact, had added to the new anthropological cache of tribal rites and ways. By the end of the nineteenth century, the mythologies known so well to Europe—those of ancient Greece, Rome, and the Aryans—were being confounded by new stories that leapt from ancient Indian texts, Middle Eastern inscriptions, and Egyptian hieroglyphics.

The human past had been found on a grand scale, and in ways it was a troubling past, eerily savage and pagan. "How to explain the cruelties and irrationalities enshrined in mythology was one of the great puzzles of scholarship in the 1870–1910 period," said anthropologist Mary Douglas, a later Gifford lecturer.[1] It was now a question for science, and to claim a scientific explanation for that puzzle, anthropologists sought ironclad laws, finally arriving at the "comparative method" to find them. What was the common core in myths, they asked, and what was peripheral? What were the earliest, and thus original, human beliefs and behaviors?

No two people began to perfect this art of comparison in early anthropology more than Müller and Tylor, each in his own way. Müller was more properly a philologist, or expert on languages, and a historian of religion when it came to discovering the ancient past. He was a world authority on Sanskrit, the basis of Hindu writings, and he studied its remarkable links to the broader Indo-European languages, which eventually stretched from Asia across Germany and on to Ireland.

Tylor was trained in a different mold, a Londoner and Quaker by background. After a short business career, he traveled widely, and in 1856 visited Mexico in the company of an anthropologist who introduced him to the topic. Having collected some artifacts and stories of ancient culture in America, Tylor began his lifelong career with his first study, *Mexico and the Mexicans* (1861). He wrote more extensively on early civilization, and by the time of *Primitive Culture* (1871) his theory on the development of religion dominated Western anthropology. But to claim such a status, it had to displace an older theory, and that was the interpretation represented by Müller.

As the sciences became major contributors to the Gifford Lectures, act two in the search for "knowledge of God" would begin. Every science—including psychology, physics, sociology, and historical criticism—would produce vying theories, and it was no different for anthropology. Like all the sciences, moreover, anthropology would raise prickly new questions for Christian belief. Yet anthropology still had a challenge of its own. It was a somewhat slippery "science," based as it so often was on the vague meanings of ancient lore and objects. Although Victorian science, in fact, looked on anthropology with suspicion, it was not to be derailed. The anthropologists dreamed of the day when their field would stand shoulder to shoulder with physics, chemistry, and astronomy.

In addition to Müller and Tylor, who both harbored this scientific dream, two of their successors in particular carried on the quest for scientific legitimacy—and they became presenters of Gifford Lectures as well. They were Andrew Lang, who was Tylor's favorite student, and James G. Frazer, the younger heir to this beginning era of British anthropology. With his literary flair, Lang explained the obstacles anthropology still faced. "To the scientific mind, anthropologists or ethnologists were a horde who darkly muttered of serpent worship," he said in the first Gifford Lectures at St. Andrews University, titled *The Making of Religion*. "Anthropologists were said to gloat over dirty rites of dirty savages, and to seek reason where there was none."[2]

Some time later, however, Frazer would herald not only the scientific ranking of the new science of man, but also its popular appeal. In the notes for a lecture he gave in 1922, Frazer wrote: "Anthropology a modern science."[3] Indeed, Frazer became the first anthropologist inducted into the Royal Academy—filled as it was with chemists, physicists, and astronomers—and the first to teach the subject at Cambridge

University. Frazer would also complete the multivolume work *The Golden Bough,* probably the greatest collection of ancient myths ever compiled.

All four men—Müller, Tylor, Lang, and Frazer—would play important roles in the early years of anthropology, and all four would give the Gifford Lectures because of their interest in the origins of religion. Not only was Müller the oldest among them (1823–1900), but he represented the older school of thought about how religious beliefs had formed and spread. It was a German-founded school called the nature mythologists, and as the old school, it was a promising target for up-and-coming rivals. Indeed, by challenging the nature mythologists with a new theory about the origins of religion, Tylor became the veritable father of British anthropology.

In Germany, the nature mythologists got their name by holding the theory that the earliest religious belief arose from human awe at natural phenomena: the sky above, the sun and moon, the day and night, or even the lightning and storms of the seasons. The nature mythologists agreed that all mythological figures, including those in religion, personified natural objects such as the sun, moon, stars, and planets, but also dramatic weather phenomena and seasonal cycles. Ancient Persians had a sun god in Mithras, the Egyptians a sky god in Serapis, and the Babylonians' chief gods were seven planets. In ancient India, Indra brought thunder and storms, as did Jupiter and Thor elsewhere, and the Phrygian god of vegetation was Attis. Looking back on these myths from the heart of liberal Protestantism in nineteenth-century Germany, this group of scholars became the first historians of comparative religion. They were also on a romantic search: to find how religious consciousness arose in every human culture.

As part of the nature mythology movement, Müller was more precisely a "solar-mythologist." His studies persuaded him that the sun was the primary natural object in forming primitive religion, especially the idea of a single and all-powerful being in charge of the universe. As a linguist, however, Müller gathered his fame for an even greater theory, a process of the spread of religious belief that he called the "disease of language." In effect, he said, primitive man tried to articulate his feelings about godlike natural phenomena and came up with images and stories not unlike what human beings experience. The dawn, personified in various mythological tales, was one of Müller's favorite examples. Storms, earthquakes, and other solar movements all became stories about the anger, sleeping, lovemaking, or waking of the gods. "These mythological expressions are by no means restricted to religious ideas," Müller said in his Gifford Lectures. "There is a period in the growth of language in which everything may or must assume mythological expression."[4]

Yet these primitive ideas were inarticulate and based on rudimentary words and images. These rudiments spread and evolved, much as the contagion of a disease; thus the "disease of language" idea was born. The proof was in the ancient languages themselves, Müller argued. In Sanskrit, scholars had found similarities to Latin, and as they compared more ancient tongues, the entire fabric of Indo-European languages seemed to be connected by threads of words and meanings. The connections, of course, suggested that there had once existed a single primitive language, and therefore a very early and original thought-world of religion. In his Giffords, Müller relished giving examples of the linguistic connections. When God had identified himself to Moses on Sinai as "I am that I am," for example, it was a name also found in Hinduism, where one of twenty Vedic titles for God is "eis Ahmi yet ahmi."[5]

Like all thinkers of the time, Müller had imbibed the evolutionary mood of science. He also viewed human history as a progression, a gathering up of higher and greater glimpses of God. Well before his Gifford talks he had lectured "On Missions" at Westminster Abbey to help Protestants understand the origins of religious consciousness in other lands, a consciousness that he also agreed had reached its peak in Christianity. "The human mind, such as it is, and unassisted by any miracles except the eternal miracles of nature, did arrive at some of the fundamental doctrines of our own religion," he later told his rapt Gifford audience in Glasgow, some of whom attacked his lectures as "subversive of the Christian faith."[6]

Like perhaps Gifford himself, Müller was part of the project to salvage as much Christian belief as possible in the face of the new skepticism, natural sciences, and intrusions of the non-Western world. Even though some of the first anthropologists operated out of a Christian background, as did Müller, their probing into the savage origins of belief, myth, and custom would often boomerang and result in questioning of the primitive origins of Christian culture itself. Despite those discomforts, their exploits prepared the public for an increasingly exotic world outside church walls. Try as it might, however, this new "science" of religion invariably upset as much conventional belief as it reinforced.

Two things were upset the most: the story of the world found in the Bible and the claim of some Christian believers that their faith and doctrines were based on an exclusive revealed truth never before seen on earth. From that point of view, beliefs not found in the Bible or the church were simply false. "Everything which could not be justified by reference to one or the other of the two Testaments was untrue; it was a 'fable,' " observed Mircea Eliade, the historian of religion.[7] Yet now, in the Victorian age, "fables" of a bewildering variety and number poured into the West. And it was hard not to notice some similarities to Christian belief. The most positive assessment was that God had given the world sundry myths to prepare it for a final Christian revelation. Anthropologists, though, often argued a more troubling case. They suggested that Christian belief itself was a descendant of primitive superstitions.

Although Müller did not take such a heavy-handed approach, it was not long in coming. It came also with the theory that finally pushed aside the nature mythologists, and that was the anthropological theory of Edward B. Tylor. It was a close-quarter battle, though never personal, for Tylor taught at Oxford, the very school at which Müller too had made his scholarly residence after immigrating to Britain. Tylor was a dashing figure in his day. Tall, good-looking, and well built, he also sported a long beard. After his travels to Mexico and the publication of his brilliant *Primitive Culture* in 1871, Tylor settled down as curator of Oxford's museum. He became the cataloger par excellence, perfecting what one student called "the method of the dragnet," which is "to gather first and sift afterwards."[8] The mass collecting gave anthropology the semblance of a science, and Tylor's concise and dispassionate way of explaining the data ensured that he soon dominated the field. What came with that dominance, of course, was Tylor's new theory about the origin of religion.

Drawing on the evolutionary themes of his day, Tylor argued that the earliest human mind developed by gradual steps of rational interpretation of the world. This gradualism, it seemed, was built into the physical laws of human learning. The result was that every human culture, no matter how separated, had gone through the same stages of rational development. By taking this scientific view of the seeming cacophony of cultural artifacts—flints, fetishes, masks, amulets, and myths—the anthropologist could discern the upward pattern, a "development" of human culture that seemed a law of nature itself. "Development depends on similar operations of mind," Tylor said in his Gifford Lectures.[9]

For all the early anthropologists, the object of interest was "natural religion," a different subject than the idea of religion being revealed by a particular deity. Natural religion was a product of human thought. It arose by human interaction with the mysteries and powers of nature. Yet if the nature mythologists said that God was conceived, for example, by an experience with the sun, Tylor had a contrary finding: he said that the first idea about gods arose from primitive man seeing spirit forces in every root and branch of nature, a force that Tylor described with the Latin term for spirit, *anima*. Hence, Tylor became founder of the "animistic theory" of the origins of religion. Already, English explorers of the Pacific had brought back the tribal concept of "mana" as a spirit power, but now Tylor latinized the idea and made it a science.

Tylor tried to show the clear steps by which "savage philosophers" first conceived of a spirit force, and it had nothing to do with the sky. Far more significantly, Tylor said, the savage philosopher had witnessed the phenomenon of human death and experienced phantoms in dreams. From these, the primitive mind concluded that a spirit force existed behind all things. Then, the savage philosopher took the next logical step: he tried to control the *anima* for human benefit and thus invented magic and religion. According to the evolutionary view, nature moved from simple to complex, and so it must have been with human development. In the human record, the more bestial and ugly remnants came earlier and the more refined later. First came the "savage" food gatherers and later the "barbarian" farmers. The lower and early mind was content with animism, but the later mind rose to polytheism. Next, when social hierarchies developed, so did the idea of a top god and, later still, a belief in monotheism.

Fortunately for the anthropologist, remnants of this animist past were still around in every culture in the form of "survivals." Even as the mind developed, remnants of its early beliefs stayed around, putting the most primitive notions alongside some of the most advanced. The theory of survivals captured the imagination of anthropologists for a generation. Looking for "survivals" was a serious science, and Tylor was its master. By amassing evidence of animism, he argued that it was the ultimate origin of religion, tracing it to the root of such practices as fetishism, ancestor worship, and the adoration of nature. For the willing Christian theologian, Tylor's theory of survivals also offered a helpful filter: in the grab bag of human religion, the theologian could evaluate which "survivals" were primitive beliefs that were still sound, which were crude beliefs that were sufficiently updated, and which were simply barbarous beliefs, retained by the dead hand of custom.

By the time Tylor had published *Primitive Culture,* his main argument and a catalog of evidence were in place. When, in the harsh winter of December 1889, he traveled from Oxford to Aberdeen University to deliver his Gifford Lectures, he went "over much the same ground as *Primitive Culture,*" a student said, and was remembered in particular for showing the audience a lot of his artifacts, the flints, fetishes, masks, tools, and amulets that were beginning to turn anthropology into a kind of museum and laboratory science.[10]

For the "animistic theory" of religion to gain predominance in England, it had to push aside the influence of the nature mythologists, people such as Müller. Although Tylor was not involved in expediting such a palace coup personally, his student Andrew Lang relished taking the anthropological battle to the enemy. Yet Lang's battles in the field would not end with the apparent defeat of the nature mythologists. After that, he changed his own mind, and then turned his considerable and acerbic wit on his old allies, the animists. When that day came, in fact, Andrew Lang and James G. Frazer went down in history as the final combatants over the topic of the savage origins of Christian belief itself—a topic that Lang took up in his Gifford Lectures.

After studying at St. Andrews in his native Scotland, Lang went to Oxford and became a literary virtuoso. His essays filled journals and newspapers, and he produced book after book, touching on everything from "the ballads and lyrics of old France," to the *Odyssey,* "rhymes a la mode," and fairy tales and myths. As Tylor's best student, Lang used the animist theory to launch a campaign on the wrongheaded nature mythologists. The first prong of his attack came in 1884 with the book *Custom and Myth,* which argued the weakness of the linguistic and solar theories. Barely taking a breath, he followed with a second prong in 1887 in a work titled *Myth, Ritual and Religion.*

On the eve of the latter book, however, troubling new evidence was trickling in from abroad. The fact, laid out for the first time at an 1884 meeting of the Anthropological Institute over which Tylor himself had presided, was that ethnologists had found monotheism, the worship of a single "high god," among southeast Australia's aborigines, the most primitive group. This god was not a ghost of a chieftain or spirit in a rock, but, as some aborigines put it, a being who "can go everywhere, and do everything." Animists knew the evidence as much as anyone, but as Lang would argue, their theory was such "a widely preached scientific conclusion" that anthropology had cynically "simplified her problem by neglecting or ignoring her facts."[11]

When Lang arrived in St. Andrews to deliver his Gifford Lectures, he was a changed man. In the world of anthropology, he had stirred a great intellectual scandal by defecting from the animist school of thought. What Lang had turned to in opposition to both nature mythology and animism was hardly a new theory: he argued that the first belief of ancient humanity was in a single high God, a god of power and goodness and also of ethics. Part of the scandal, of course, was that Lang seemed to be talking about the biblical God. Although Lang delivered his Gifford Lectures somewhat obscurely in 1889, like a faint echo within the stone walls of ancient St. Andrews, their published version, *The Making of Religion,* was called "epoch-making" for identifying a primitive high god and characterized as the "crucial break" from the evolutionist orthodoxy of animism.[12]

As an animist insider, Lang used his erudition to ask the right questions. Did primitive man, for example, have to derive the spirit idea from death and dreams? He could just as well have had a psychic experience, much studied then in Victorian police departments, or arrived at the "argument from design"—the world looks made, so perhaps there is a Maker. "The conception of God, then, need not be evolved out of reflections on dreams and 'ghosts,' " Lang said.[13] In fact, many primitive views lacked the very idea of physical death, so ghosts and ancestor worship were not even a tool to devise "savage high gods."

In short, Lang said, "Certain low savages are as monotheistic as some Christians." These savages, moreover, did not fearfully propitiate the god, as animism had theorized. They acted in awe and appreciation. They had even drawn up high-god ethics, a fact denied by animism: "Anthropology had simplified her task by ignoring that essential feature, *the prevalent alliance of ethics with religion,* in the creed of the lowest and least developed races."[14]

For this, Lang was hardly propitiated in return. Mostly ignored, he was also attacked as a closet Christian and for dragging out the old degeneration theory, which he had. Lang identified "two moods" in history, one religious and the other mythological. If religion came first, myths arose tangentially. He argued that religion began "in a kind of Theism, which is then superseded, in some degree, or even corrupted, by Animism in all its varieties." Later in history "mythology submerged religion." Why? "Man being what he is," Lang said, "man was certain to 'go a whoring' after practically useful ghosts, ghost-gods, and fetishes which he could keep in his wallet or medicine bag."[15]

Anthropologists should give far more credit to the savages than animism had once allowed, Lang seemed to say, for "if savages blundered (if you please) into belief in God and the Soul, however obscurely envisaged, these beliefs were not therefore necessarily and essentially false." But he rested his case against animism on the best available facts, which the Gifford Lectures had allowed him to compile for posterity: "We found a relatively Supreme Being, a Creator, sanctioning morality, and unpropitiated by sacrifice, among people who go in dread of ghosts and wizards, but do not always worship ancestors."[16]

By now the outlines of a successful Victorian anthropologist were becoming clear: a master of ancient languages such as Müller; a skilled cataloger like Tylor; or a literary savant like Lang. When James G. Frazer came on the scene, he seemed to embody all three of those talents. In the end, that made Frazer the favorite anthropologist of the educated public. Frazer took Tylor's theory of the rational evolution of the human response to animistic powers and expanded it into a sweeping understanding of the human story: the rise of man from magic, to religion, and finally to science. He would back up his claims, moreover, with the most voluminous collection of ancient acts and faith yet compiled. Titled *The Golden Bough,* it captured the Victorian imagination with evocative tales of ancient savage life, idyllic hideaways, secret love affairs, and human sacrifice.

Frazer was just over five feet tall, a shy, scholarly Scotsman. He spent most of his life at Cambridge University, living in an apartment whose floors bowed with the sheer weight of full bookshelves. Whereas someone such as Tylor had taken at least one rugged and youthful trip to Mexico, and had written on it as an ethnologist, Frazer's wildest adventure was a tour of modern Greek cities. Yet in the Scottish, Calvinist home of his upbringing, Frazer's imagination had wandered around both history and the globe. The family read Bible stories and heard accounts of maternal relatives who had trodden the exotic soils of the Caribbean and Tibet.

We have an 1883 oil painting of Frazer when, after studies in his native Scotland, he had settled in Cambridge as an expert in the classics of ancient Greece, Rome, and northern Europe. The year of the painting, in fact, marked his conversion to anthropology. A seated Frazer, age twenty-nine, reads an open newspaper in the painting, his thin face framed by dark sideburns and a handlebar mustache, and his "piercing, sapphire blue" eyes gazing out behind spectacles. During an Easter walking tour in Spain that year, a colleague handed Frazer a copy of *Primitive Culture* to read. Tylor's magnum opus inflamed Frazer's imagination with what the comparative method of anthropology could do for elucidating the strange past of the ancient Mediterranean.

As expected of him, Frazer had embraced the rationalist Enlightenment project of debunking ancient mythology. When he found in Tylor's work the idea that the Catholic Mass arose from magic, a kind of primitive "survival," for example, the broader implications gripped Frazer's imagination. He thought of all the strange rituals of the classical Greek and Roman world. With "survivals" of animism in mind, Frazer began a great search, as his biographer offered, for "those aspects of antiquity that resembled the behavior of lesser breeds."[17]

Frazer's strict Calvinist upbringing made him a perfect rebel, a scholar who clearly reveled in finding out about primitive appetites and superstitions, and displaying them with sarcastic humor. "He was always fascinated and delighted by the 'misbehavior' of his savages," according to one study of Frazer's life and work, an idea also confirmed by Frazer's most famous student, Bronislaw Malinowski. Frazer enjoyed the primitives' "pranks and pleasures" with "an almost maternal concern," as Malinowski relates the story.[18]

Yet classics remained Frazer's forte, and that was why in 1886, during his sixth year as a Cambridge fellow, his first major project was to translate *Description of Greece,* the monumental travelogue of Pausanias, an Asia Minor physician. At Cambridge also, he would come under the fateful influence of William Robertson Smith, the young Scottish scholar of the Old Testament. Smith was holed up in Cambridge teaching Arabic after his articles on "Angels" and "Bible" in the *Encyclopaedia Britannica* had prompted a heresy investigation that finally ousted him from the Free Church College faculty in Aberdeen.

Thanks to two of the most fascinating discoveries of the new anthropological age—the primitive concepts of "taboo" and "totem"—the work of Smith and Frazer would intersect closely and finally have a dramatic impact on Western thought.

The idea of taboo, brought back from the Tonga Islands by the adventurous Captain Cook, stood for the tribal rule against touching what is sacred or profane. Ever after, the taboo was the forbidden thing, whatever its modern Western incarnation, and this would range from crossing between social classes, to the topic of sex, to the sacred symbols of nations, religions, or clubs. The idea of totem, meanwhile, had been found among the Algonquin tribes of North America and brought back for English inspection. The totem was an animal figure that contained a higher power and that represented a family or tribe. As Smith himself would write in 1886, "Totemism is a subject of growing importance, daily mentioned in magazines and papers."[19] He lobbied for its inclusion in the *Britannica.*

Like the idea of taboo, totem also became a popular Western notion.

As every athletic team in the West would come to understand, the totem embodied the very life and power of the group, and woe to the person who defiled or denigrated such an important symbol. The taboo and totem would also take on overwhelming significance in the new and exotic field of psychiatry, which was inaugurated as a "science" by the Austrian doctor Sigmund Freud. Based on his belief that the taboo and totem were basic to human origins, Freud developed an entire theory of the human psyche, families, and religion that hinged on the fears and pleasures surrounding these magical objects. In 1913 Freud put down his bizarre theory, which took science by storm, in the earthshaking little book *Totem and Taboo*.

Long before those rumblings, however, William Robertson Smith, a Protestant scholar of ancient biblical literature, had hoped that taboo and totem would explain activities of the ancient Hebrew people. One of their propensities, for example, was animal sacrifice, followed by the priestly act of eating the ritually slaughtered animal. More significant, Smith felt that the totem idea helped make sense of the "scape goat" of Hebrew scripture, an animal that carried the people's sins into the oblivion of the wilderness. Such totemized animals, he declared in his 1890 book, *The Religion of the Semites*, stood for no less than a god dying for the sake of his people.

A few years before this, in 1887, Smith had become well enough established at Cambridge to ascend to the editorship of the *Britannica*. In that post, he commissioned Frazer to write an entry on "Pericles," the Athenian statesman. He next urged Frazer to write entirely new *Britannica* entries on "Taboo" and "Totem." Frazer spent seven months on totem alone. He also began sending surveys to travelers, foreign officers, and missionaries, seeking reports of customs seen abroad. What Frazer produced was the first definitive monograph on totem, bristling with examples. A short abstract went into the *Britannica*, and the rest became his first book, *Totemism*.

With this new kind of anthropological radar turned on, Frazer began to see survivals everywhere. They were in Pausanias, whom he laboriously translated, and throughout the classical literature. At some point, a fantastic theory came to Frazer's mind; it became his life's work, and his

strange legacy. Although it took some time to emerge, Frazer finally declared in his magisterial work, *The Golden Bough,* that the past revealed
countless stories of gods taking on human form and being killed for their
people. To the horror of many, of course, this sounded like Christian
belief, and as far as Frazer was concerned, people had to make up their
own minds on that matter. Published in two volumes in 1890, *The Golden
Bough* then began to expand. It became encyclopedic, going through
three editions and growing into thirteen volumes with a total of 1.3 million words, what Frazer called a "moving panorama of the vanished life
of primitive man all over the world."[20] One theme seemed to tie the
work together, anthropologist Mary Douglas said of Frazer: "He never
wanders from his theme of the slain god."[21]

For Frazer, the whole idea for writing *The Golden Bough* began with a
topic very familiar to both classicists like himself and the public who read
the classics. In an ancient book titled the *Aeneid,* the Roman poet Virgil
told the story of a sibyl who was poised at the gates of the underworld.
The gates opened near a sacred grove with "a mighty tree, that bears a
golden bough." The golden bough had inspired both poets and painters
in Victorian times. It was a topic taken up by the British impressionist
Joseph Turner, who limned the scene of the sibyl in oils in 1843, a
panorama of cliffs, vines, temples, and lakes with a maiden in the foreground, bearing the golden branch.

Yet a classical scholar like Frazer knew there was more than just the
pretty scenery going on in this story. Like a detective, he followed the
trail of other investigators and came upon two important clues regarding
the origins of the golden bough story. First, there was Servius, a writer
who commented on Virgil's lines four hundred years after they were
written. Servius said the lines spoke of a grove tied to the cult and temple
of Diana, the fertility goddess, and that here was "a certain tree whose
branch none could disturb," except a runaway slave who might "challenge the fugitive priest to single combat, and so become priest himself."[22] Second, there was Pausanias, who reported a famed Diana cult
near Aricia. According to ancient Roman legend, moreover, the deity associated with the cult of Diana at Aricia was Virbius, a god-man who
was the first king and priest of the grove where Diana was worshiped.

Just as Frazer was feeling confident that he had identified a primitive rite and some of the characters involved, its location would also become clear. The British ambassador to Rome, an amateur archaeologist, had unearthed figurines of Diana at a little town called Nemi. It was near Aricia, just seventeen miles south of Rome on the old Appian Way. The pieces of the plot were now in place: a golden branch, a priest named Virbius, and a haven for exotic rituals at a place called Nemi.

Putting aside his work on the Pausanias translation, Frazer turned his powers of research and writing to *The Golden Bough,* which became a lifelong project. The key, he believed, was finding survivals that paralleled the story of the priest at Nemi and the struggle, succession, and death that was his apparent fate. The priest became so well known in folklore and Frazer's work, for example, that a historian such as Thomas Macaulay would speak of the Nemi grove with its "ghastly priest." What Frazer finally hit upon was a parallel myth in Norse mythology: there, the god-king Balder was slain, and Frazer was convinced that both stories—of the Norse king and the Nemi priest—were "survivals" of the *same* primitive event. He rapidly expanded his search for royal succession accounts, scouring history for stories of kings who were required to step down after fixed periods of rule, or who defended their authority in a contest of strength.

Although Frazer would publish multiple versions of *The Golden Bough,* they all opened like a macabre detective story: who was that "mysterious priest of Nemi," this priest who was also a god named Virbius? Behind this pretty picture of idyllic groves, springtime renewal, and the waters of Lake Nemi lay a forbidden tale with its naughty allure, suggesting fertility cults, sexual affairs with gods, and finally human sacrifice. "Who does not know Turner's picture of the Golden Bough?" Frazer began quaintly enough. What follows are chapters on "killing of the divine king," taboos, totems, magic, the slaying of Balder, and finally "death and resurrection" itself.

As he considered the ancient facts and survivals, Frazer also applied the theories of the taboo and totem, both of which were related to the power of the anima, or spirit, in a person or object. In a general way, it seemed, when primitive peoples viewed that spirit as alive and active in an object, the taboo was in effect: they would not touch or violate the object. When the spirit had left, the object could again be approached. As Frazer viewed this system, primitive man made these objects totems, which were vehicles of the spirit power. Primitives then set up a protocol of taboos for controlling the spirit power, usually for some kind of human benefit.

And the most obvious benefit, as Frazer saw the case, was the food needed by early man. Food became central to human beliefs about the supernatural. Religion thus had its origins in agricultural magic, a rite of destroying one life to guarantee another life. Nature's cycles had required this, for to bring about bountiful crops or children, a magical birth-death-rebirth cycle had to be acted out. The old vehicle—whether a withered crop or an aging king—had to be destroyed to free the anima to enter a new vehicle. By killing the totem, according to Frazer, primitives believed they had stopped the divine power from leaving the animal god, and then by eating it, they took the power into themselves.

Because the king came to represent a totemic god, what could be more naturally fruitful than a royal succession, even by ritual sacrifice? To show that this was a primeval belief, Frazer compared the killing of the priest-king Virbius at Nemi, by use of the golden branch, to the slaying of Balder in the Norse myth; Balder was killed by an arrow made of golden mistletoe. Frazer had a feeling of great discovery, for the two stories seemed to describe the same drama: a god was assaulted, his soul was taken from him (in the form of mistletoe), and then it was used as a weapon to slay him. Here was the totem belief in full, for a god was killed in order to magically snatch his divine power.

Publication of *The Golden Bough* began a prosperous forty-year relationship between Frazer and George Macmillan, whose London publishing house bore his name. In presenting his manuscript to Macmillan, Frazer said that, using the comparative method, he could "make it probable that the priest represented in his person the god of the grove, Virbius, and that his slaughter was regarded as the death of the god." And it matched a "widespread custom of killing men and animals regarded as divine."[23]

Frazer would say in the preface that his "intention merely was to explain the strange rule of the priesthood of sacred kingship of Nemi." But for his readers, there was much more. Douglas, the anthropologist, says that *The Golden Bough's* "full ambition" was to bring Christian doctrines, and particularly the crucifixion, "into the same perspective as totemic worship."[24] Frazer's letter to Macmillan suggested as much. "The resemblance of many of the savage customs and ideas to the fundamental doctrines of Christianity is striking," he wrote. "But I make no reference to this parallelism, leaving my readers to draw their own conclusions, one way or the other."[25]

Although *The Golden Bough's* hints about Christian origins were not at first so explicit, that changed in its second edition, published in 1900 with expanded interludes on Christian doctrines. Frazer subtitled the second edition *A Study in Magic and Religion* and articulated in full how magic had evolved into religion, and then how science had eclipsed both.

Frazer presumed that the savages had acted rationally, with an "instinctive curiosity concerning the cause of things," but particularly out of need to obtain food.[26] So once primitives believed in anima, they tried magic, either by mimicking nature or by manipulating it in replica (such as a voodoo doll) to produce a desired outcome. The magician rose to kingship. But as magic exposed its own fraudulence, the people turned to the gods with pleas for help, and the magician became the priest who led those worshipful rites. "When men find by experience that magic cannot *compel* the higher powers to comply with their wishes," Frazer said, "they condescend to *entreat* them."[27]

Finally, when human curiosity doubted even the gods, the next step was science, a rational manipulation of nature just like magic, minus the ghosts. "We in this generation live in a transition epoch between religion and science," Frazer declared. "Those who care for progress [must] aid the final triumph of science."[28] As a lad studying at the University of Glasgow, Frazer had been part of a generation caught between physical sciences and philosophical idealism. The first had clearly intrigued him most, although he kept the evolutionary philosophy that even idealism espoused. Clearly, he disliked the Hegelian abstraction of the Cairds at Glasgow, as he joked to a friend, the future Gifford lecturer Robert Marett (1931), when they were both senior men in anthropology. "I have just reconciled my three theories of totemism in a higher unity, as

Edward Caird would have said in his Hegelian jargon, which used to make me sick at Glasgow."[29]

During the first decade of the 1900s, *The Golden Bough* had made Frazer famous, and he was invited to move to Liverpool to oversee that city's impressive university museum. Once he arrived, however, the urban hubbub did not suit him. So he and his wife returned to Cambridge, nearly broke from the cost of their one-year escapade. He was eager to expand on *The Golden Bough* and reluctant to ask for a loan. Then in February of 1911, St. Andrews University offered him the appointment for a Gifford Lecture, with its payment of 800 pounds.

It was a godsend for Frazer. He not only accepted immediately but wanted to perform at the earliest possible date, which came in the fall of that year. Once again he put aside the voluminous *Golden Bough,* and to do the Giffords gathered up his years of notes on belief in immortality and the worship of spirits and ghosts. Although Aberdeen had kept Frazer on a short list for the Giffords from the very beginning, it was St. Andrews—Lang's own alma mater—that awarded him that final honor. The readability of his *Golden Bough,* however, was quite the opposite from Frazer's typical public talks, which at St. Andrews risked no polemic or rebuttal of Lang's argument for early monotheism. As Malinowski said, "He was a poor public speaker." Prone to stage fright, Frazer "preferred to read his lectures rather than deliver them *extempore."*[30]

Frazer delivered his first series of Giffords in October 1911 and returned for the second set in 1913, the published result being *The Belief in Immortality and the Worship of the Dead*. There was no whodunit here, no ghastly glamour of the priest of Nemi, but the "dragnet" style of information for which Tylor and he were renowned. The same was true when Frazer returned to give the Giffords at the University of Edinburgh in 1923 and 1925, laying out with encyclopedic precision how the "worship of nature" had busied primitives around the globe.

The Golden Bough would keep tumbling off the Macmillan presses, but between the second and third editions a new mystery would arise.

Why was the third edition, out in 1913, so different from the second in 1900? The second had been the veritable "forbidden text" because of how Frazer had drawn Christianity into the debate about totems and sacrifice. "Victorian Christians did not like to be told that other people have religions, that many of these religions are sacrificial," the critic Robert Fraser would say. "Nor did they relish being instructed that sacrifice could sometimes be explained as mere magic, or that magic might lie at the tap-root of much they themselves held dear."[31]

When the third edition of *The Golden Bough* came out, that controversial material—from a commentary on the crucifixion and church feast days to a look at ritual prostitution—had been excised or moved to an appendix in the back. There, Frazer said that his crucifixion theory was "in a high degree speculative and uncertain." In place of these "risky paragraphs," the critic Fraser said, the Victorian anthropologist instead explains "the principle of taboo at laborious length, and entertains us to a lengthy discussion of tree worship in Northern Europe, a safe enough subject at any time."[32]

What had happened? Like a rival who came to confront Virbius or Balder, the intrepid Lang had resurfaced to fire off a barrage of attacks on Frazer's new assertions in the second edition, ensuring that the tenuous friendship between the two men would now come to a complete end. Figuratively, says Frazer's biographer, "Lang was driven half-insane by the book."[33] First, Lang criticized the second edition in excruciating detail in the *Fortnightly Review.* Then he rapidly packaged this and more into a counter book, *Magic and Religion. The Golden Bough,* Lang wrote to a friend, was "the most learned and the most inconceivably silly book" of the era. "To criticize it is really too like hitting a child. And the gifted author thinks he has exploded all of Christianity."[34]

Lang's wittiness had not dulled, and he lampooned Frazer for his "vegetable" theory of the gods in which Frazer believed human sacrifice and resurrection were seen mimicking the growth cycle of nature. "The solar mythologists did not spare heroes like Achilles; they too were the sun," Lang said of his earlier foes in Müller's camp. "But the vegetable school, the Covent Garden school of mythologists, mixes up real human beings with vegetation." In other words, Lang was poking fun at Frazer's claim that primitives likened the breadth of human affairs—from priest-hoods to politics and successions of rulers—to the growing and wither-ing of nature. A master of mythology himself, Lang showed the utter disparity between the Norse Balder myth and the ancient Roman myth of Virbius. "Balder was not sacrificed, but cremated, and the 'huge ship,' of course, is a late Viking idea," he said. To account for a Persian myth about a mock-king hanging, he went on, Frazer "has to invent the hy-pothesis that real kings, in olden times, were annually *sacrificed.*"[35]

Regicide did exist, Lang offered, but usually because kings were dic-tatorial. "We do not kill a commander-in-chief, or an old professor; we pension them off," he joked. "But it is not so easy to pension off a king." While giving credit to the sheer "ingenuity" of the vegetation-god theory, Lang gasped continually at the cavalier claims about annual king sacrifices, from Babylon to Italy. "The idea is incredible," he said. "There is no evidence, or none is given, to show that a man has ever been sacri-ficed for the benefit of a god whom he incarnates."[36]

Since his Gifford Lectures, Lang had been quietly ignored by his former allies in the animist camp. Lang's prominence rose this one last time, and then he died in 1912. It was the year in which Frazer was giving his own first Gifford Lectures at St. Andrews University. Frazer took a moment to make a kind remark about his friend-turned-rival. "He was essentially a man of letters rather than a man of science, though no doubt by his writings he did an immense deal to popularize and extend the study of primitive man," Frazer said in a letter. Yet it may have been better, Frazer concluded, if Lang had done more with literature "and less of what I would call adulterated science."[37] Still, the next year, in 1913, Frazer had rid the third edition of the parts Lang had ravaged.

For the anti-Christian rationalists of later years, Frazer's softening in the third edition was chalked up to his self-interest—namely, "to keep his beautiful rooms at Trinity" College at Cambridge. Malinowski, the student of Frazer, intimated that he suffered a breakdown under Lang's attack; he was so "deeply upset and irritated" that he stopped work and thereafter "never read adverse criticisms or reviews of his books." Biogra-pher Robert Ackerman rebuts that claim, noting that Frazer had openly complained of Lang's "customary assaults on me" but acclimated to them. Ackerman says they "certainly did not (then) crush him."[38]

In his final word on *The Golden Bough,* published as *Aftermath* in 1936, Frazer stuck to his theory, "which on the whole I have seen no reason to change," he said. "But now, as always, I hold all my theories very lightly, and am ever ready to modify or abandon them in the light of new evidence."

After the relative excitement of the early anthropological debate, later Gifford Lectures on the topic would naturally seem tame. They were typically delivered in the catalog style, their themes including Asian myths in Greek culture, image worship by pagans and Christians, the artifacts of Rome, primitive altruism, and the ethics of islanders, Indians, and aborigines. On the Gifford circuit, Müller, Tylor, Lang, and four others with anthropological themes had come before Frazer. But his own claim to study only "natural religion," and not theology, spoke for the six others who were to follow. Anthropological topics seemed to peak around 1920 and peter out by 1950.

As the Victorian age gave way to the twentieth century, anthropology had come forward with fresh and bold confidence. It had claimed discovery of the very laws of human development, including belief in God, and it was a splendid drama in act two of the Gifford century. To do their science, the anthropologists often distinguished between natural religion and natural theology. Frazer, for example, separated the two in his opening Gifford series. Natural theology, he said, was "reasoned knowledge of a God or gods." But anthropology aimed far short of that, of course. It talked only of natural religion, which was the realm of the human imagination. While the spirit of anthropology inescapably challenged Christian belief—suggesting it was evolved superstition—the science of anthropology also claimed scientific neutrality. Yet its association with debunking religion is indelible, and for many that is the true legacy of the "science of man."

Frazer had ended *The Golden Bough* on a sentimental note. In the final chapter, "Farewell to Nemi," the sun was setting, the ship's sails were furled, and a "long voyage of discovery" was over. This quest of the early anthropologists also had a sentimental, if overly idealistic, tinge. They "felt no embarrassment in claiming to explain the entire phenomenon of religion," an ambition that "astonishes us by its naive overconfidence," says historian of religion Daniel Pals.[39] Yet even today, the street-corner philosopher or the typical college student who says religion is just "a lot of hocus-pocus" is heir to Frazer's theory.

Today, the Frazer volumes, thick books bound in black or green, still line university stacks. The volumes testify to anthropology as a quest to compile lists, data, and catalogs. But Frazer had always made a two-part distinction about his field. In the 1922 lecture for which his notes say "Anthropology a modern science," Frazer argued that the interpreter of the data (the "comparative ethnologist") must stay separate from the field collector (the "descriptive ethnologist"). It was a handy distinction that gave the ivory tower pride of place. That tower was mostly pulled down by the next generations, for anthropologists who wanted to claim that name had to dirty their hands and shoes in the field.

The fieldwork began with a stage of expeditions and grand surveys, typified by American anthropologist Franz Boas. A physicist and geographer, Boas ventured out from Columbia University to survey the Kwakiutl Indians in the Baffin Islands (of northern Vancouver) and adjacent British Columbia. A next stage was akin to full-immersion baptism. Malinowski, for example, spent the early 1940s among the Trobriand Islanders of Melanesia, a pioneer of living among a native people and using their indigenous language all during his research.

This new, prudent anthropology—characterized by Boas, Malinowski, and others—rejected Tylor's evolutionary view, which held that every human culture went through the exact same gradual stages. The great catalogs of Tylor and Frazer would have perennial value, but their ideology of evolution, once taken as a kind of scientific talisman, has been dismissed as a starry-eyed attempt to find a total theory. "Instead of trying to understand the whole thing, anthropologists now tend to chip off a bit and to develop refined tools for interpreting that bit," says anthropologist Douglas. "The first job is to understand a cultural system as a rational way of behaving for people who know each other and who make the same assumptions."[40] A group's myths and customs, in fact, can be so mixed up in symbolic, practical, and arbitrary uses that a "final theory" of origins seems fantastic.

Andrew Lang, remembered mostly for fairy-tale collections, is still a namesake at St. Andrews University. A street bears his name, and the medieval St. Salvator's Chapel his bronze memorial plaque. Lang never systematized his theory of the "savage high god" but had left behind a foothold for others. The Austrian ethnologist and Catholic priest Wilhelm Schmidt rallied collaborators for a decades-long search for primitive monotheism, cataloged in a twelve-volume tome called *The Origin of the Idea of God*, published from 1912 to 1954. They found monotheism among pygmies, Bushmen, and primitives in Tierra del Fuego, Australia, the Arctic, and North America. If Tylor and Frazer had mounted up the "proofs" for animism, Schmidt said, "I have collected all the counterproofs."[41]

Anthropology, like all sciences, rapidly specialized. It focused on the particular, accentuating intense observation and minute data. But the sweeping generalization, an art that made Frazer so popular, still tantalized the public most. Such was the case with Margaret Mead, a student of Boas. She had lived with Polynesian teenage maidens for six months and produced an ethnological study on the joys of free sex in a tropical Eden with few taboos. The 1928 book, *Coming of Age in Samoa*, was like a scientific homage to an erotic Gauguin painting, and it became a scientific best-seller (although her data were rebutted before she died, leaving a cloud of fibbing girls hanging over the supposed Pacific idyll).

As before, for every rationalist view of a religion in a primitive culture, there was a counter theory that emphasized human awe and emotion. Whereas Schmidt, for example, said the savage had rationally contemplated a first cause, the Italian anthropologist Raffele Pettazzoni argued that awe-inspiring phenomena in the sky were taken emotionally to be God. This was echoed by Mircea Eliade of the University of Chicago, who became a celebrated historian of world religions. Eliade amplified the idea of a human encounter with "the sacred." Reared in Eastern Orthodoxy, he was amenable to monotheism and presumed that the unique symbols of "autonomous" cultural groups all pointed to the same ineffable Creator.

The final leap for myth-hunting anthropology would be into the depths of pure psychology. In Tylor's footsteps, Frazer had expanded on this as well. He attributed common myth to the common evolving psyche of early man. Myths were "the effect of similar causes acting on the similar constitution of the human mind in different countries and under different skies."[42] This was the Frazerian adage so beloved by the last great popularizer of psychological mythology, Joseph Campbell.

Campbell was reared in a Catholic home in New York and his favorite boyhood pursuit was looking at Indian totem poles, and reading their stories, at the American Museum of Natural History. He studied at Columbia University and went into comparative literature, translating Hindu texts, annotating Irish prose, and compiling the works of Carl Jung, who in the age of "depth psychology" presented the case for hereditary archetypes—namely, common mythological images—in the biological unconscious of humankind.

Beginning his work in 1949 with *The Hero with a Thousand Faces,* Campbell would later announce that his comparative mythology offered the "lineaments of a new science," reminiscent of Müller's unveiling of the "science of religion" in 1870 London. And like Müller, Campbell generated public interest in anthropological universals, becoming "the most famous and in some circles the most esteemed writer on myth" by the time he died in 1987.[43]

Campbell was not in search of God but rather the psychic unity of myths, looking for their similarities rather than differences. He looked at all past definitions—Müller's poetic fantasy, Tylor's animism, Frazer's primitive fumbling, and Jung's group dream, and said: "Mythology is all of these."[44] That was possible because the modern quest for mythological origins had become psychological, if not mystical. Every myth, if not taken literally, can be true, Campbell said.

If Campbell had been romantic about archaic man, Frazer had looked upon him as barbarous. It was perhaps an understandable sentiment to come from a man in a Cambridge ivory tower in the midst of an imperialist age. A humorous anecdote expands on that point. During the Christmas season of 1900, Frazer and his wife had traveled for a vacation to Rome, just north of ancient Nemi, where the god Virbius vied for his life against a golden bough. In Rome, Frazer met William James, the American psychologist. James asked Frazer whether he had any hands-on experience with the tribes he wrote about. "Good God, no!" Frazer replied. The response may be apocryphal, but the meeting was not.

The Frazers had checked into the same Italian *pensione* that housed James, who had fled with his wife to the same Mediterranean climes. James, whose famed *Principles of Psychology* came out the same year as *The Golden Bough,* was on his way to delivering the Giffords at Edinburgh, but a heart ailment had delayed him in Europe for another year. Having met Frazer, James wrote home about the encounter. "He, after Tylor, is the greatest authority now in England on the religious ideas and superstitions of primitive peoples." Frazer's humility, he remarked, was colored by "unworldliness and molelike sightlessness to everything but *print.*"[45]

Frazer had come out of an old school of psychology—the empirical school of Locke and Hume, the famed "way of ideas" in which simple ideas from sense impression multiplied into more complex ideas. By this rational "association of ideas," the human mind created its reality, and for Frazer this was the process that explained human development through magic, religion, and science. For James, the biology of the mind was more basic than its rationality. His *Principles of Psychology* launched a frontal assault on associationist theories.

The mind could be anything but rational. James spoke of the human tendency toward neurotic beliefs and feelings of "morbidity," a fearful worry about death or meaning in life. Frazer and James talked about how the mind works. "I have been stirring him up so that I imagine he will now proceed to put in big loads of work in the morbid psychology direction," James recalled.[46] But no, Frazer preferred the rational. Primitive rationality had led to human sacrifice; how much worse might the irrational mind, with its dark inner wellsprings, behave?

When James finally reached Edinburgh, his lectures, *The Varieties of Religious Experience,* dismissed the theories of Tylor and Frazer. He said they were yet another failed attempt to understand the psychological present by reaching into the past. The theory of "survivals," moreover, was another tyranny of the abstract, whereas science needed to look at the immediate, concrete, and real. "So I unhesitatingly repudiate the survival-theory of religion," James said. "It does not follow, because our ancestors made so many errors of fact and mixed them with their religion, that we should therefore leave off being religious at all."[47] It was a message the public wanted to hear.

Alongside anthropology, psychology was another burgeoning science that offered its services where religion and philosophy had once been dominant. As the science of mind, psychology set out to probe the material functions of the brain, and by experiment it linked perceptions and behaviors to the senses and the nervous system. In act two of the Gifford legacy, psychology, like anthropology, offered itself as an alternative to much Christian explanation, and in doing so took the Western public by surprise. When William James began his Gifford series, his opening lesson was on "Religion and Neurology," a topic that seemed only slightly less shocking than Frazer's stories about human sacrifice and boughs of gold.

FOUR

STREAMS of EXPERIENCE

Psychology

W HAT TURNED OUT to be a milestone in the psychology of religion, and perhaps psychology itself, began prosaically enough. When William James arrived at Edinburgh University to begin his Gifford Lectures, it was a sunny spring morning in 1901. Outside the great quadrangle, teams of white horses pulled trolley cars up cobblestone streets and men walked about in bowler hats. The scene was intensely urban compared to the town around James's home campus, the almost bucolic Cambridge at the edge of Boston.

About noon, James took his seat at an oak table set on a small stage, which was painted white and enclosed by a waist-high wall. Given the lecture hall's design, with James down in front, he might have been speaking in a medical theater. The hard benches swept up to the back of the white-walled room, known as the English classroom. The eyes of nearly 250 people stared down that opening day. The numbers would grow. James would keep Edinburgh's attention for two springtimes, in 1901 and 1902. "Crowded," said an organizer's notes.

If the Edinburgh setting *had* been a medical school, James would have felt quite at home. Not only was his opening topic "Religion and Neurology," much of his young manhood was spent in a world of chemistry, dissections, and biological specimens. He earned a medical degree at Harvard, did duty in hospitals, and ran a little psychology laboratory that tested frog nerves and studied calf brains, which look a lot like human brains.

But James was there to tell his audience not to be intimidated by the "medical materialism" of his day. Medical materialism wanted to reduce all human experience, including love, hope, and charity, to what people had for breakfast or what the liver secreted. Religious experience did indeed have parallels in the nervous system, James told his audience, but it was more than that, more than just vibrating matter. His words were sympathetic, but they did not arise from a desire to espouse some spiritual system of his own. James was in Edinburgh to explain "the empiricist principles I profess," and how those principles challenged the dogmas of materialist science, and that included the science of psychology.[1]

Act two of the Gifford legacy introduced much troubling "empirical" science to the Christian mind. For the field of psychology, however, James made the encounter as friendly as possible. He was a man of the laboratory who never belonged to a church or adhered to a religious doctrine. Yet he assured the public that religious instincts were as profound as scientific tests. The musings of James, meanwhile, hinted at a new movement of the twentieth century, one that viewed reality as a flowing stream of consciousness. Reality, in this view, was made up of organic processes, streams of thought, and drops of experience connected across time. This "process" view finally arrived at metaphysical conclusions about God, but in many ways it began that journey in the mental science of William James.

James arrived in Scotland in 1901 as a Yankee mystic, an open-minded skeptic with a large forehead, graying beard, and sparkling blue eyes. On the opening day of his lectures, he wore a dark suit just purchased in London. Not in good health, James sat while he spoke and kept a very

light social schedule while in Edinburgh. He had only nine more years
to live because of a weak heart. By the time James was done with the
Giffords, they had lifted his spirits and then left him exhausted. To complete the task, he had steamed across the Atlantic four times. When *The
Varieties of Religious Experience* was finally published in 1902, James called
the draining project "that terrible book."[2]

Since then, *Varieties,* with its twenty lectures, has been published again
and again. The occasional pilgrim has also made his way to the urban Edinburgh scene. Over the years, the lecture hall in which James spoke had
been lost to memory, at least until 2002. Using old floor plans and a detective's instincts, a member of the William James Society finally pinned
it down—in time for a world centennial of *Varieties.* The room, now
serving the law school, retains its Victorian ambiance. On the stone wall
outside, a bronze plaque now honors James: "Philosopher, Psychologist,
Teacher, Gifford Lecturer."[3]

Of the four titles, James was best known in 1901 for the second.
"Psychology was the only branch of study in which he was particularly
versed," the *Scotsman* reported him saying. He wanted to offer "a descriptive survey of the religious propensities of man." Still, "it was hard for a
psychologist not to be also, in a way, a little of a philosopher."[4]

Only a few years before the Giffords, James had made his debut as a
published philosopher in *The Will to Believe.* For both science and philosophy, his book of essays and lectures espoused two of his own world-
destroying ideas, the first being "radical empiricism." Empiricism studied
concrete events. It allowed people "in the course of future experience" to
change their ideas, be flexible, be free of dogmatic theories. This empiricism cared nothing of ultimate causes, making it radical. The empiricist
relished "the crudity of experience" over gleaming systems of thought.[5]
Truth was out there for the radical empiricist, but that search found the
world in a permanent state of "pluralism"—and pluralism was James's
second world-destroying idea. It was James who introduced the term *pluralism* into Western philosophy.

In a word, pluralism recognized the limits of human perception, arguing that people, in effect, had their own universes when trying to grasp
ultimate things. Pluralism was a commonsense interpretation of the

world and eminently practical; it suited the needs of human thought far better than a single abstract system, which had to be forced onto so many unwilling minds. In his own age, James faced two such abstract, all-or-nothing systems, Absolute Idealism and scientific materialism. With pluralism, James had hit upon a weapon to break apart such great consuming systems—which he called "block-universes"—so that individuality, unpredictability, creativity, diversity, and radical empiricism could all find room to breathe.

As part of act two in the Gifford legacy, James began to do this with the science of psychology, but in time he moved on to philosophy and even metaphysics. The Gifford Lectures were a high point in the Jamesian assault on idealism and materialism, and the fallout has been felt ever since. In time, James and his allies against the block-universe proposed a different kind of cosmos, one that was organic and made up of moments, experiences, processes, and emerging realities. The revolution began with biology and psychology, with nature and the human mind. But as will be seen, in time it moved all the way up to God.

James began delivering his Giffords in 1901, at about the midpoint in the rise of modern psychology, a time when it was shifting from metaphysics to the "brass instrument" approach in research. Two decades earlier, the first laboratories had been opened at Harvard (1877), the University of Leipzig in Germany (1879), and Cambridge in England (1879). By 1920 psychology was a diverse and professional field, groomed by its ability to help shell-shocked soldiers after the First World War. Yet the early push to make psychology an experimental science had not been universal. Oxford had delayed the project for a generation. As one faculty member said, it might "insult religion by putting the human soul in a pair of scales."[6]

When James published his two-volume *Principles of Psychology* in 1890, it established his reputation internationally and became a benchmark in the field. *Principles* laid out basic problems that psychology still faces today. But for all its brilliance, James was apologetic. Psychology was "hardly more than what physics was before Galileo," he said. "It is a mass of phenomenal description, gossip and myth, including, however, real material" that may one day become a real science. "I wished, by treating Psychology like a natural science, to help her become one."[7]

In the next few decades, psychology would swing in one direction, then another, and finally end up remarkably diverse. The first swing was to introspection, whether in personal testimony or in Sigmund Freud's analysis of the subconscious. In reaction to introspection, behavioral psychology arose soon after James's death in 1910. Its adherents desired an approach that viewed mind and body as biological machines conditioned by pain or pleasure. At midcentury, computers gave psychology yet another model, the computational theory of cognitive science. By then, the approaches in psychology were so varied that no school dominated the field, and soon enough psychopharmacology—the use of one chemical or another to make the brain work better—was on its way.

In James's Giffords, starting on that sunny and cool day in Edinburgh, the topic of the hour was the psychology of religion, which meant religious experience in the concrete. Over the weeks, he would treat his audience to a smorgasbord of spirituality, beliefs in the unseen, ecstatic experience, conversions, reformed lives, healed bodies, and eccentric saints, a world he ended up calling "piecemeal supernaturalism." Using testimonies, accounts, and letters, he carried his audience into that strange universe, with its fabulous claims, delirious swoons, flesh torn in abnegation, catharsis, defeats of addiction and anger—indeed a world of people waking up and loving everything as if their hearts would burst.

James's lectures were a catalog of cases both modern and ancient. There were religious ecstatics such as George Fox, the founder of the Quakers, and Teresa of Avila, the Catholic saint. They were unusual people. Their psyches and behaviors pushed the extremes. The medical materialists had an explanation, James said, but it should not worry ordinary believers too much. Materialists said these religious figures showed "symptoms of nervous instability." They called Saint Paul's revelation on the road to Damascus "a discharging lesion of the occipital cortex." Everyone knew Paul's story and this pedantic materialist claim brought a laugh. Yet even if true, James said, such medical arguments were self-defeating for atheists. Following medical materialism, "we should doubtless see 'the liver' determining the dicta of the sturdy atheist as decisively."[8]

By making fun of materialism, James leveled the playing field for believers. After that, though, he was not going to make religion out as a picnic, for it was the extremes of religion, James said, that gave insight into psychology and human nature. His next point surely stirred his well-bred, churchgoing audience: religious people often *were* neurotic. The "morbid"-minded person (who felt that all was not right with the world) *was* the typical searcher after salvation. Two generations after James, modern psychology, in its humanistic and "positive" approaches, tried to understand the integrative role religion played in individual lives. It did not compensate for neurosis, but added to the normal, stable life. Religion led to physical health and mental moderation.

James would not have disagreed, but the purpose of his Giffords was to explore the wild frontiers, the place where experience turned lives inside out, founded religions, and revolutionized others. The "morbid" person was a likely place to look for such extremes in religion, but James told his audience that almost everyone is morbid to some degree. It was a natural state, given how universal were the evils, disappointments, and ills of life. Yet in this psychology, the psychology of William James, the cardinal point was the outcome in one's life, not the origin or fact of morbidity. "You must all be ready now to judge the religious life by its results exclusively, and I shall assume that the bugaboo of morbid origin will scandalize your piety no more," he said.[9]

The morbid person, however, was not the only human specimen that James would cover in his Giffords. There was a second, if rarer, kind of human temperament, which James called healthy-minded. He was fascinated by healthy-minded souls. By a trick of mind they expel evil from the world. They are carefree, athletic optimists. They say to the disconsolate, "Stuff and nonsense, get out into the open air!" With a slap on the back, they offer, "Cheer up, old fellow, you'll be all right ere long, if you will only drop your morbidness!"[10]

James had always taken a scientific and personal interest in the "mind cure" movement, known today as faith healing or positive thinking. He was persuaded that such a mustering of mental attitude, whether in thoughts, readings, prayers, or denial of evil, could ease physical woes. But for others a simple mind cure could not work. For them, another kind of experience came in, which James called being twice-born. Optimists needed to be born only once. For the morbid soul, religious experience gave a second chance, a second birth.

As his audience could well imagine already, the healthy-minded and the morbid-minded are not two groups that easily get along. James asked them to consider "how great an antagonism may naturally arise between the healthy-minded way of viewing life and the way that takes all this [morbid] experience of evil as something essential." To the morbid-minded way, "healthy-mindedness pure and simple seems unspeakably blind and shallow. To the healthy-minded way, on the other hand, the way of the sick soul seems unmanly and diseased."[11] Yet for all the virtues of healthy-mindedness, James concluded, it was transient in the face of the ultimate realities of life, the palpable and grasping hands of age, disease, and death. "We are all such helpless failures in the last resort," he said. "The sanest and best of us are of one clay with lunatics and prison inmates, and death finally runs the robustest of us down." On this great slope of life, "religion comes to our rescue and takes our fate into her hands."[12]

And so goes the origins of human religion, James argued. It could not have been more different from the argument of the anthropologists. James found religion in the psychological present, Frazer and Tylor in the primitive past. James concluded with his pithy definition of religion: it was not doctrine (or survivals) but deliverance. "The warring gods and formulas of the various religions do indeed cancel each other, but there is a certain uniform deliverance in which religions all appear to meet," said James. They all resolve human uneasiness. "The uneasiness, reduced to its simplest terms, is a sense that there is *something wrong about us* as we naturally stand," he went on. "The solution is a sense that *we are saved from the wrongness* by making proper connection with the higher powers." The deliverance could have both its physical and spiritual sides, James said, like a gift of grace—"a gift of our organism, the physiologists will tell us, a gift of God's grace, the theologians say."[13]

Throughout his Giffords, James argued the usefulness of religion. He extolled the search for fruitful *ends,* not vague *origins.* But he did not exactly declare the philosophy for which he became so well known, a view called pragmatism. That declaration came five years later in a set of published lectures titled *Pragmatism*. Pragmatism was the idea that what is useful and effective is also what is true; the "truth" is proven by its successful application. At Edinburgh, he saved "pragmatism" for the end, the eighteenth lecture, given April 5, 1902. The common man uses a "thoroughly 'pragmatic' view of religion," he said. He fills the world with miracles, beliefs, and hopes beyond the grave, whereas materialists and metaphysicians traffic in abstractions. "I believe the pragmatic way of taking religion to be the deeper way," James said. "It gives it body as well as soul, it makes it claim, as everything real must claim, some characteristic realm of fact as its very own."[14]

Engaging his listeners so often with the "we" persona—"*we are saved,*" for example—James still retained his typical ambiguity. What did James truly believe? He was so good at stating other people's theologies and theories that it was often hard to know where he left off.

After *Varieties* was published, James added a postscript clarifying that he put himself in the "supernaturalist" camp if the choice were between that and brute materialism. He could not believe in popular Christianity or scholastic theism. But he nevertheless preferred the hypothesis of "piecemeal supernaturalism," the "crasser" theism of ordinary folks. He would take their concrete experiences over the abstraction of philosophers any day. He wanted to believe in God, one biographer said. But he seemed unreconciled to a Supreme Being who needed ritual, worship, or prayer. Personally, he never found the spiritual source of the mind cures.[15]

James may have held to some kind of whim or hope to the last. If James was anything, he was open. The dedication page for *Varieties* noted "filial gratitude and love" to his mother-in-law, Eliza Putnam Gibbens. She had never doubted God or the afterlife, and by the dedication, James may have hinted at his own lifelong desire to have a deep-felt theistic belief.

Since the delivery of *Varieties,* it has been remarked upon how Lord Gifford and William James helped each other. Without the bequest, James would not have written the book. Without the book, the Giffords would not have become so widely known so quickly. James's journey to Edinburgh, an event that still inspires pilgrimages in his footsteps, had an array of causes, with the Gifford invitation the most immediate. Further back was James himself, and the story of his unique family and the pregnant age of psychology in which he lived.

James was born into a warm and eccentric family in New York City. His father and mother were independently wealthy and reared their children while traveling between America, England, and Europe, educating them at home under tutors. William mastered several languages, French and German in particular. The spiritual backdrop of life was father Henry's enthusiasm for the spiritualist teacher Emanuel Swedenborg, a Swedish engineer and metallurgist who as a philosopher argued for a universe governed by a divine will. William's younger brother by a year was Henry James, who became a famous novelist and migrated to England. Their sister, Alice, was inclined to illness, but the whole family was prone to hypochondria, and Mrs. James worried excessively about her children's overexertion. With disposable income, though, they could always travel to a European hot spa to "take the cure," as William often would.

Physical health aside, William's two darkest moments in life were psychological, and both came in his youth. Once he contemplated suicide during his medical studies. Another nadir came after he graduated. To escape the first depression, James fled to the spas in Germany, where he became enamored of the psychological field after attending lectures in Berlin. As he wrote his father: "There are two professors there, [Hermann] Helmholtz and [Wilhelm] Wundt, who are strong on the physiology of the senses."[16] Studying the mind as a physical mechanism was new, a step beyond the old metaphysics. It was a search for thought and perception as products of brain and nerve operations. Measurement by "brass instrument" eclipsed the old way of "introspection," which asked people about their inner psyches.

Although James was influenced by German experimentalists, he drank in the outlooks of France. When he traveled there he saw a bedside psychology that still valued personal testimony, what hard-science critics dismissed as "folk psychology." In this tradition, in fact, it was not uncommon for the researcher to sit down and study his own mental introspection and processes. James was not beyond personal application either. In France he learned of an up-and-coming philosopher, Charles Renouvier.

After James had returned to New England, the ideas of Renouvier helped him resolve a deeply personal crisis, the nub of his second malaise. James struggled with a depressing philosophic problem: was there free will, he wondered, either in a mechanistic universe as taught by science or in the predestined one espoused by his Swedenborgian father?

Renouvier provided a solution. The French philosopher defined free will as "the sustaining of a thought because I choose to when I might have other thoughts." As James famously wrote, "My first act of free will shall be to believe in free will." A new phase in life apparently began for James. It was conceptual, no doubt. Perhaps it was a mind cure. But it was surely linked to a stage of final independence. He would soon leave his parents' household, where he had lived until 1878, when he was married. Having endured such a range of personal experiences, James shifted his focus in psychology. He wanted to find its practical solutions for coping with real life.

After his informal study in Europe, James returned to Harvard as a lecturer. He eventually launched a novel course, "Physiological Psychology." In those days, university psychology was usually taught in the philosophy department. At Harvard, it was taught by the head of philosophy, Francis Bowen, an idealist who followed Sir William Hamilton, the Edinburgh don who had melded common sense with Kant. Bowen tried, but failed, to block James's promotion to professor of psychology. Across America, tensions were growing between philosophers and experimental psychologists—housed in the same philosophy departments—until a solution came: independent psychology departments emerged to do their own kind of science.

Textbooks on psychology had been around since the 1820s. By James's period of study, they ranged from the materialism of Herbert Spencer to the Christian common sense of James McCosh and the Hegelianism of the young John Dewey. When New York's Henry Holt was seeking an author for yet another version, James was circuitously recommended. The two-year deadline turned into twelve, allowing James to build a tower of expert papers that were finally amassed as his monumental "science of mental life." Rapidly enough, the two-volume *Principles of Psychology* (1890) overshadowed all other works and remains in print today. A single-volume "short course" also came out, and for the next few generations, students spoke of "the James" and "the Jimmy" to keep them straight.

Once that was accomplished, James, now forty-eight, would move on to his more philosophical and metaphysical work. Yet *Principles* represents the kind of overall legacy he left behind. "He founded no school, unearthed no verity, solved no problem," said philosopher Daniel Robinson, but he "traced out the boundaries of thought and supplied the terms with which an entire generation would discuss and understand Psychology."[17] The failures and frustrations that James met in his own laboratory and speculation, moreover, anticipated the puzzles psychology still faces: defining the self, locating perception in the nervous system, and understanding the nature of experience.

Whether it was James's *Principles* or his just-published *Will to Believe* that brought him to the attention of the Gifford committees in Scotland is unclear. *Will to Believe* had gained quite a following with its argument that people had the "right to adopt a believing attitude in religious matters," especially when it speaks to a personal background and is beneficial; for many the religious choice amounts to the "sole chance in life of getting upon the winning side."[18] When the book sold so well, James realized that the public had an insatiable appetite for "religious philosophy that is both unconventional and untechnical."[19]

James was actually first invited to give the Giffords at Aberdeen. *Will to Believe* was published in the spring of 1897, about the time its academic senate invited him to deliver the lecture beginning in 1899. Yet there were reasons to turn it down, and not only ill health, according to his biographer. "He decided that Aberdeen was not the better choice: Edinburgh would be able to offer him the lecture series in 1900, giving him more time to prepare; equally important, Edinburgh paid more." His wife, Alice, said turning down Aberdeen was a "gamble." It would "probably result in our missing the Gifford lectureship altogether."[20] Of course, they did not. Meanwhile, James had recommended his Harvard colleague Josiah Royce to take the Aberdeen platform.

Edinburgh was a "great honor" but also "very highly paid," said James, who despite family wealth always worried about making a living.[21] Both he and Royce paid another kind of price—the sheer workload. Said

Royce before his deadline, "I have no time to do more than breathe and work, and sleep between times."[22] At age fifty-seven, James felt there was no more time for "postponed achievements." His "message to the world" would be empiricism and pragmatism combined with a pluralist metaphysics.[23] Then James took an ill-fated summer break in the New York Adirondacks, damaging his weak heart in a climb to the top of Mount Marcy.

Still, he worked and worried up to the last, bundling up all his books and notes to carry with him to Nauheim, Germany, near Hamburg, to take the heart cure at the hot springs. "Perhaps if he had gone to bed and rested instead of working for the two weeks before we sailed," his wife lamented, "we might have escaped Nauheim and many subsequent woes."[24]

After visits to Switzerland and then a stay with his brother, Henry, who was writing his novels in London, James was too weak to meet the deadline. Both Harvard and Edinburgh gave more time. The lectures were pushed back more than a year, to May of 1901. Until then, James was mobile, fleeing damp London for Rye in the south and then a vacant chateau near the French Riviera. Back and forth the strained family of three (including their daughter) went, including two stops at the spa. James opened the new century, January 1, 1900, plagued with his lifelong ailments. "My nervous system is utter trash, and always was so," he wrote to his brother, Henry.[25]

Four of William James's close friends died during these long months in Europe. "I find myself in a cold, pinched, quaking state," he put down, "when I think of the probability of dying soon with all my music in me." Some interpreters have looked back on James's Giffords as a final chance for him to reconcile his own struggle over science, doubt, and religious belief. Whatever the true case, he rose to the occasion. He summoned new strength and in March, two months before delivery, he was on the road again with his daughter and wife, stopping at Rye before going on to Edinburgh, where their son, Harry, would also join up with them.

Each afternoon in Edinburgh, James and his wife took a carriage ride in the countryside. "The green is of the vividest," he noted. "The air itself an object, holding watery vapor, tenuous smoke, and ancient sunshine in solution."²⁶ At the Old College, the Victorian lecture hall, so much like a medical theater, was always full, and available space grew smaller as word got round.

His second lectures were a lighter burden. James simply read them from page proofs for *Varieties* (cutting back the long quotes seen in the book version). With the publication of *Varieties,* James received the rare acknowledgment that his work had made a difference. "No previous book of mine has got anything like the prompt and thankful recognition," he said. He had merely explained what "everybody knows," that "the real life of religion springs from what may be called the mystical stratum of human nature."

But James felt that something more still had to be done. He wanted to write an "epoch making" systematic work of philosophy, not the "squashy, popular-lecture style" of his others. "I actually dread to die until I have settled the Universe's hash in one more book."²⁷ That project never came. On the day in 1910 when James breathed his last, his head cradled in his wife's arms at their summer home in Chocorua, New Hampshire, he was still working on *Some Problems of Philosophy.*

By thinking about "epoch making," James might have been comparing himself to the idealists and Hegelians, whose lengthy and systematic works had swamped Western philosophy. Or perhaps he was thinking of Josiah Royce, who had produced the lengthy and systematic *The World and the Individual.* James had been waging a "battle for the Absolute" with idealists everywhere, even in psychology.

In Hegelian psychology, which held sway in German and American schools for decades, the focus was on the metaphysical ego of the person. Hegel presented human personality as an individual ego realizing the Absolute ego. The force behind individual development was the dialectic of the self and the development of rationality, the Absolute becoming incarnate. To James it was intellectual witchcraft. Driven by a penchant to make psychology a science, James saw Hegel as the enemy. As long as Hegel's dialectical philosophy influenced the psychology departments of Europe and America, he believed, scientific research on the human psyche—driven by experiments, interviews, and measurements—would not advance.

The lighter side of this battle revealed itself in his friendship with Royce, whom James had brought to Harvard. During the Gifford episode, James and Royce crossed paths in England, on the way to their respective lecture appointments. James was staying at brother Henry's house in Rye when Royce arrived by ship in December, so he paid James a visit. Back in their Cambridge neighborhood, Royce told James by letter, a cat had been seen chasing squirrels daily, which Royce likened to James's pluralistic world of conflicting interests and individuals. Royce joked that he had tried in vain "to persuade them all to take a monistic view of the situation," to unite the many into the One, a theme of Absolute Idealism.[28]

When Royce arrived in Rye, he gave James a copy of *The World and the Individual,* its first volume already in print. After reading it, James encouraged his younger compatriot. "I don't see how you can fail, from now onward, to be recognized as the champion of Absolute Idealism for Anglo-Saxondom, and the one thoroughly original leader of that whole school of thought," he wrote Royce. Playing off the doctrine that the Absolute is manifest in individuals, James said to his friend: "The Absolute himself must get great fun out of being you." After his own lectures, Royce sent a note back from Aberdeen. "Everywhere they ask about you, & regard me only as the advance agent of the true American Theory. *That* they await from you."[29]

James and Royce carried on the friendly duel in a Harvard course they jointly taught on Royce's book, with James as the antagonist and dissenter. But James saved his biggest attack on the block-universe of the Hegelians for the Hibbert Lectures he delivered in 1908 at Oxford, still a seat of British idealism. His sustained argument was published as *A Pluralistic Universe.* He criticized Oxford's Francis H. Bradley for an "intolerance to pluralism" and Absolute Idealism for the "vice of intellectualism."[30] Although James had no impact in Europe to speak of, he is credited with chasing Hegelianism out of psychology at Harvard, Chicago, and elsewhere. The probing of Hegel moved to religious studies.

In the formative period of psychology, not only James but two other pioneers in the field gave the Gifford Lectures: James Ward and Conwy Lloyd Morgan, both Englishmen who also looked askance at medical materialism. They were part of act two in the Gifford legacy, when materialist science confronted popular beliefs about God, free will, and the soul. Under the purview of the Giffords, in fact, psychology had become a kind of natural theology. For example, while James dismissed the "argument *ex consensu gentium*"—that widespread belief is proof of God—he cited so much piecemeal supernaturalism that mental contact with something higher seemed very common. Ward and Morgan openly grappled with questions of immortality, morality, and why human beings are different from animals, despite similar brains, sense organs, and nervous systems.

These were topics that crossed the line between science and theology, and Ward and Morgan seemed even more willing than James to commit that trespass. Ward was the leading psychologist at Cambridge at a time when the field was just as much philosophy as "brass instrument" research, which he had undertaken in Germany with great distinction. He was known for his celebrated 1886 "Psychology" entry in the ninth edition of *Britannica*. Ward was a clear defender of the spiritual. He gave the Gifford series twice, in 1896 and 1907, the first being an assault on the agnosticism of Herbert Spencer and the naturalism of Thomas H. Huxley, both of whom explained the mind as having purely material causes. As Huxley said, the mind was merely the whistle on the steam engine.

At the time Morgan gave his Gifford Lectures, delivered at St. Andrews in 1921, he was a senior statesman and philosopher in British science. His lectures popularized "emergent evolution," an idea that was in the air but which he formalized. A noted psychologist and zoologist, Morgan wrote books on psychology for teachers and published seminal studies comparing instinct and learning in animals and human beings. As a former student of Thomas H. Huxley, though, Morgan dramatically diverged from his mentor's materialism. "For better or worse, I acknowledge God as the Nisus [impulse] through whose Activity emergents emerge, and the whole course of emergent evolution is directed." His study of animals convinced him still more that the human mind emerged almost abruptly, suggesting "the direct presence of God."[31]

In his *Principles,* James had already rejected the "automata," or reflex, theory of the mind espoused by Huxley, and in a similar way Ward and

Morgan fought against psychology being reduced to mere physiology. The two Englishmen were also foes of the new behaviorism when it began rolling into Britain from the United States. But if Ward and Morgan were nominal churchmen, and in their Giffords tried to explain not only the *origin* of the mind but also the possible role of God, James was the outsider looking in, having no religious affiliation or ken for theology. Yet no psychologist in the period could escape that tantalizing yet intractable problem of human "consciousness" and "the self." On these too James would make an influential contribution by finally speaking of the "self" as a stream of consciousness.

In Ward's *Britannica* entry in 1886, the question of the self, subject, or pure ego was handled delicately. "For empirical psychology this notion is ultimate; its speculative treatment falls altogether—usually under the heading 'rational psychology'—to metaphysics," Ward explained.[32] By "rational" he meant that the "self" was a logical construction. A self could not be put on scales, poked with brass instruments, or probed by experiments. Yet there were two ways scientists could approach this mystery, Ward said. They could view the self as a single substance, the proverbial "soul," or as a process, an entity formed by so many small mental events like pearls on a string. The self was either a thing or a process—an eternal stuff that never perished, or a medley of experiences that finally went away. The two options had been around for some time, but James infused the debate with new life.

When James wrote *Principles,* he concluded that the idea of the self as a singular entity—indeed as a soul—seemed natural enough. The soul was a plausible hypothesis. "A soul influenced in some mysterious way by the brain-states and responding to them by conscious affections of its own, seems to me the line of least logical resistance," he wrote. Still, James had his serious doubts, especially if the pursuit was to build a *science* of psychology, to which a soul-substance offered no material to study and no "verifiable laws." So he asked his readers whether the self as a "succession of states of consciousness" was not the better option. What followed was James's famous chapter nine. He introduced "*the stream of thought, of consciousness, or of subjective life.*"[33]

Before James, the British empiricists had tried to build mind and the self around an "association" of ideas. The person was a "bundle of perceptions." Idealists imposed an ego on sense perception in order to create the self. James said that all such attempts failed for the same reason. They could not see the person as an unbroken stream. Mental activity was not chopped up but continuous. *Principles* argued this with diagrams and cited experiments. "Thought from the outset is a unity," he said, "a single pulse of subjectivity." The best guess for how this happened was the mental phenomenon of attention. Attention seems to define the self at any moment. But what was being attentive? Was it the soul, or a moment of perception itself? James again sided with something like a soul, which he called a "spiritual force." But he chose this for ethical reasons and not scientific ones.[34] After he wrote this in *Principles,* James gradually changed his mind.

James and his wife were no more immune than anyone else to the great mystery of the self, the soul, and indeed what happens after one dies—what remains of a person. Just before her husband's death, Alice James wrote plainly, "I believe in immortality." James was perhaps more philosophical in his last years. Writing to a colleague, he said he "didn't think death ought to have any terrors," though "the great thing is to live *in* the passing day, and not look farther."[35]

As a psychologist and philosopher, James would finally collapse both the self and consciousness into one perishable process. Empiricism was a flexible approach after all, and its employer could change his mind almost with the weather. Already, James had changed once on consciousness. By the time he gave the Giffords, he had accepted the idea of a subconscious, a second level of awareness hidden deep in the brain. Religious experience, he proposed, happened in that "subliminal self," a place where biology touched an unseen world. "*If there be* higher spiritual agencies that can directly touch us," James said, "the psychological condition of their doing so *might* be our possession of a subconscious region which alone should yield to them."[36]

In his dedication to empiricism, James had apparently grown weary of the elusive soul, an unseen substance that was such an easy answer to so many problems. So he swung to the opposite side and opted for the self as a stream of experience. James came out most dramatically in his 1904 essay "Does Consciousness Exist?"

For twenty years, he wrote, he had distrusted the concept of consciousness as a *thing,* like a fly suspended in amber. Then he taught against it for eight more years. Now "the hour is ripe for it to be openly and universally discarded." Yet it was not so nihilistic as it sounded, James said. He jettisoned consciousness as an "aboriginal stuff or quality of being" but kept it as a "function." Drops of experience linked mind and world. The history of philosophy was "one long wrangle" over how this happened, and James proposed "pure experience" as the bridge between the two. "Consciousness connotes a kind of external relations, and does not denote a special stuff or way of being," he said. "There is no general stuff of which experience at large is made. There are as many stuffs as there are 'natures' in the things experienced."[37]

Pluralism and process reigned. Woe to the block-universe. A small revolution had begun.

Yet revolutions in conceptualizing the physical world seemed common enough in the first half of the twentieth century. The upheavals, all fair game for the Giffords, shared a concern about human experience in a world known so well by its two basic coordinates, time and space. In his *Principles,* James had analyzed the psychological perception of space and time, a world in which drops of experience had their continuity. For the average person, of course, time and space were practical daily matters, a practice of watching the clock (time) while traveling to work (space) in order to arrive at the appointed hour. Time and space were to be conquered, not questioned.

Yet the scientist and philosopher could not let it be. Questions arose in science about cosmic measurements and in religion about a timeless God's "presence" amid time and space. These two coordinates, now a metaphysical question, needed understanding in the deepest possible way, and that meant on the scale of the universe itself. At this level, two great solutions were offered: the organic continuum of experience (as James had emphasized) and Albert Einstein's proposals in 1905 and 1915 that matter and energy were the same thing and the universe had a four-dimensional reality called "space-time." The two views—which might be simply labeled organicism and relativity—were major approaches to explaining the universal order and the human place within.

In brief, organicism came out of biology and psychology and saw the universe as a flow of interrelated events, like a complex and growing organism. From this organic reality, creativity emerged and the whole was greater than the parts. Relativity was a product of mathematics (and Einstein's thought experiments) and it too described new connections in the universe. In the classical physics of Isaac Newton, space was an eternal receptacle in which objects moved. Like the man traveling to work, motion was measured by space and time. Then came Einstein's theory. According to relativity, as objects approached the speed of light, their space and time were changed (rods shortened and clocks slowed, in Einstein's favorite example). In this reality where space-time was altered, motion and time no longer had absolute measures: the measures were "relative" to each frame of reference in the universe. In Einstein's universe, a man going to work at nearly the speed of light and a man traveling by bus could not use the same ticking clock to chart their progress.

With organicism and relativity to consider, the God-believing metaphysicians of the day had a strange new world to work with, and some of them used the Gifford Lectures as the place to complete that exercise. Although every Gifford lecturer of this era had to keep Einstein's bizarre new claims in the mind, it was the organicist worldview that was by far the most predominant. Organicism rejected a mechanistic view of the universe and yet wanted to adopt a scientific approach to existence. That required it, in the scientific mode, to begin with empirical facts and then move to generalizations, such as universal laws or metaphysical principles. Through the use of this approach, organicist ideas earned the label of "realist" metaphysics, contrary to the dialectical "idealist" metaphysics of Hegel. Early in the twentieth century, organicism was called the "new realism," and decades later the "new natural theology."[38] Friends and enemies gave it other names as well: vitalism, holism, and emergentism, since it argued that higher qualities of life "emerged" from brute matter. In the long run, it was called process philosophy, for it emphasized not a static world, but a world in the very process of becoming.

In all of this, William James had played an inaugurating role, though he was never a leader or joiner of intellectual movements. After his passing, however, three other men represented an expansion of these new intellectual concepts and forces: Henri Bergson, the French philosopher, Samuel Alexander, a philosopher at the University of Manchester, and Alfred North Whitehead, the British mathematician who went to teach philosophy at Harvard. All three gave the Giffords, Bergson's and Alexander's overlapping the First World War and Whitehead's coming in 1928.

They were part of a wide movement that was linking together human psychology, biological evolution, and the new physics. They put an emphasis on immediate "drops" of experience, the flowing interrelatedness of things, and the great question of how people live in space and time. After James, the earliest on the scene was Bergson. He exchanged letters with James and met the Harvard psychologist, who was seventeen years older than the talented young French professor, mathematician, and philosopher. In the new realm of organicist thinking, James and Bergson would both influence and encourage each other. But who influenced whom the most—and thus was the most original—is lost in the mists of their correspondence and of history.

James had written about "streams of thought" and "pure experience" earlier than 1890, when they appeared in *Principles*. A year before that, Bergson had published *Time and Free Will,* which attacked the old associationist psychology (that mind was an "association" of simple experiences of the physical senses) and helped speed its demise in scientific circles. To explain the human mind, Bergson emphasized the ancient philosophical notion of "duration." Duration explained the nature of time; it lasted. It continued, from one point to the next. Bergson said that time was the essence of life and mind, an unbroken flow. Without time as the connecting tissue, a real life of activity would not come into being. This was contrary to how mathematics, mechanics, and natural science tried to divide up the world, putting time over here and space over there. Indeed, the mechanistic curse on modern man, said Bergson, was to make the measure of space his god.

For real human experience, Bergson went on, time was everything while space was an abstraction that followed in its wake. By failing to recognize this, science had fallen into dark geometrical pits of ignorance. In a world in which time truly defines reality, the deepest kind of human knowledge, Bergson said, comes from a "practical" intellect that operates on intuition, sympathy, action, creativity, and freedom. He had the zeal of the convert. As Bergson had written to James, he had once been a devotee of science "wholly imbued with mechanistic theories," especially those of Herbert Spencer. But after pondering how time worked in mechanics and physics, Bergson had a great conversion. "I saw, to my great astonishment, that scientific time does not *endure.*"[39] Time was sacrificed to space, and space was just a jumble of pieces. Something else glued life together.

That something, according to Bergson's 1907 work *Creative Evolution,* was the *élan vital,* or vital force, of the cosmos. It was undirected, but when it encountered brute matter, its fountainlike proclivity produced nothing but freedom and creativity. Bergson, an original thinker for sure, may have borrowed his ideas here and there, including from Ralph Waldo Emerson. In Emerson's essay "Experience," he had italicized *vital force.*[40] Emerson was struggling with another kind of rigidity in the universe, not scientific mechanism but Calvinist predestination. His solution was the free power of intuitive knowledge. Bergson would use the concept of "sympathy" as interchangeable with intuition.

By the time James gave his Hibbert Lectures in 1908, he dedicated a session to Bergson, a "young man, comparatively, as influential philosophers go." He was young and original enough to trouble European rationalism and materialism. In a world that said permanent things were nobler, Bergson espoused an intellect that, far from being static, grasped "the inner flux of life," James explained. "Bergson alone has been radical." His writings over the years must have affected James, for he said the Frenchman "had led me to renounce the intellectualist method," with its strict use of logic on what can or cannot be.[41] As a final kindness, James helped with the English translation of *Creative Evolution,* but he did not live to see its publication in 1911.

In a mechanistic age, Bergson's almost poetical ideas and writings enticed a remarkable number of biologists, many of whom felt stuck in their science. Before the First World War, when the English translation of *Creative Evolution* came out, Bergson took a triumphant lecture tour in England. Later, in 1914, he delivered the Giffords at Edinburgh on "The Problem of Personality," a set of springtime lectures that defined personality as a "continuity of movement" in time; the person was a spiritual experience that projected the past into the future by a life of will and action. The war intervened, so those lectures were never published and the second series was abandoned. Other matters were never finished as well. At the end of his life, Bergson tried to reconcile Einstein's space-time with his own emphasis on duration, creativity, and freedom, but his powers were dimming and it was a task probably too great for any mind. In the end, many consider Bergson the father of process philosophy, although, as will be seen, the mathematician Whitehead's Gifford Lectures finally stamped that doctrine on history.

Reared in a Jewish family, Bergson lived until 1941, seeing Europe consumed by the Nazi onslaught. Four years before he died, he declared his Catholic convictions in his will. Yet when France's Vichy government, under Nazi coercion, began to register Jews, Bergson resolutely stood in line, a shy and noble Frenchman with a Nobel Prize in literature. Bergson may finally have adopted the beliefs of a traditional theist, but in *Creative Evolution* he conceded only an impersonal cosmic fountain. "God, thus defined, has nothing of the already made; he is unceasing life, action, freedom," he said. "Creation, so conceived, is not a mystery; we experience it in ourselves when we act freely."[42]

In *Creative Evolution* Bergson urged "the collective and progressive effort of many thinkers" on this new system.[43] Yet he was no more an organizer than William James, and so other forces such as the Gifford Lectures would assist in gathering in and connecting up the new thinking about nature, the mind, and God. Early comers were such figures as Emile Boutroux, a French philosopher at the Sorbonne, whose 1903 Glasgow lectures, published as *Science and Religion in Contemporary Philosophy* (1908), foreshadowed Bergson's assault on the mechanistic worldview. Similarly, at Aberdeen in 1907, the German biologist and philosopher

Hans Driesch argued for directedness in biological life (reviving Aristotle's word *entelechy*). Driesch's title, "The Science and Philosophy of the Organism," set a tone for the movement. Reality was nonmechanistic, for it harbored vital forces, connected drops of experience, and pulsed creatively and purposefully, very much like organisms in the world.

When the new organic thinking streamed into the British world of science and theology, it produced the "New Theology" in the nonconformist churches, the "Broad Church" movement among Anglicans, and an acquiescence to "Modernism" among Roman Catholics. All three movements stood for the same Christian attitude, a willingness to adapt ancient church doctrines to the new findings of science. The literal belief in miracles, for example, was replaced by seeing God's redemptive scheme in an evolutionary, holistic, and creative world, a journey from Genesis through Christ to a new future. Anglican traditionalists, of course, scoffed at their Broad Church brethren. Rome took an even firmer stand with Catholics; the pope issued an encyclical against Modernism in 1907. The encyclical, which required priests to take an oath against Modernism before ordination, cited two heresies in particular: modernists rejected metaphysical reasoning, which proved God, and they replaced supernatural causes with scientific and psychological ones.

Despite such protests, however, the modernist desire to harmonize God, the Bible, and immortality with sciences drove many of Britain's brightest to the Gifford rostrums, and John A. Thomson was typical. A former Methodist minister, he became professor of natural history at Aberdeen and wrote books such as *Gospel of Evolution* and articles such as "Biological Philosophy." Oxford biologist John S. Haldane lectured in Glasgow in the late 1920s and went on to declare that biology would supersede physics as the true way to understand the universe. Anglican Bishop Ernest William Barnes, a former mathematics professor at Cambridge and leader of Anglican Modernism, lectured in Aberdeen in 1927. But by then, Modernism was fast becoming a spent force.

As Modernism ran its course in the churches of the West, empirical philosophers such as William James and Henri Bergson made their contributions and then gradually disappeared from the scene. Others would now take the torch of empiricism and organicism, and the two most prominent would be Samuel Alexander and Alfred North Whitehead, both of whom gave the Giffords. First came Alexander, a famed speculative philosopher of his day. On the intellectual front, Alexander inherited Bergson's leadership role as the Frenchman receded into the war years, a leadership status Alexander gained primarily by his Gifford Lectures. Delivered at Glasgow during the First World War, the lectures were published in 1920 as *Space, Time and Deity.* Although Whitehead was born only two years after Alexander, Whitehead's Giffords came later in his own career, presented at Edinburgh in 1928 as *Process and Reality.*

As British thinkers, neither Alexander at Oxford nor Whitehead at Cambridge escaped the influences of Absolute Idealism, as James had dodged it in America and Bergson eluded it in France. The rise and fall of idealism may have been applauded by many, but as alternatives, the pluralism of James and the spontaneity of Bergson still left the universe in an unseemly state of affairs. Alexander and Whitehead wanted to piece it back together again, making it an organized system of some kind. They both kept Einstein's new theories of space-time and relativity in mind, but they finally set off on a different path from him. What Alexander and Whitehead did mostly was turn Hegel, with his idealistic metaphysics, on his head. As the "new realists," they began with empirical facts (not with Hegel's rational whole). They began with the facts offered by science, and only then moved to metaphysical speculations on God's place in the universe.

This was finally a speculative quest, of course, and so it had its critics, some of whom called the "new realists" the same ilk of fuzzy thinkers seen in Absolute Idealism. James might have been horrified at this attempt by the "new realism" to create a system. It would create a block-universe that, while softer than most, was still a monstrous whole that defied radical empiricism and pluralism—or so James might have protested. On the other hand, he might have been pleased. Some new realists said they wanted merely to "complete" the work of James.

Samuel Alexander, who read and quoted James, may have been one of those completers. He was born and reared in a Jewish family in Australia. Moving to England, he studied philosophy under the Oxford idealists. When his interests shifted to science, he never abandoned the theism of his Jewish heritage. But he drew on Spinoza as a chief inspiration for trying to create a metaphysical system. A large man with a long beard, Alexander was a bachelor and teacher who was hard of hearing. He became one of the most beloved figures around Manchester during his years of teaching, especially after years of opening his home, with his mother and sister also as hosts, to students and others for Wednesday evening discussions in his large study upstairs. A philosopher's philosopher, he was elected president of the Aristotelian Society three times, and discussion about a single paper of his could dominate several society gatherings.

In 1915, the year Alexander received the invitation to give the Giffords, Einstein produced his "general" theory of relativity, which expanded the implications of his 1905 "special" theory of relativity. The general theory, born in the midst of world war, reached Britain by stealth about 1916 but was not widely known or understood for a few more years. In the general theory, Einstein described the largest scales of the universe as having a curved "space-time," the curvature producing the effect of "gravity." Whereas Newton had described gravity as an attracting force between two objects, Einstein said it resulted from objects moving into each other's curved space.

As a good metaphysician, Alexander also wanted to unite space and time. In his Giffords, in fact, he coined his own use of the term *space-time* to describe the nature of God. As a first step, physical reality was made up of points, and each point was an intersection of space and time. Each point was an instant of duration at a particular location. Sounding much like Spinoza, Alexander said that time was God's mind and space his body. He might have sounded like Bergson also, except that Bergson had said that *only* time was reality whereas Alexander demanded both sides of the coin, time *and* space. Motion was the force of God's mind that moved the universe forward, creating new qualities, according to Alexander's system.

All these arguments grew from Alexander's pledge to be empirical, which meant beginning with concrete things and experiences. He took the same approach in observing human religion, with its belief in God. In this empirical sense, Alexander said, human worship of God showed that God was for all practical purposes a real being. Next was to speculate on how God fit into the physical universe. Alexander concluded that God, being Space-Time, emerged along with physical reality. Emergence created a higher and higher "quality." This quality, in fact, was even higher than God, and Alexander called it Deity, the infinite ideal of the universe. His system seemed to be a kind of theism, but even Alexander's friendly critics were not sure.

Alexander's theism had a kind of hierarchy. Because God transcended man, God was worthy of worship. This produced human religion with its personal and communal virtues. Yet there was a still higher stage in the hierarchy. While the emergent God was higher than man, the highest of all was Deity. "Deity is a quality, and God a being," Alexander said. But for religious man, God was quite enough. "God is the whole universe engaged in process towards the emergence of this new quality," Alexander explained. "And religion is the sentiment in us that we are drawn towards him." God was thus immanent *and* transcendent: he was in the world as Space-Time and beyond the world as an eternal ideal. Alexander insisted this was not pantheism. "I believe that, in the end, a theistic conception of God's deity is demanded by the facts of nature," he said.[44]

In Alexander's being given the Gifford invitation in 1915, the world of British philosophy, in effect, wanted to see what he would do with a major project. The result, delivered as lectures two years later, was Alexander's "realist" alternative to idealism. The influences of the new organicist thinking were seen everywhere. Alexander's *Space, Time and Deity* made reference to Bergson and Whitehead, and later the Giffords by Conwy Lloyd Morgan credited Alexander's influence. As a psychologist and zoologist, Morgan said his *Emergent Evolution* borrowed from Alexander's metaphysics.

Alexander delivered the lectures in winter during the war, a time when railway travel north was hardly easy and when few students populated any university campus. Despite the bleak overhang, however, he drew crowds of one or two hundred and, by one account, charmed them in the process. The listeners were "touchingly" attentive, recalled Alexander, by showing an interest in his metaphysics of God as the world seemed bound for destruction. For a few weeks after the first lecture series,

Alexander remarked that he was still trying "to sleep them off," so intense was his presentation.[45]

Alexander did not hold out for an afterlife, however; it did not fit his metaphysics nor was immortality a tenet of his Jewish heritage. Memories were enough. Alexander died in 1938, not imagining the plight of his people in Europe, and his ashes were buried in a Reform Jewish cemetery. Well before then, however, Alexander had introduced his friend, the Manchester chemistry professor Chaim Weizmann, to a visiting Arthur Balfour, the future prime minister (and a Gifford lecturer as well). In 1917, the year of Alexander's Giffords, Balfour declared Palestine a "national home" for the Jews, and after much catastrophe Weizmann, in 1948, became Israel's first president.

Alexander presaged another seminal event in history, and that was Pierre Teilhard de Chardin's *Phenomenon of Man,* one of the most influential popular works of metaphysics in the twentieth century. A Jesuit paleontologist, Teilhard envisioned God evolving with humanity toward the Omega Point, a cosmic reality made possible by the incarnation of Christ. Whatever the specific formulations, whether Spinozistic in Alexander or christological in Teilhard, some believers worried that such ideas were merging God with the universe. A Jewish admirer of Alexander told him "you do walk humbly with your funny God."[46] In London, the Anglican dean William Inge spoke for Christian theologians who fretted that the "new theology" was confusing God with physical matter. Inge gave the Giffords on Christian Platonism at St. Andrews in 1917, but elsewhere mocked the futility of measuring God like a physical property. "Phrases like 'emergent evolution,'" Inge said, "only cover up an attempt to assert and deny change in the same breath." He said Alexander's philosophy seemed "to be entangled in a subtle materialism."[47]

In his published Gifford Lectures, Alexander lamented that Einstein's relativity theory on space-time and gravity had come out so recently that he could not incorporate it into the universe proposed in *Space, Time and*

Deity. Alfred North Whitehead, however, had a different vantage. He had followed Einstein's work since 1905—and would famously challenge it later.

In many respects, Whitehead summed up the generation of thought from James and Bergson to Morgan and Alexander. By the time of Whitehead's 1928 Giffords at Edinburgh, he had credited all four as opening the way for a more complete doctrine known as "process metaphysics," or what Whitehead called "the philosophy of organism." In this, Whitehead put together drops of experience, streams of thought, sympathy and intuition, time and space, and an organic world evolving into the future. When he added God to this recipe, moreover, process thought became a remarkable attempt at comprehension. It also became one of the hardest metaphysical systems to understand, but nevertheless, the most fruitful for a long-term encounter of science and theology.

Whitehead was no small player in the world of Western mathematics. At Cambridge he was the teacher of Bertrand Russell, who went on to philosophical and atheistic fame himself. Together they had produced *Principia Mathematica* (1910–13), the mathematical tome of the twentieth century. Whitehead then moved to an academic post in London where, from 1910 to 1924, his interest shifted to the philosophy of science. He rejected science's most mechanistic approaches and would challenge even Einstein's theory of relativity, offering "an alternative rendering" in his 1922 book *The Principle of Relativity.*

By ending the absolute measurement of space and motion, and making them "relative" to different frames, or coordinates, in the universe, Einstein had fragmented nature, according to Whitehead. For Whitehead, this was an image of people living by false psychological "appearance" while nature was doing something entirely different in time and space. He said Einstein divided the world into the real and unreal, putting everyday human perception in the latter. "We must reject the distinction between nature as it really is and experiences of it which are purely psychological," Whitehead protested. For him, human experience is part of nature so it knows the real space and time: "Our experiences of the apparent world are nature itself."[48]

In this rebuke of Einstein, Whitehead was no crank. With a different mathematics, and quite esoteric terminology, Whitehead explained the same phenomenon as Einstein. As the British astronomer Arthur Eddington said in his Giffords of 1927, Whitehead and Einstein were "tunnelling" from opposite sides of the mountain to describe the same wonders of nature.[49] In that battle, Whitehead lost, yet the contest prepared him for a lasting contribution to the God question. Rather than a relative universe, Whitehead proposed a unified cosmos of shared "events," each event a point in unified space and time. As the points overlapped, they produced a continuous reality: the natural world. After Whitehead left London for Harvard University in 1924, he put this idea into religious terms, a kind of echo of his youth in the English countryside.

Born in 1861, Whitehead was the son of an Anglican priest, a "thoroughly countrified vicar," as he recounts. Whitehead finally turned to agnosticism after losing patience with the rival claims of Canterbury and Rome and their belief in God as a reigning monarch. Life's experience, however, can change prevailing beliefs, and this is what apparently took place with Whitehead, a loving father of two sons and a daughter. The younger boy, Eric, was killed in action in 1918, the last year of the First World War. Such a loss, said colleague Bertrand Russell, promoted Whitehead to "want to believe in immortality." Whitehead's wife agreed that Eric's death probably was "behind" her husband's conversion to a kind of theism.[50]

After arriving in America, Whitehead openly addressed science and religion, putting a mild but friendly pox on both houses. Materialist science had great limits, he said. In turn, religion needed updating from its medieval past. Two series of his popular American lectures were published as *Science and the Modern World* (1925) and *Religion in the Making* (1926). Although Whitehead's rejection of mechanistic science—and of relativity—was well known, he now became known for speaking of God. Yet this was not a conventional Creator at all: not a deity who was monarch, moral force, or rational proposition. In his own modernist way, Whitehead valued God as simply a loving being—and a being who was both absolute *and* in process, growing with the universe.

In *Science and the Modern World,* Whitehead announced that James's theory of consciousness as a "stream" of experiences was the "inauguration of a new stage in philosophy." Later, in his Giffords, titled *Process and*

Reality, he said he wanted to complete the work of James, expanding what that psychologist had said about the mind to what science could say about a metaphysical universe. In doing this, Whitehead also confessed that he wanted to give the Jamesian outlook a new credibility, for European thinkers in particular had seen the musings of both James and Bergson as celebrations of irrationality. In defense of the American genius, Whitehead said in *Process and Reality* that one of his "preoccupations" had been to rescue James's philosophy "from the charge of anti-intellectualism."[51]

Yet to do that, Whitehead had to break one of James's rules—that philosophy shall not make a "system," a dreaded block-universe, out of human experience. With apologies to James, of course, Whitehead went forward anyway. His plan was to turn idealistic metaphysics into realistic metaphysics: both sought a rational whole and tried to explain all "internal relations" within a single reality. But now Whitehead would begin with empirical data, and the scientific world of experience, in an attempted "transformation of some main doctrines of Absolute Idealism onto a realistic basis."[52] The Gifford Lectures gave Whitehead, at age sixty-seven, that final opportunity.

The invitation came quite by surprise, thanks also to Eddington, the Cambridge astronomer. Eddington became too busy to deliver a second series of Giffords in 1928, as planned, so the host committee fired off a telegram to Whitehead, who was making a name at Harvard. Whitehead happily said yes, adding that the invitation "gives me an opportunity to put out a systematic work on the metaphysical notions which are occupying my mind." With the lectures still a year away, Whitehead offered up as a title "The Concept of Organism." He began work on them in the summer of 1927, ensconced at Caspian Lake in Greensboro, Vermont, and at some point, he wrote his son, enjoyed "a complete holiday and at last have got rid of the metaphysics buzzing round and round in my head." By April he had a new lecture title to send to Edinburgh: "Process and Reality."[53]

Preceding Whitehead, Eddington had given one of the most successful Gifford series yet. Eddington was a tall and expansive public speaker and storyteller. When Whitehead began his talks on June 1, 1928, filling the slot intended for Eddington, the audience was just as large and enthusiastic. They expected more of the same, perhaps, but here before them was not a tall Cambridge astronomer but an avuncular, bald mathematician. In the United States, Whitehead had drawn large student audiences and published his books to popular acclaim. But now in Edinburgh, thinking on his feet before a general audience, he rehearsed one of history's most complicated metaphysical systems. In his opening lecture, "Speculative Philosophy," he explained its role to a general audience. He brought up "things which the man on the street did not imagine could possibly exist," the *Glasgow Herald* reported. Yet even his most educated listeners, sadly, found the topic unintelligible. "It is painfully obvious," recounts a biographer, "that as public lectures, Whitehead's Giffords were bound to be a fiasco."[54]

Although not successful as lectures, their published version as *Process and Reality* was far more successful in identifying Whitehead's goal: to create a metaphysical system that relied neither on mechanistic science nor on Einstein's relativity. Whitehead had chosen the "organism" as his model. It was a model, in fact, that he found in an array of philosophers from Plato, the ancient Greek, to John Locke, the seventeenth-century Englishman. Plato saw a transience of worldly matter, and Locke speculated about simple impressions adding up to produce more complex experiences. For Whitehead, all these outlooks bespoke the reality of an organism, which was not a machine but a processing whole. By being a whole, moreover, it was not the fragmented reality of Einstein's relativistic universe. In this way, Whitehead conceived of the universe as organismic, not biological to be sure, but as integrated at every level of being, within space and time, and finally between God and nature.

In this cosmos, Whitehead described the basic units of reality as transient "actual entities," which he also likened to "drops of experience." Every level of existence was an actual entity, from molecule to God, and they were all in a process of existing for a moment and then becoming something new or different. In this sense, all things were constantly perishing and being rejuvenated. Whitehead sounded very much like James or Bergson on this topic, for while each actual entity came and went, it carried memories of the past into the future, as if handing on a baton in a relay race. If both traditional Christian thought and Absolute Idealism had made permanent "being" its most important category, Whitehead proposed that process was the fundamental rule. " 'Being' is constituted by its 'becoming,' " he said. "This is the 'principle of process.' "[55]

What held the process together was a contact made between the entities that Whitehead called "prehension," a momentary grasping of all other entities. By making these contacts, entities clustered as societies, and these formed higher levels of existence. Yet Whitehead realized that God had to be something more than just the highest conglomeration of societies; he should transcend them in some ultimate way. To make this separation between what was, in effect, spiritual and material, Whitehead said that all actual entities had a mental pole and a physical pole, a capacity for mind and for material laws. By having poles, each entity bridged the mental and physical realms, though in a hierarchy: a rock did not have the mental polarity of a bird or, even higher, a human being. Accordingly, nothing had a mental polarity like God—the highest order of mind.

In this formulation, Whitehead proposed a solution to the great question of God's immanence in nature and God's transcendence beyond the material world. Yet the most striking contribution Whitehead made was in talking about the character of the transcendent God. This God was a being who, while above the world, was in the world to "lure" the creation forward. The lure was cosmic love itself, experienced at all the mental levels. By plunging God into the physical world, Whitehead also made him an emergent being, much as if the world embodied God—an idea not too dissimilar from those espoused by Bergson and Alexander. Yet God had to be above the world as well, and Bergson, Alexander, and Whitehead offered different solutions. Bergson, in effect, put God at the root of reality in the *élan vital*. Alexander placed Deity, the highest ideal, ahead of God as God emerged with the growing universe. Whitehead did something similar. Although God's mental pole was his highest aspect, a still higher and deeper impulse in the universe was creativity, the production of novelty. Even God, as part of nature, developed toward this creative end.

While the God of Whitehead's metaphysics was quite different from a monarch, he was still above the world, setting out creative possibilities and infusing the world with love. At the same time, God grew with the creation and promised a kind of salvation, even if salvation was only higher forms of creativity and goodness. For all its optimism, however, process metaphysics did not offer a final "heaven" or even immortality, the very thing, reportedly, that Whitehead had wished for his son Eric. In a system in which actual entities perished, the idea of an immortal soul seemed out of sync, though there was one other option: a God of love could certainly retain that person as an object of love. Still, that was a mystery, Whitehead seemed to say, but in a cosmos of love it could certainly be hoped for.

The other important side of that love—and this had implications for science as well—was that God gave freedom to all things, whether under the laws of physics or in the moral choices of human beings. God could only lure human beings toward what was good and right, but never coerce. Lacking either coercion or persuasion, in fact, many actual entities go astray, producing the world's chaos, suffering, and evil. Nature itself, for example, has laws but also contingencies, and it is these that explain destructive outcroppings such as disease and natural disasters. Human beings are also free to reject God's persuasion, perhaps suggested most in the impulse of human self-destruction, otherwise known as war. Yet in all of this unhappiness, no entity in the universe is finally alone. The universe is moving through a great evolutionary ordeal, both holy and demonic, and God is sharing in every experience. As Whitehead explained, God is "the great companion, the fellow-sufferer who understands."[56]

Back in Edinburgh in 1928, the future influence of Whitehead's process thought could not have been anticipated. By his second lecture that week, the audience had dwindled to a half dozen. He would soon be bogged down in detail, despite a fairly clear syllabus. "It will be impossible for me to deal with the complete scheme of my Giffords in the ten lectures," he informed his hosts. So he adjusted and trimmed. What probably came across were his many difficult and sundry technical distinctions, while his broader view of God participating in the instants and process of the universe may have been lost.

After a summer in England, he and his wife returned to Harvard, where he now faced a publishing crunch. When *Process and Reality* was completed in January 1929, it was still imperfect; a well-edited "corrected" edition was produced years later so that Whitehead scholars could keep track of his entire system. At completion of his own imperfect draft, however, Whitehead lovingly wrote his son. "It is the biggest piece of imaginative work which I have attempted," he said. "Whether it will be a success I cannot have any idea. It is rather an ambitious book, of the sort which may be a dead failure."[57]

For a few decades, in fact, the book did seem to dissolve into oblivion, although by the 1950s it was being talked of again as presenting the most formidable metaphysical system of the century. The talk arose primarily in theological circles, in fact, where the quest to reconcile theism with an increasingly expansive science became more urgent. What Whitehead had provided was an alternative to both atheism and a mechanistic universe. Yet his idea about God, and his categories of "process" and "creativity" seeming to overshadow those of "being" and "immortality," brought down both praise and condemnation in the world of Christian thought.

For modernist Christianity, Whitehead explained how God could be outside the world and yet inside the world as well. For this group, process thought became a powerful alternative to a monarchial God who simply intervened in nature, and was thus hard to reconcile with science. It also helped solve the problem of why God, if all-powerful, did not swoop down and eradicate both evil and suffering. The process solution, in a word, was that God was *not* all-powerful, though ultimate in directing the universe.

A God without omnipotence was offensive to Christian orthodoxy. Yet the intractable problem of evil—and why God allowed it to endure— moved some evangelical theologians to consider what they called the "openness of God." While God might be omnipotent, these theologians said, he suspends his absolute monarchial power to allow nature to follow its own rules (and thus produce earthquakes or disease), and God puts himself in a situation even to negotiate with his creatures (as when Old Testament prophets prompted the Lord to change his mind). Here was a God of love who communicated and saved but did not mandate every disaster in the world, an idea that classical theology, with awkward and strained explanations, had to accept.

In the view of more orthodox Christian thinkers, Whitehead's metaphysics had something to offer, but the main complaint was that creativity claimed the pinnacle of the universe, not the person of God himself. With God as a second-tier player, theism was attenuated, according to the more orthodox, and belief in God could be bloodless indeed, despite the ethic of love that Whitehead's system espoused. After Whitehead, a group that came to be known as "process theologians" firmed up a more classic theism—with a personal God, immortality, and free will—working within the designs of Whitehead's metaphysics. As in generations past, here was a "new theology," and the doctrines of creation, love, soul, and redemption could all be reconciled somehow in a metaphysics of process, drops of experience, and an evolutionary universe.

The snags, however, were always two things in particular: Whitehead seemed to deny the individual soul and the power of God to bring ultimate salvation, as dramatically promised in the Bible. These two snags were often enough a topic of future Gifford Lectures. "The notion of the soul as a society of occasions rapidly succeeding one another seems . . . inappropriate for an understanding of human reality," the theologian John Macquarrie, somewhat modernist himself, said in his 1983 Giffords.[58] A decade after that, John Polkinghorne, an Anglican priest and physicist, gave due credit to process metaphysics for allowing an openness to science not seen before in theology. But could people die for this faith? "I do not find the God of process theology to be an adequate ground of hope," Polkinghorne said. "To put it bluntly, the God of process theology does not seem to be the God who raised Jesus from the dead."[59]

Yet the era of James and Whitehead, one that spanned early psychology and a new organic metaphysics, did offer much to theological reflection—and thus an invitation to keep God in an age of science. The Gifford psychologists—Ward, James, and Morgan—rejected medical materialism and offered openings for the spiritual life of the mind. Ward seemed to back theism and immortality, while James at the least verified the reality of religious experience. Either way, this was psychological music to the ears of theologians struggling with the assaults of modern science. Religion, after all, was geared to individuals, their experience and their destinies.

After the mind was conceived as a stream of consciousness, that idea expanded to the wider universe. It gave rise to realist metaphysics that began with science and experience but did not shun arriving at a concept about God. In Bergson, Morgan, and Alexander—and especially in Whitehead—God was deemed a creative force, emerging within the natural world and giving it spiritual qualities. In some of these views, God also existed beyond the world, a deity of goodness, power, and love on high.

One reason Whitehead received the Gifford invitation, it will be recalled, was that Eddington, the astronomer, was unable to present a second series of lectures. Like Whitehead, Eddington also had come to terms with Einstein's revolution in physics, but very much on Einstein's side. Eddington was the first to popularize relativity in the English-speaking world. When Eddington gave his 1927 lectures, published as *The Nature of the Physical World,* he was more interested in the nature of matter itself, marveling at how the hard little atom was finally being understood. For physics, the atom—the "indivisible," as the Greeks called it—was always the greatest question.

That question would also enamor a young Danish physicist named Niels Bohr, who would give the Gifford Lectures many years later. The son of a prominent biologist in Copenhagen, Bohr had grown up amid all the organicist debate of the early century, and his father had been quite favorable toward the belief in emergence, vitalism, and entelechy. The young Bohr, in fact, had befriended Alexander, the philosopher of emergence. From 1914 to 1916, Bohr was in England working at the Manchester University laboratory and was invited to the most exclusive discussion group at Alexander's home, where Alexander surely mentioned his forthcoming Giffords on natural theology.

Yet it was not for a decade more that modern physics seemed to reach the Gifford rostrum, primarily in the 1927 lectures by Eddington. It was a time, Eddington said, in which physics had been thrown into "anarchy," especially in the controversy over the atom. It was a year that well represented the start of a new debate in science and religion, a year when Eddington gave his lectures and Bohr and Einstein, over dinner in Brussels, argued about whether God played dice with the physical universe.

FIVE

DOES GOD PLAY DICE?

Physics

W HEN THE YEAR 1927 came around, the stock market still boomed, kings and presidents ruled in relative peace, and new fashions in clothes, dance, and jazz distracted Western society. The average citizen could hardly have noticed the great turmoil in physics, which was witnessing the "downfall" of the classic understanding of matter.

Amid the social and political calm that pervaded the West before the Great Depression, two events in physics were emblematic of its cutting-edge debates. One was the Gifford Lectures delivered by the Cambridge physicist and astronomer Arthur Stanley Eddington. The other was the Fifth Solvay Congress on Physics, which gathered the best minds in the field in the storybook city of Brussels, Belgium.

In Edinburgh, Eddington's task was to bring his large and enthusiastic audience up to date on the world of physics, and especially to apprise it of the new concepts in which the "greatest philosophical significance are to be found."[1] By philosophical, he also meant religious, for his 1927 talks went down as a rare case in which a big name in English science hinted that the changing concepts about matter itself might allow an opening for the God hypothesis, which had been losing ground in physics for more than a century. When his talks were published as *The Nature of the Physical World,* the book was perhaps the most widely read science text of the decade, something to think about after the world's stock markets came crashing down.

Eddington's talks also had a stop-the-presses quality, for as would be seen at the Solvay Congress several months later, a great new debate had arisen on whether matter—as it operated *inside* the atom—followed the strict laws of mechanics or revealed a new world of "uncertainty." The idea was already out, but it would come to a head at the Fifth Solvay. Since 1911, Solvay congresses had been the great family gatherings of leading physicists, made possible by the funding of the Belgian industrialist Ernest Solvay. Every few years, the sessions gathered amid Brussels's ornate hotels, grand palaces, and cobblestone streets. By the Fifth, the names spanned two generations, with Max Planck, Ernest Rutherford, and Albert Einstein among the older and Niels Bohr, Werner Heisenberg, and Erwin Schrödinger among the younger.

In the preceding year, Bohr and his student, Heisenberg, had argued that an innate "uncertainty" was attached to physical particles as they moved within the atom. Either a particle's velocity or its position could be measured at a given time, but not both: a situation that defied classical mechanics. If there was uncertainty, or "indeterminacy," at this level of matter, the implication might be staggering. Could the entire universe have this kind of ambiguity? The very idea struck horror in some scientific circles. At the Fifth Solvay, this was the big question, and Bohr and Heisenberg were about to defend their stance before the greatest minds in physics—especially Einstein himself.

Many stories have come down from the events in 1927, but perhaps no event resonates quite like the evening on which Einstein, Bohr, and Heisenberg had dinner together in Brussels and the uncertainty topic arose. It was already a pregnant year for such a debate, as 1927 was England's bicentenary celebrating Isaac Newton. Newton had given physics its bedrock principle: all matter operated in "strict causality," with all actions having reactions that were determined and measurable. With the debate on uncertainty brewing, Einstein had evoked Newton against the new theory of indeterminacy in matter. "The last word has not yet been said," he wrote in the spring of 1927. "May the spirit of Newton's method give us the power to restore unison between physical reality and the profoundest characteristic of Newton's teaching—strict causality."[2]

Over dinner in Brussels that fall, that debate had continued. Outside the Solvay's formal sessions, the meals were opportunities for the thirty or so physicists attending to talk over food, smokes, and drinks. On this particular night, amid the clink of dinner plates, Einstein lost his patience with the idea that the electrons moving inside the atom followed paths of chance, not Newtonian certainty, and so he said, "Yes, but do you believe that Almighty God plays dice?" It was not the first time he had used the phrase, but perhaps the most famous time. Bohr replied with equal resonance that Einstein should stop telling God what to do.[3]

Although Einstein never gave the Gifford Lectures, his theme of God playing dice captured the central issue for physics in the second act of the Gifford legacy. As with all materialist science, the question for religious belief was whether it left any room for the activity of God. Newton certainly had, but for a God who acted on the outside of a world-machine, while impersonal laws worked within. The God who acted outside, of course, finally became less relevant in a scientific world, and this explained why physics seemed to be a synonym for atheism. Yet the God question was always below the surface, and the Eddington lectures and Solvay debate illustrated how it arose again. In his Giffords, Eddington spoke of a new "anarchy" in physics, a time in which vying theories were trying to "seize the throne." The world, he said, might be seeing "the final overthrow of strict causality by Heisenberg." If that were the case, "religion first became possible for a reasonable scientific man about the year 1927."[4]

Before this time, the Victorian world of science had offered a useful but simple explanation of matter. In the late Victorian age, Lord Gifford had dedicated his body's "enduring blocks of materials" to the earth. A little way into the twentieth century, William James had asked whether human experience was merely "blind atoms moving according to eternal laws, or . . . the providence of God."[5] Looking back in his Giffords, Eddington said that "the Victorian physicist felt that he knew just what he was talking about when he used such terms as matter and atoms."[6] Atoms were like tiny billiard balls, each one a positively charged gel embedded with negative charges. Quaintly enough, it was called the plum pudding model of the atom, with electrons being the plums. Here was the simple electrical theory of matter. In the Victorian worldview, the cosmos was composed of charged atoms, attracting and repelling each other. But most interesting of all was that everything from atoms to planets was believed to move through a universal substance, never discovered but theorized to exist. This was the cosmic ocean of ether.

The Giffords began, in fact, when "ether physics" still reigned. The first physicist to give the talks was George G. Stokes, who had held Newton's former chair at Oxford University. His 1891 Giffords at Edinburgh, published as the two-volume *Natural Theology,* came just as ether physics was about to be debunked, yet it was still part of his description of the physical world. Stokes specialized in optics, or the physics of light. He had discovered fluorescence, turned waves and wind into mathematics, and helped answer "why the sky is blue." With his high forehead, sharp features, sideburns and wavy hair, Stokes cut a stately figure in the halls of British science, and his memorial plaque lines a crowded wall in Westminster Abbey along with those of Newton, Darwin, and others. Yet as he gave his Giffords, his world of science was fast becoming obsolete.

In previous centuries, light had been conceived of as both a wave and a particle, a controversy finally settled in Stokes's time in the theory of electromagnetism, which gave light a wave quality in some kind of medium. Light was the visible part of electromagnetism, a family that included all kinds of other wavelike forces, from magnetism to heat. To describe how light traveled and refracted, science had proposed ether as a universal medium, perhaps gel-like, in which light moved. As Stokes told his Gifford audience, without the ether hypothesis "the splendid edifice of modern optics could never be erected." Thanks to ether, science has "what we have every reason to regard as the true theory of light."[7] That theory, however, along with ether, was soon headed for the grave.

Stokes was both a rigorous scientist and an evangelical Anglican, so while he talked about God and the world, he was careful in making any claims of proof from the transient world of scientific theory. For his own faith, science was not needed as verification of God, but he gave his audience hints at what physics might suggest in God's favor, from a world of design and order to a substance such as ether, which suggested how invisible realities can pervade the universe. Some physicists of that day, in fact, had pushed ether physics to theological conclusions about God and the paranormal, though Stokes did not.[8] Instead, he used the complexity of the eye to present the classic argument from design: the eye could not have stumbled into existence by Darwinian evolution but must have had a designer. Yet his was the old Newtonian universe, and there was little use in citing the atom, as he did the eye, for pondering the Creator.

Thirty-five years later, when Eddington took the Edinburgh rostrum, the world of physics had dramatically changed. Like his predecessor Stokes, Eddington was an establishment figure in British science. He had been chief assistant at the Royal Observatory and from 1914 headed the observatory at Cambridge, where he held the chair in astronomy. His theory that nuclear reactions (the annihilation of matter) produced the "shine" in stars was the beginning of modern astrophysics, but perhaps most significant, he was the first physicist in the English-speaking world to grasp Einstein's theory of relativity, first presented in 1905 and then expanded in 1915. Reared a Quaker, Eddington remained active in the Friends society his entire life, a life spent mostly as a quiet bachelor. As time went on, however, he also showed other sides of himself: an adventurer, a public expositor of science, and finally a philosopher in search of a "final theory" about the universe.

Soon after arriving at Cambridge, Eddington, as secretary of the Royal Astronomical Society, received Einstein's 1915 paper on relativity, smuggled from Berlin across the battlefronts of the First World War. By 1918 Eddington had produced a summation, *Report on the Relativity Theory of Gravity,* which explained Einstein's proposal that the continuum of matter and energy at the largest scales of the universe produced a curved space-time, causing the effect of gravity. Einstein predicted that the curved space would bend a distant ray of light, and Eddington's decision to test that made him the Indiana Jones of physics. In 1919 he led a rugged expedition to Principe, West Africa, defying reports of storms to observe a solar eclipse, which allowed the viewing of passing starlight. His finding was international news: the light, indeed, bent. Before the expedition, Eddington's assistant had asked a senior scientist about the consequence of failure. "Then Eddington will go mad and you will have to come home."[9]

Quite sane, however, Eddington had mastered Einstein's relativity, but his abiding interest, it seemed, was in the tiny atom. Later in life he argued that "atomicity," the accumulative effect of all atoms, provided the real clue to the grand scale of the cosmos. Eddington's career came into full bloom during the "heroic years" of physics, roughly between 1900 and the 1920s. Part of that was Einstein's work on relativity, but a great deal more had to do with opening the book on the inscrutable atom. The atom, which meant "indivisible" when the Greeks first named it, had a longer history of speculation and mystery than perhaps any other object in the science of physics. It was viewed, simply enough, as the basic unit of matter.

When the ancient Greek philosophers—the first physicists, in fact—speculated about matter, their solutions were many. Some viewed it as containing earth, air, fire, and water; others said it was a continuous substance behind all things. The third idea was matter as particles, the smallest being the atom. Aristotle gave matter *potentia,* or potential, saying this was what determined that plants bore fruit and stones fell to the ground. For Christian thought centuries later, *potentia* was a kind of God-given design or, if the Greek philosopher Plato was preferred, matter was the raw material that God molded into forms that reflected his perfect ideas. In Christian doctrine, God in fact created matter from "nothing," the famous doctrine of ex nihilo, important to early theologians in order to show that God was entirely separate from matter—and its master.

Every interpretation of matter, it seemed, was open to some kind of philosophical interpretation. When Eddington delivered his Giffords in 1927, the two features in which he saw the most "philosophical significance" were fairly recent discoveries: the emptiness of the atom and the indeterminacy of the movement of electron particles around the atom's nucleus. To drive home the first point, Eddington opened his Gifford Lecture series with the story of two tables, one considered by the common man and another by the scientist. In the first case, the table was an item of hard wood, very solid and useful as a result. The scientist saw it quite differently: the table was made up of electrical charges with incredible amounts of emptiness between them, and thus anything but solid. As Eddington said about the atom—and thus the table—"it is 'really' empty space."[10]

Befitting the Giffords, Eddington presented this emptiness, and the indeterminacy of matter, as openings for a transcendent world spoken of by religion and metaphysics. The old physical universe—what Eddington called a "self contained world in which God becomes an unnecessary hypothesis"—had considerably softened up, if not evaporated in some sectors. "Having gone so far physics may as well admit at once that reality is spiritual," he said, characterizing one response to the new findings. Yet Eddington was no mystic, and he reminded his listeners (and readers) that taking apart the new physics was not a blank check to prove particular religious doctrines or put God in scientific terms. "We must go more warily," Eddington said, for no one should "reduce God to a system of differential equations." In 1927 bringing back the God hypothesis to physics seemed quite enough.[11]

In a strange way, for that to happen, so friendly a doctrine as universal ether, a godlike substance, finally had to be thrown out by the materialist research of science itself. This was the story of modern physics beginning in about 1900, and the stage was set somewhat earlier by the discovery of "fields" in matter, the space between particles where the electromagnetic waves—magnetism, heat, and light—spread. Einstein said the field concept was the greatest discovery in physics since Newton's laws. The operation of fields, however, contradicted universal ether, and it was Einstein's 1905 theory of "relativity" that resolved problems in field theory so that ether was no longer necessary. Because matter and energy were the same thing, Einstein said, particles and fields were also the same substance, just in different concentrations. This continuum of matter and energy turned space-time into a four-dimensional reality, which curved at large scales and created the effect of gravity.

As this upheaval descended on physics, the understanding of radiation and light—both forms of electromagnetism—took a dramatic turn as well. In 1900, while studying how heated material changes color, the German physicist Max Planck found that the energy was emitted in discrete amounts, which he called quanta. Quanta were so small that they seemed to be a continuous stream. In addition, part of Einstein's 1905 "miracle year" was to show that, similarly, light behaved like a quantum (a particle absorbed and emitted by atoms) as well as behaving like a wave (going around corners). Einstein called the light quanta photons. As will be seen, this particle-wave duality in nature was not restricted to light. It also seemed to be the dual personality of electrons inside the atom, and this was the basis of the "uncertainty" principle that so troubled Einstein.

Eddington, however, was not troubled at all. When he gave the Giffords, in fact, he conceded that Einstein's world, consisting of space-time, relativity, and the gravitational curvature of space, was a bizarre new discovery that had made science stand up and cheer like nothing else ever had. In physics, he suggested, the concepts that most boggle the mind—and are incomprehensible to ordinary people—are usually worthy of the greatest awe and praise. Yet when it came to the modern "downfall of classical physics," the new ideas about space and time were thin gruel compared to one other event: the discovery in 1911 that the atom was mostly empty space. Remarkably, he said, physicists suffered "no great shock" at this experimental revelation.[12]

That discovery had been made by Ernest Rutherford, who first revealed that the atom had a nucleus around which electrons orbited. For this, Eddington said, Rutherford deserved a place alongside Einstein, especially on the question of who in modern physics had made the world a stranger place. "The most arresting change is not the rearrangement of space and time by Einstein but the dissolution of all that we regard as most solid into tiny specks floating in [the] void," Eddington said, in effect casting his vote for Rutherford. He said that the empty atom was "more disturbing" than even the vastness of interstellar space, or the strange curvature that bends rays of light, as revealed by the revolutionary Einstein. "And yet," Eddington said, "when I hear today protests against the Bolshevism of modern science and regrets for the old-established order, I am inclined to think that Rutherford, not Einstein, is the real villain of the piece."[13]

The story of Rutherford's great "villainy"—said in loving jest by Eddington—opens the way to the modern understanding of the atom and, finally, the great debate on the uncertainty within matter. During the "heroic years" of physics, Rutherford had begun making his name in research on radioactivity, a new area of study of the inner workings of the atom. Radioactivity was the outcome of small charged particles leaving the atom's core, flying out at various rates, sizes, and velocities. Rutherford latched on to one particle in particular, which he called the alpha particle. It was one of the most common radioactive emissions, a positively charged, large, and fast-moving thing, and it was the tool by which he began a revolution in physics.

A sturdy and jovial New Zealander, Rutherford had studied at Cambridge University's Cavendish Laboratory, a leading center of physics. Over his long career, those around Rutherford remembered him in two different ways, first as "the one who could swear at his apparatus most forcefully," but also as a cheerful and boyish man of science: "When things went well with Rutherford he would bound upstairs singing 'Onward Christian Soldiers.' "[14] By way of Canada, where he had begun his work on radioactivity, Rutherford went to the University of Manchester, which had Britain's second-best physics laboratory. It was at Manchester that Rutherford used the alpha particle, probing the atom by pummeling it with particle blasts to see what happened.

In the experiment, Rutherford's team bombarded various metal foils with alpha particles. To their surprise, very few particles bounced off, and those that did angled away at extremes of 90 to 180 degrees. "It was almost as incredible as if you fired a fifteen inch shell at a piece of tissue paper and it came back at you," Rutherford said.[15] He had found an atomic "nucleus," which held a positive charge and around which electrons seemed to "orbit," much like a tiny solar system. For this, which revealed the atom to be mostly space, Eddington gave Rutherford the greatest compliment by calling him the true "villain" in the upheaval of early-twentieth-century physics.

Yet Rutherford's discovery was hardly the final word on the inner ways of the atom. The orbits had created an entirely new problem for physics: why did the electrons not spiral into the nucleus, destroying the whole thing? In 1911, the year in which Rutherford announced his theory of the nucleus, he met a young man from Denmark who would solve the problem of electrons keeping their orbits. His name was Niels Bohr and he was a novice in science who had recently come to England to study physics at Cambridge University.

From the days when he earned his doctorate in physics "a pale and modest young man," Bohr followed the mechanistic and experimental model, observing, forming hypotheses, and testing.[16] At the annual Cavendish banquet, Bohr met Rutherford and through a family connection was able to work in Rutherford's laboratory in Manchester during 1912. It was there, over the course of a few months, that Bohr solved the problem of the unstable solar-system atom. He took Planck's theory of quanta and argued that electrons had "steady state" orbits. The orbits changed only when a quanta of energy was absorbed or released. When the theory was published in 1913, it argued that inside the atom, the electrons that orbited farthest from the nucleus, at higher quanta, acted according to smooth Newtonian physics. On the other hand, particles at low quanta and nearer the nucleus jumped between orbits.

This apparent unevenness in the quantum states would remain a mystery and give rise to the debate over the indeterminacy of matter. Yet Bohr's steady-state theory solved a central puzzle in the atom's behavior: the spectrum of light it seemed to emit. Bohr showed that each quantum state of an electron represented a color on the well-known spectra, a series of bright-colored lines produced when light bounced off elements. Bohr thus began a revolution in chemistry, which used the spectra to understand how outer electrons produced chemical bonds. With this knowledge, Bohr and others completed the modern periodic table in 1922, the year Bohr won the Nobel Prize in physics. The quantum atom also changed astronomy, for now the atomic activity inside stars could be deciphered by spectral lines. At age twenty-eight, Bohr had put a new kind of atom on the map. He had also begun his collision course with Einstein over the nature of the universe.

Bohr returned to Copenhagen after his short stay in Manchester in 1912, but was back at Rutherford's laboratory, newly married, for a two-year stint from 1914 to 1916, a journey he and his bride made across the war front. When they returned to Denmark, Bohr stayed in Copenhagen to build his own legacy, despite overtures by both Rutherford, who would become head of Cambridge's Cavendish Laboratory, and Max Planck in Berlin, to recruit him to their respective institutions. Instead, Bohr opened his own institute in 1921 in Copenhagen, and before long it became a mecca for physicists from around the world. By the time Bohr died in 1962, his institute had been a research stop for seventeen people who went on to win Nobel Prizes.

One of them was the German student Werner Heisenberg, who first heard Bohr speak at his university in Göttingen. Senior scientists there had once called Bohr's quantum theory "awful nonsense, bordering on fraud."[17] But now his Göttingen lectures became known as the Bohr Festival. During that occasion, the twenty-year-old Heisenberg asked Bohr a penetrating question. The Danish physicist took him for a walk to talk it over. The next year, Heisenberg arrived as a researcher at Copenhagen.

What brought Bohr and Heisenberg together, in fact, was their mutual recognition of a major flaw that still went along with Bohr's quantum atom. For Bohr's theory could explain neither the perplexing way that light interacted with atoms nor why orbits and quantum jumps did not always correspond. Every scientific theory has a honeymoon, but because of these anomalies, Bohr's ended in 1922. The quantum atom was true as far as it went, but the idea of crystal-clear orbits of electrons began to face problems of measurement and description.

In its early life, the orbit concept worked well because it became a visual model that aided such projects as the periodic table. As a scientist, Bohr always preferred to describe physics in plain language. For all its plainness, though, the orbit concept could not explain the seemingly irregular behaviors in the atom, and that is where Heisenberg, the arch-mathematician, came to the rescue. He devised a mathematical equation to describe why some electron orbits were predictable but others were not. Yet he did so at the cost of throwing out a visual model of the atom. It was now a mathematical equation. As one historian said of him,

"Heisenberg, who—with the abandonment of the orbital conception—brought mathematics to Bohr's theory."[18]

Although mathematics may have saved the Bohr atom, Bohr himself was not at all pleased. His visual model had been turned into a mathematical abstraction, a thing that he abhorred. He and Heisenberg descended into a heated debate about taking such a drastic step, but in the end Bohr had to come around. As Heisenberg recounts, "We finally concluded that the formulae were correct, and I felt that we had come a good bit closer to the atomic theory of the future."[19]

The year was 1925, the birthday of "quantum mechanics," also called matrix mechanics for the kind of mathematics that Heisenberg had used. The mathematics linked the smooth action of outer electrons with the "discontinuous" action of inner electrons. The consequence was a move from Newtonian certainty at one level to quantum uncertainty at another. According to this formula, it was impossible to measure *both* the position and the velocity of an inner electron. Measuring one required a fuzzy approximation of the other, and vice versa. Here was Heisenberg's "uncertainty relations" in atomic matter. Heisenberg published this conclusion in a famous 1925 paper, and after that, Wolfgang Pauli, a Bohr associate, said, "Heisenberg has learned a little philosophy from Bohr in Copenhagen."[20]

While Bohr and Heisenberg shared a philosophy that recognized some kind of indeterminacy in the measurement of matter, they still had to decide how, for science, they would describe the remarkable claim—now famously known as the Copenhagen interpretation of the atom. Einstein had already openly rejected this apparent loss of strict causality in the quantum approach. But it was a younger physicist, Erwin Schrödinger, who tried to prove Bohr and Heisenberg wrong by coming up with a deterministic "wave" theory that explained the anomalies in electron orbits, and thus the inner nature of the atom. Using mathematics, Schrödinger came up with the same predictive results as Heisenberg but with a smooth, Newtonian "wave mechanics" to connect the behaviors of inner and outer electrons. Opponents of uncertainty rooted for Schrödinger.

When Eddington gave his Gifford Lectures, he noted the determinist bandwagon. "Schrödinger's theory is now enjoying the full tide of popularity," Eddington said in early 1927. He ascribed the acclaim to the merit of wave mechanics, but also to its relative ease of understanding, for it described a mechanical world so beloved by classic physics. As Eddington put it playfully, wave mechanics began with a "something," which then drew itself together, sketched an outline, became an orbit, and then congealed into an electron. "The electron, as it leaves the atom, crystallizes out of Schrödinger's mist like a genie emerging from his bottle," he said.[21]

Months earlier, in the summer of 1926, Schrödinger had made his famous journey to Copenhagen to confront Bohr and Heisenberg and persuade them of his wave-mechanics solution. Just as famously, the young physicist failed to make Copenhagen blink. From the moment Bohr met Schrödinger at the railway station, they began their friendly debate, but in the end Schrödinger left Copenhagen despondent, although not totally without hope. He did not solve the quantum anomalies in the atom. But his method of calculating the mechanical wave traveled by a subatomic particle won him a 1933 Nobel Prize in physics, awarded to him and England's Paul Dirac for "discovery of new productive forms of atomic theory."

Finally, Bohr and Heisenberg, working together in Copenhagen, were left to settle their own personal disagreement: how were they to describe their findings to the world of science *and* the world of the common man? Their strong personal preferences—Bohr for visualizing theories and Heisenberg for reducing them to mathematics—already divided them on how to talk about the Copenhagen interpretation. In February of 1927, according to one story, the two men came to loggerheads. Bohr left on a skiing trip in Norway, and there arrived at the conclusion that the quantum atom could be described as yielding to two kinds of measurement, though it was one reality. What was wrong, Bohr seemed to say, with having two pictures of the atom, a kind of dualism—Bohr's future "complementarity"—of viewpoints about a single thing?

Heisenberg, who stayed at home that month in 1927, decided on a different description, and he would upset Bohr greatly by writing it down in a paper for publication. Heisenberg emphasized the technical impossibility of making a measurement of the electron, simply because to see it, a photon of light had to be introduced, and this would deflect its current orbit. In a way, it was a very technical argument—human intervention interfered with nature—but the technical problem had deep philosophical implications. In effect, Heisenberg was suggesting that nature itself is finally indeterminate for human ways of knowing. Hence, it might be called indeterminacy in reality itself.

When Bohr returned, they argued over the paper that Heisenberg had already put into the publication process. The two men, as a team, really wanted to present a united front, it seems, but they still had their two personalities and views of science. So as Heisenberg declared a new "uncertainty relations" in the measurements and reality of physics, Bohr added a note to the paper describing a new perspective called "complementarity." Either way, Bohr said, the quantum effect "requires the abandonment of the causal connectedness between physical phenomena—previously the very basis of all descriptions of nature."[22]

Later that year, Bohr and Heisenberg attended the Fifth Solvay Congress in Brussels, and while their findings were a topic of a formal session, they also matched wits with Einstein over dinner. The question was, of course, did God play dice with the physical universe? For this Solvay and the next one in 1930, Bohr and Einstein engaged in a series of "thought experiments," stretching their physicist minds to the utmost in a debate on whether the location *and* velocity of an atomic particle released from a box—known thereafter as "Einstein's box"—could be measured precisely. Each time that Einstein set up the experiment that would work (a deterministic measurement), Bohr showed it would not (an indeterministic measurement). For the history of science, much is unresolved, but Bohr has generally been recorded as the victor. In 1928 he was signing letters to friends with "Your complementary devoted Niels Bohr."[23] In that year also, a few other problems with the quantum atom were tidied up and the "standard model" was completed. Rutherford called it "a triumph of mind over matter, or rather—over the rays."[24]

Decades later, when Bohr and Heisenberg gave the Gifford Lectures on natural theology, the "heroic years" of physics were a thing of the past, a subject for museum exhibits or historians of science. Yet the philosophical debate raised by the new conceptions of matter lingered on. In some ways, the Giffords by Bohr at Edinburgh in 1949 and by Heisenberg at St. Andrews in 1955 were an attempt to summarize what the Copenhagen interpretation of the indeterminacy of matter was all about. Despite the two men's apparent victory over Einstein, the ideas of complementarity and uncertainty still lived in a shroud of ambiguity, especially for those who expect science to give hard-and-fast answers. Indeed, in his own lifetime, Bohr had gotten fed up with professional philosophers, for that group, in Denmark at least, had never stopped complaining that they could not understand his principle of complementarity.

The puzzle was this: when Bohr spoke of complementarity and Heisenberg of uncertainty, were they talking about the limits of human perception or about the nature of reality itself? Bohr gave the Gifford Lectures first, and in some ways they supported a general verdict about his philosophy among students of his life: for Bohr, complementarity was about human perceptions. As he said of quantum laws, they are "responsible for the properties of matter on which our means of observation depend."[25] In philosophy, the technical field that studies human perception is called epistemology, a term that Bohr put right at the top of his Giffords. He gave them the title "Causality and Complementarity: *Epistemological* Lessons of Studies in Atomic Physics" (emphasis added). Bohr was *not* stalking the very stuff of reality, said Aage Petersen, his last assistant. "For Bohr, philosophical problems were neither about existence or reality, nor about the structure and limitations of human reason," he said. "They were communication problems." As Bohr used to say, according to Peterson: "We are suspended in language in such a way that we cannot say what is up and what is down. The word 'reality' is also a word, a word which we must learn to use correctly."[26]

Accordingly, Bohr emerged from the discussions with Heisenberg in early 1927, well before the Fifth Solvay, preferring the noble idea of "complementarity," a plain-language, reconciling viewpoint, a unified description. Both the wave and the particle, the position and the velocity, could be pictured. But the two "cannot be comprehended in a single picture" at the same time.[27]

An optimist and lover of poetry, Bohr was captivated by concepts of harmony and unity in nature. He did not like the somewhat more jagged, disorienting viewpoint of Heisenberg, who spoke openly about "discontinuity" in the life of the atom. As some philosophers viewed Bohr, he was in the line of a Hume or a Kant, who worked on a theory of knowledge but believed a theory of existence itself was a futile project. To understand how people know reality is enough, in this line of thinking, which sounded very much like Bohr's central interest. When questions of ultimate reality came up, however, Bohr liked to remain ambiguous at best. To such queries, he would tell a little story, as was his method over the years, and then leave the inquirer amused, and even touched, but usually not any clearer.

To deliver the Giffords, Bohr set off for Scotland in the spring of 1949, accompanied by his assistant Stefan Rozental and loaded down with manuscripts, notes, rough drafts, and even a wire recorder, forerunner of the tape recorder. Both he and Rozental hoped the ten lectures would amount to a long-awaited overview of Bohr's philosophy. Bohr was famed for speaking at a chalk-smudged blackboard, and he did the same in Martin Hall at New College, the divinity school in Edinburgh. The blackboard collided, however, with the lectern used for morning devotions. "With the touch of boyishness which remained in him, he got Rozental to help him to move the lectern—but only a few centimeters each day, so that it was not noticed," goes the story.[28] The blackboard finally stood where he wanted it. Despite the best intentions, however, Bohr never published his lectures, and although he made efforts at a draft, only some rough notes remain. The hoped-for overview would never materialize.

By this time, Bohr was a veritable world diplomat, not a book writer. His legacy comes down to 150 concise papers. His audiences often came to bask in his celebrity, not the obtuse physics. A student at the Edinburgh lectures spoke of "boring Bohr," and the man who helped Bohr prepare the Gifford talks, Abraham Pais, agreed that he was "a divinely bad lecturer," and not only because his ideas jumped around. "He spoke too softly."[29] Bohr also made frequent references to "the box," and in fact, just before he died, his last chalk illustration on a blackboard was of Einstein's box experiment—he was still arguing with the master.

Outside of physics and his difficult lectures, however, Bohr himself was a magnetic personality, drawing the world to his residence in Carls-berg, Denmark. On a trip to the United States in 1938, he told American scientists that, based on rumored experiments in Germany, he believed a uranium atom could be split in half, a fact that was verified at Columbia University, and was the first glimmer of the American atomic program. The Allies later brought Bohr in on the atom bomb project briefly, if somewhat blindly. For his part, Bohr would rue the day of the atomic bomb and in 1950 call for an "open world" of shared science for the sake of global peace.

Relations between Bohr and Heisenberg sadly chilled after the Second World War. Heisenberg had stayed with the German physics es-tablishment, as did Planck and others. Yet Heisenberg, who became the director of the Max Planck Institute in Göttingen, the leading center of physics in postwar Germany, was on par with Bohr as a candidate to give the Gifford Lectures. He delivered them in 1955 at St. Andrews, and when published, *Physics and Philosophy* became one of the popular sci-ence books of its time, much as Eddington's *Nature of the Physical World* had seen remarkably good sales among the reading public.

By the time Heisenberg gave his Gifford Lectures at St. Andrews, his many interpreters took his "uncertainty" to apply to reality, to the stuff of existence itself. His lectures corroborated that view, even though it was never stated so flatly as that. What Heisenberg added to the picture was the ancient Greek concept used by Aristotle: the *potentia,* or potential and direction, of any given part of nature. Thomas Aquinas had taken Aristotle's *potentia* and put it in God himself. Revising Aristotle in his own way, Heisenberg said that quantum reality had substituted *possibility* for neces-sity at the foundation of matter. "Scientists have gradually become accus-tomed to considering the electron orbits, etc., not as reality but rather as a kind of 'potentia,' " he said.[30]

While not taking a stance on the role of God in the quantum uni-verse, Heisenberg said its uncertainty had allowed a new appreciation of the "natural language" of religion. Science itself, for example, was forced to use plain words for elusive realities, which was not too different from religion after all. "After the experience of modern physics," Heisenberg said, "our attitude toward concepts like mind or the human soul or life or

God will be different." The words are not scientific, but "we know that they touch reality."[31] If Heisenberg was essentially mute on physics and theology, he had other forums after the Giffords to express his views on religion, as will be seen.

By the time Bohr and Heisenberg gave their Giffords, the exciting era of relativity and quantum physics had become part of the background, not the center of an earthquake. In the days of Einstein's relativity, for example, Western culture tried to mirror what science said was physically real. Modern jazz threw out convention to meld creative freedom with an underlying order, and poetry shunted aside metrical rhyme to do its own version of the same. European painting was the most vivid expression of that age, a burst of abstract design and cubism, embodied most famously in the work of Pablo Picasso, who turned ladies of the night (*Les Demoiselles,* 1907) and musicians (*Accordionist,* 1911) into geometric shapes seen from shifting frames of view. Such paintings, Bohr would later testify, had indeed caught the spirit of complementarity. At St. Andrews, where Heisenberg gave his Giffords, a bit of cultural legacy also carries on. A leading musical group at the school is the Heisenberg Ensemble, an orchestra of about thirty instruments played mostly by professional musicians that often accompanies the St. Andrews Chorus. As a volunteer orchestra, however, the ensemble's exact composition of players at each performance can be very "uncertain" indeed.

While Western culture had swooned under the popularized influence of the new physics, the theological world was probably more cautious. In time, however, relativity and quantum uncertainty would become props for theological speculation. But as theologians stood back, it was the scientists themselves who seemed willing to engage in "God-talk"—and the public had certainly hankered for such commentary from the great names in science. Such was the case, of course, with Eddington, Einstein, Bohr, and Heisenberg.

Perhaps Eddington, a Quaker, and Einstein, from a Jewish background, would do the most to inject God into the modern physics discussion.

Eddington did so by the prominence of his Giffords in 1927. Einstein did so by the off-the-cuff comments he frequently made about the deity.

By speculating about God in his Gifford Lectures, Eddington may have become popular with the general public, but at some cost to his scientific reputation. In response to *The Nature of the Physical World,* for example, a cottage industry of atheist writers sprang into action, and others in his field frowned on such exploits. Eddington had good company in suffering this outrage, however. The astronomer James Jeans, a former president of the Royal Astronomical Society, also viewed mathematics as the mind of God (and in 1931 the University of Aberdeen appointed Jeans to give the Giffords, though he withdrew later). Jeans said that with the new physics, "the Great Architect of the Universe now begins to appear as a pure mathematician."[32] Also in Eddington's corner was a small group of British theologians working on "empirical theology" who argued that the cumulative order in the physical world—not just an object such as the human eye—bespoke design and purpose.

The closest Eddington came to theology in his Giffords, however, was his argument that "mind-stuff," not brute matter, seemed to hold the universe together. With so much emptiness in the physical world, that world is "entirely abstract and without 'actuality,'" Eddington said, except for the order given to matter by consciousness. In a universe of empty space, he went on, "we restore consciousness to the fundamental position." Science had thus made an opening for a higher consciousness that could be spoken of as "a universal Mind or Logos," what others called God.[33]

He also addressed the classic theological question of free will, which a deterministic universe such as Einstein's, for example, technically ruled out. The theory of uncertainty in matter, however, gave free will a new opening. In fact, Eddington had pushed the idea as far as it could go in science. While Heisenberg spoke carefully of the uncertainty "relations" in matter, Eddington felt emboldened to make it a law of nature: he called it "the *principle* of indeterminacy" in his Giffords. "Physics is no longer pledged to a scheme of deterministic law," he said. At the least, he continued, "we may note that science thereby withdraws its moral opposition to freewill."[34]

Eddington stood roughly in the Christian tradition as a Quaker, but he had "no impulse to defend" religious doctrines in his Giffords. In fact, he warned against using the latest scientific findings to try to prove one doctrine or another. As he put it bluntly, "I repudiate the idea of proving the distinctive beliefs of religion either from the data of physical science or by the methods of physical science." Ordinary people had every right to adopt one doctrine or another, and in his published lectures, he spoke to believers who were ambivalent about science interfering in belief. In fact, Eddington had taken a rather modest route in his Gifford talks, pointing only to a transcendent realm beyond ephemeral matter. As he said, he had chosen "the lightest task by considering only mystical religion."[35]

Before his career was over, Eddington would help open up another area of physics that gave natural theologians much to talk about: the idea of an expanding universe. In that fateful year of 1927, this revolution had begun quietly when the Belgian priest and physicist Georges Lemaître, a student of Eddington's, published an obscure paper that built on Einstein's large-scale universe. Back in 1917 Einstein had formulated the first quantitative model of the universe, calling it his "cosmological considerations" and showing how the push-and-pull of energy and gravity ended in a state of cosmic equilibrium. Lemaître was among a few physicists who wondered whether the Einstein universe was truly as stable as it seemed. Lemaître finally decided that it was expanding, and that the expansion may have begun at a point Lemaître called the first fireworks, or the primeval atom.

For Eddington, the Einstein universe of eternal equilibrium was best of all, until the equations began to show instability and at least one astronomer was finding in distant nebulae a major "red shift," a shift in the light wave that suggested the universe was spreading apart. When Lemaître sent Eddington his paper, it spurred the British astronomer with the zeal of a convert, at least on the expansion idea. In 1931 Eddington said the expansion theory was "so preposterous" that it demanded hesitation, adding coyly: "It contains elements apparently so incredible that I feel almost an indignation that anyone should believe in it—except myself." Two years later Eddington's *Expanding Universe* laid out the entire scheme in popular-book form, bringing the idea to the masses. "The original unstable Einstein universe might have turned into an expanding universe or into a contracting universe," he said. "Apparently it has chosen expansion."[36]

Eddington knew what genie had been let out of this bottle, for the expansion, and especially Lemaître's model, challenged the eternal universe of Newton and Einstein for the first time in a millennium. The challenge also included the whiff of a creation event, a sort of "primordial nebula," as Eddington described Lemaître's theory. "Lemaître does not share my idea of an evolution of the universe from the Einstein state," he wrote. "His theory of the beginning is a fireworks theory." In the beginning was a "violent" projection. It began from a "point or atom." And it was "strong enough to carry it past the Einstein state."[37]

As was Eddington's custom, he addressed his readers personally, humorously, and candidly when the topic became delicate for a scientist. "I cannot but think that my 'placid theory' is more likely to satisfy the general sentiment of the reader," he said. "But if he inclines otherwise, I would say—'Have it your own way. And now let us get away from the Creation back to problems that we may possibly know something about.' "[38] As a science book, *Expanding Universe* did not cite God, and the word *creation* was used but once. Between the 1960s and the 1990s, however, the theory of the big bang gained increasing empirical evidence and, as will be seen, became eminently fair game for Gifford Lectures of that era.

While Eddington may have gotten in some trouble for his God-talk, Einstein was usually let off the hook. His use of the God word was typically in a humorous vein. In 1919, when Eddington had gone to Africa to view a solar eclipse and try to verify Einstein's theory of relativity, Einstein was asked how he would have felt if the experiment failed. "Then I would have been sorry for the dear Lord," he said.[39] The "dear Lord" was Einstein's code word for the laws of the universe, and those around him picked up on the nomenclature. A few months before the Fifth Solvay, Heisenberg wrote to Einstein, saying that despite uncertainty, cause and effect could still be real in the utter beyond. "We can console ourselves that the dear Lord God would know the position of the particles, and thus He could let the causality principle continue to have validity."[40]

Einstein spoke about God in a variety of forums, including the Solvay Congress, where he apparently confounded some of the younger physicists. "Einstein keeps talking about God: what are we to make of that?" one Fifth Solvay participant asked.[41] While his fellow physicists usually understood Einstein's God allusion, the slightest mention in public could become an international incident. In 1929, for example, when Einstein told an inquiring rabbi that he believed in the God of Spinoza, the rabbi enthused that Einstein might "bring to mankind a scientific formula for monotheism."[42] If true, it would have been for Spinoza's monotheism, or perhaps pantheism, according to which "All things are conditioned to exist and operate in a particular manner by the necessity of divine nature," as Spinoza made the case in his *Ethics.* Einstein also dreamed of the day when a "unified field theory" could explain such a necessary system of matter. In the footsteps of Spinoza, he equated the deterministic universe to God, a deity who did not play dice because the laws of nature did not operate on chance.

Einstein's comments could stir other kinds of uproars, quite the opposite from the foregoing rabbi's enthusiasm. In 1940 in a speech of Einstein's that was read to a New York conference on religion, science, and philosophy, he urged clergy to give up a personal God who was interested in human affairs. Now came a negative storm over Einstein's atheism. For one historian of physics, in fact, it was the question of God that loomed large in the 1927 Solvay debate between Einstein and Bohr. "The actual subject was not physical theory, but God," Gunter Stent, a molecular biologist who studied at Copenhagen, has argued. Einstein operated out of ancient Middle Eastern monotheism, which clashed with Bohr's ancient Asian precepts that defined reality as a godless flux. "Bohr's atheism, the counterpiece of Einstein's monotheism, . . . was more affined to traditional Far Eastern philosophy," according to Stent. Thus, no argument by Bohr could persuade Einstein, who could not abandon Spinoza's God.[43]

Throughout his career, Bohr had been viewed as among the most philosophical of the physicists, an interest that went back to his church upbringing. He was reared in a prosperous Copenhagen family, his father a professor of biology with a Christian background and his mother born into a culturally active Jewish family. The young Bohr thus lived in two worlds, but mostly the cultural Christianity of the Danish middle class. As a young man, he had read Søren Kierkegaard, a fellow Dane and a Christian existentialist from the nineteenth century, with some enthusiasm. But he finally faced a religious crisis, and by the time he went to England to study physics, the idea of God had lost its appeal. The aim of life was happiness, he wrote his fiancée, making it impossible "that a person must beg from and bargain with fancied powers infinitely stronger than himself."[44]

What stayed with Bohr were the philosophies that talked about how the human mind knows the physical world. They were ideas that leaned toward a subjective interpretation, a view learned from Kierkegaard, from Bohr's own teacher, a Danish philosopher named Harald Høffding, and apparently even from William James's writings on pluralism. While in Manchester working at Rutherford's laboratory from 1914 to 1916, moreover, Bohr had been invited to speculative sessions with the philosopher Samuel Alexander, who had also proposed ideas about human knowing built on the "double aspect" theory of knowledge—the mind can know two realities, physical and mental, that work at different levels and are not always reconciled.[45]

In his only published paper on the topic of religion, Bohr spoke not of deities and doctrines but of psychological experience. He identified consciousness, creativity, and free will as the spiritual matters at stake in physics. Complementarity, he said, helped reconcile these three human activities with the seeming determinism of organic life. In the past, he said, the religious idea of free will flew in the face of "the mechanical conception of nature." But after frustrated attempts by science to explain "our own consciousness as a causal chain of events," Bohr said, free will had gained a new "recognition, in modern psychology."[46]

As a final revelation of his outlook, Bohr turned to the joys of modern art and Eastern thought. Cubist painting intrigued the physicist-philosopher, for in its theoretical core it was about attempting to show the viewpoints of several observers in one scene, the "single picture" that was denied by the atom. Similarly, when his coat of arms was designed, the yin-yang symbol of Taoism took the center. "Complementarity, the elucidation of quantum mechanics, was his contribution most precious to him," Pais said. "There lay his inexhaustible source of identity in later life."[47] The celebration of complementarity was religion enough for Bohr, as perhaps Spinoza's metaphysics had been for Einstein.

In contrast, Heisenberg seemed far more willing to discuss the God question in public. Indeed, he was happy to divulge what other physicists were saying during the heroic years of physics. At the Fifth Solvay Congress, Heisenberg reported later, Bohr said to him: "The idea of a personal God is very foreign to me." For Bohr, religious language spoke of reality in the same way as poetry. Still, Heisenberg reminded Bohr, many people were using indeterminacy as "an argument in favor of free will and divine intervention." Bohr called it a "simple misunderstanding." Human decisions and deterministic biology could be complementary. Bohr reportedly said, "When we speak of divine intervention, we quite obviously do not refer to the scientific determination of an event, but to the meaningful connection between this event and others or human thought."[48]

Heisenberg also recalled events that disclosed his own beliefs. In 1952, for example, he was among the physicists who gathered in Copenhagen to discuss building an atom smasher, or particle accelerator. Being summer, it was a season of long nights, when the sun skirted the horizon, and during a walk with his old friend Pauli, Pauli asked Heisenberg, "Do you believe in a personal God?" Heisenberg said yes, but only after redefining what "personal" meant. For Heisenberg, God's "existence seems beyond doubt," but as a "central order" who can be reached "as directly as you can read the soul of another human being."[49] Of the quantum revolutionaries, Heisenberg and Planck (though Planck sided with Einstein on strict determinism) ended up the most explicit in their theistic beliefs.

In later years, Heisenberg was described as a Neoplatonist, a moniker that Eddington and Jeans had also received in their day. In 1973, three years before he died, he as much as declared Plato's victory, for science now had only mathematics to conquer the last frontiers. "Mathematical forms . . . , if I can express it in a theological manner, [are] the forms according to which God created the world," Heisenberg said. "Or you may also leave out the word God and say the forms according to which the world has been made. These forms are always present in matter, and in the human mind, and they are responsible for both."[50]

In time, with careful and tentative steps, theologians began to apply concepts in the new physics to problems faced in Christian theology, typically problems of paradox, such as God being infinite and finite, a unity but also a trinity. One person inspired by Einstein's relativity was the Scotsman Thomas Torrance, a professor of Christian dogmatics at the University of Edinburgh. In the past, he said, theology was reluctant to allow an infinite Creator into the "finite receptacle" of space and time. But with Einstein's relativity, space and time become "relational," thus open to new dimensions. By the incarnation of Christ, he said, God "is at work within space and time in a way that He never was before." In his *Space, Time and Incarnation,* Torrance even proposed a "theological geometry" that pictured God as a vertical dimension intersecting space-time: "We must not think of the Incarnation as an intrusion" or a miracle, he said, but as the "chosen path of God's rationality in which He interacts with the world."[51] In a similar vein, the theologian Wolfhart Pannenberg has used "field" theory as a way to talk about the Holy Spirit moving, like an electromagnetic wave, through physical creation.

The concepts of complementarity and quantum uncertainty also gave new handles to Christian thought, again helping solve the paradox of dualities in various doctrines. How could Christ, for example, have two natures, divine *and* human? Or at least, how could human believers conceptualize such an apparent contradiction? In a Gifford lecture, the Oxford theologian Keith Ward said such seeming opposites in God—his mercy and justice or his trinity and unity—can be resolved by the analogy of a quantum world. "They are not contradictions, but inadequate attempts to articulate the Divine nature, which we cannot grasp in itself," said Ward, speaking of God as Bohr and Heisenberg might speak of the electron. "Like the wave-particle duality in physics, they may seem like contradictions to the uninformed, but they have a consistent application in fact."[52] Perhaps the modern Christian believer had to approach God from several different vantages—and not seek a single picture.

For all such projects applying physics to God, there was an equally cautious acknowledgment of what Eddington had said to the "religious reader" of his *Nature of the Physical Universe.* That reader should be happy, Eddington said, "that I have not offered him a God revealed by the quantum theory, and therefore liable to be swept away in the next scientific revolution."[53] After Eddington died in 1944, the theories of relativity and quantum mechanics generally stood the test of time. Yet there were still new levels of the physical universe to probe, and this has produced even more particles and forces. Eddington's lifetime saw the discovery of the neutron, a particle that joined the proton to form the nucleus of the atom. It was revealed by attempts to "smash" the atom into smaller pieces.

From the days when Rutherford fired alpha particles at the atoms that made up various metals, atom smashing expanded dramatically. The Cavendish Laboratory began operating its first atom smasher in 1932, sending protons down a long glass cylinder to break apart a lithium nucleus. The year before, in Berkeley, California, Ernest Lawrence had constructed an eleven-inch circular "cyclotron," which generated a million volts to propel protons. By the time Heisenberg spoke at St. Andrews in 1955, atom smashers had confirmed about twenty-five new subatomic particles—and the number surged to more than two hundred by the end of the twentieth century.

This proliferation seemed, Heisenberg said, to undermine belief in the unity of matter. "But this would not be a proper interpretation," he said in his Giffords. "The experiments have shown the complete mutability of matter [and] final proof of the unity of matter. All the elementary particles are made of the same substance, which may be called energy or universal matter; they are just different forms in which matter can appear."[54] The matter took its forms, as Heisenberg suggested later in life, by the guidance of mathematical truths, which was also his way of saying by the mind of God.

Whatever the "universal matter" was, it was manifesting itself in a veritable "zoo" of particles. What became increasingly clear, moreover, was that particles had a range of different behaviors, though they were presumably made of the same energy and matter as everything else. Some particles existed in their own right. Others grouped themselves together. Most surprising, when particles "jumped" from one state to another, they created the forces between still other particles and a variety of energy fields. Fortunately, by the 1960s scientists began to organize this zoo of particles into three kinds of categories. The heaviest particles (protons and neutrons, for example) were made up of quarks "glued" together by gluons. Thus quarks and gluons accounted for two basic constituents, and the third was the lightweight particles (such as electrons, photons, and neutrinos); as a group they are called leptons (Greek for "light").

In their days, Einstein and Eddington had looked at the particles of the universe and begun a kind of religious quest to unify them in a single mathematical equation. It was the search for a "final theory." For Einstein, the mechanics of "unified field theory" held out the most promise. He envisioned a day when matter and energy fields would be understood as one unified phenomenon, a deterministic universe in the mold of Spinoza's metaphysical God. For Eddington, the "atomicity" of the universe impressed, and he spent his late career in search of a "fundamental theory" that accounted for all the atoms in the universe and the proportional sizes of objects, small and large, in the cosmic architecture. Although neither Einstein nor Eddington completed this journey, the quest for a "final theory" or "theory of everything" has not been abandoned, and continues to give physics a quasi-religious ethos. The closest physics has come to that Holy Grail, though, is in reducing all relationships of particles to four fundamental forces. The task of science—yet unachieved—is to unify the operation of all four.

The first of these forces is gravity, the weakest of all forces, and a product of space-time. The second force is electromagnetism, which produces electricity and light, and scientists have understood it to arise from the exchange of leptons. The final two forces are located in the atom's nucleus. There, a "weak" force explains how the nucleus deteriorates and lets off little particles called radioactivity. Thanks to the weak force, for example, Rutherford had obtained his alpha particles to bombard other atoms and discover the nucleus. Finally comes the fourth force, which may be called the greatest of all. It is the "strong" force that holds together the nucleus, a nucleus made up of particles called quarks that are glued together by particles called simply gluons.

Although modern physics had explained how some combinations of these four forces have a mathematical unity, it has failed to describe their ultimate unity. That is why modern physics, to the bewilderment of ordinary people, has switched from particle theory to string theory, in which the smallest parts of matter are visualized as vibrating loops of string. It is the strings—which have never been seen but are imagined in highly advanced mathematics—that presumably form the particles.

Even if strings would hardly be a wise topic for the Giffords as a public lecture, the theory of everything still claims some attention. It still evokes the dreams of God-thinking Einstein and Eddington, though not necessarily of Bohr. "Our monotheistic traditions reinforce the assumption that the Universe is at root a unity," British astronomer John Barrow explained, giving his Gifford talks at Glasgow in 1988. Yet a theory of everything finally is a misnomer, he said, because particles and forces, galaxies and gravity, are *not* everything. When it comes to human life, physics has a limit. "A Theory of Everything," for example, "will shed no light upon the structure of the human brain and consciousness."[55]

By the 1970s and 1980s, however, physics in general had begun to intersect with biology on the question of how biological life arose over billions of years in a universe that is otherwise mostly cold, lifeless matter. It had been a topic of natural theology before, even in the era of Eddington, but the science matured only at the end of the twentieth century. When this time arrived, the terminology was not "particles" but rather "constants," speaking of the portions, balances, and strengths by which particles and forces worked together to create matter, elements, chemistry, and finally biological life. Thanks to the constants, it seemed, the universe was not a cosmic soup of particles, as it might have been. The constants were also mysterious. Where did they come from? As properties of physics, they did not evolve into place by trial and error, as biological life might do in arriving at its balances and forms. The constants were simply *there* from the very beginning. They produced what would be called a "fine-tuned" universe, and in later Giffords, the problem of cosmic fine-tuning brought physicists, biologists, and theologians onto the same turf where, again, the God question was unavoidable.

It might not have surprised Eddington. Yet he gave his Giffords in the second act of their legacy, at a time when physics basically suggested a mechanical universe that needed no God hypothesis. What emerged in Eddington's age, at least, was a softening of that machine-universe, a finding of enormous space within the atom and indeterminacy in matter itself. Energy and matter, like spirit and form, seemed to oscillate back and forth between each other. For the ordinary observer, the suggestion was tantalizing enough. First, the spirit of God might be in that empty space. Second, the cosmic spirit may come into the physical world in ways suggested by Einstein's equation $E=mc^2$, where invisible energy becomes visible matter, and vice versa. Third, God might indeed be playing dice with matter, though as Heisenberg said, "the dear Lord" was well in control of ultimate causation.

For all the spiritual good that the new physics may have done, from one point of view, its benign side was invariably contrasted by an evil twin—the ability to unleash nuclear destruction. Scientific research, as Heisenberg said in 1973, is always aimed at applicability, and for nations at war, superior weaponry is a desired application. With atomic physics it was "extremely disappointing," Heisenberg said, "that the first practical application was for warfare."[56] The year 1927 had seemed filled with promise, both on the stock market and in physics. Then the economic crash of 1929 changed the world, for as life became fragile, many people turned to dictators and their promises of social salvation by means of war.

Although physics provided new weaponry, warfare was ultimately a social problem, a smashing apart not of atoms but of social relations. To understand how this happened, a new science had been born—sociology, the science of society. One Gifford lecturer, the American theologian Reinhold Niebuhr, delivered his prognosis of the modern social problem just as the new weaponry went to war. When he spoke in Edinburgh in 1939, Nazi airplanes were dropping conventional bombs on the city. A year before, German scientists had discovered how to split the atomic nucleus, and soon enough the United States Army would also be on the nuclear trail. It launched the Manhattan Project and in August 1945 dropped two atomic bombs on Japan. From Scotland to Japan, the social problem was being writ large, and though Niebuhr had followed sociology for a generation, he realized that something more than just "social forces" had gone terribly wrong.

SIX

SOCIAL FORCES *and* SIN

Sociology

R EINHOLD NIEBUHR ARRIVED in Britain in March 1939. On the eve of his Gifford talks, a newspaper said he had brought "a teaching to be reckoned with." Niebuhr himself was about to reckon with the distractions of war. Could he make theology as relevant as gas masks? He carried one for a while, then put it aside. "Amidst this whole impending tragedy my little lectures seem futile and foolish," he wrote home.[1]

During his nine months in Europe, the Nazi occupation of Czechoslovakia gave way to all-out hostilities. Having signed a mutual nonaggression pact, Hitler and Stalin invaded Poland in September, and Britain declared war. By the time Niebuhr began his second series of talks in October, German airplanes were bombing Edinburgh's naval yards at Port of Leith, where Lord Gifford sometimes had gone to resolve commercial disputes in shipping tonnage.

"We had our first air attack yesterday," Niebuhr jotted in his October 17 diary entry, the day after his lecture "Man's Essential Nature," with its emphasis on Christ as the model. "I was too preoccupied with theology to hear the anti-aircraft guns, though I noticed my audience was not too attentive."[2]

Niebuhr's wife and son would soon return to their home in New York, but the nine months of mounting crisis did little to slow down the tall, midwestern social critic and German Reformed church pastor. Niebuhr had several invitations to speak, and he was reckoned with in various ways. At Oxford he drew crowds on the topic "The Relation of Christianity to Marxism, Fascism, and Liberalism." His German heritage raised some eyebrows in Scotland, and when he traveled to Amsterdam for a summer World Conference of Christian Youth, Dutch authorities detained him. They required a letter from Union Theological Seminary, where he taught in New York City, certifying that he was not a communist. Niebuhr had twice run for political office as a Socialist. His writings on Marxist sociology, namely the class struggle, had prompted the *New York Times* to report, "Doctrine of Christ and Marx Linked."

A half century into the Gifford Lectures, Niebuhr was one of its most political figures, a churchman attuned to the new tool of political analysis called social science. Besides two world wars, the political upheavals of the day between capitalist owners and organized labor had colored Niebuhr's views of the world. Sociology, the new science of society, analyzed the social forces—economic classes, political structures, age groups, occupations, and even symbols of collective "meaning"—shaping the new urban and industrial societies. Religion's embrace of the new explanatory science was a mixed blessing. In the Social Gospel movement, social science provided a practical means to apply Jesus' ethics of love and justice. But in matters of supernatural belief, sociology allowed less room to see God's mysterious activity in society.

As a science, sociology naturally fell into act two of the Gifford legacy, where theology and philosophy were pushed aside by scientific theories. From the start, sociology had relished the challenge of charting group behaviors and putting them in models, much as a chemist would organize the elements in his laboratory. By reducing social forces to a kind of mechanics, sociology would equal anthropology, psychology, and physics in their attempts to dispel mystery from the world—and thus relegate God to an afterthought in human affairs. In Niebuhr's lifetime, the encounter of sociology and religion was in full swing, especially in the Social Gospel movement and in Christianity's brush with Marxism, one of the most influential sociological interpretations produced in the West.

Born in 1892 and ordained in 1913, Niebuhr came of age as the Social Gospel movement was peaking in influence. In the spirit of that time, Niebuhr also wanted to "save the world." He had hoped that the Social Gospel could prod middle-class Christians to change society out of good-will. But when that did not work, Niebuhr decided the Social Gospel offered a more extreme option, a revolutionary upheaval in class structures to end economic injustices. A German-American patriot during the First World War, Niebuhr then became a passionate pacifist and advocate of social justice. He finally moved from his prosperous Detroit congregation to New York City where, as a writer and seminary professor, he became a Socialist political candidate vowing to take on capitalism.

Niebuhr would change his mind over the course of his career, but the use of the social sciences by his generation was never entirely absent. In his Gifford Lectures, *The Nature and Destiny of Man,* for example, Niebuhr still used such social analysis, and on a favorite topic of the sociologists: the way that people's beliefs are determined by their class interests. For Niebuhr's purpose, this explained why Western philosophical rivalries were so fierce. Wealth, rank, privilege, and power were at stake, not just beliefs. "The real dynamic of the struggle must be explained in socio-economic terms," he said.[3]

To illustrate this point, Niebuhr contrasted the belief systems of Christianity and its two great rivals in the nineteenth century, naturalism and romanticism. Naturalism rejected the supernatural and located all life forces in nature itself. Romanticism, in turn, offered myths of a utopian past, and its vision of a human brotherhood would often play on European racial memories as well. According to Niebuhr, social classes adopt beliefs that advance their class interest. Since the European aristocracy had based its privilege on Christian belief, the rival middle class adopted naturalism to topple those aristocratic oppressors. Next in line were the lower-middle-class and industrial workers. When they felt oppressed by the middle class, now the property-owning bourgeoisie of Europe, romanticism was adopted as a unifying belief for the underclass revolution.

Although naturalism produced its modern problems in secularization and the hegemony of science, Niebuhr found the worst excess in the legacy of romanticism. Romanticism became a seedbed of fascism and Marxism. Fascism viewed the racial state and its dictator, as manifest in Nazi Germany, as the primary unit of human loyalty, while Marxism advocated a workers' revolution that broke national boundaries and distributed property equally. By the time Niebuhr delivered his Giffords in Scotland, the ruinous nature of both ideologies was clearer than ever before, and Niebuhr was having second thoughts about the usefulness of both social analysis and revolution.

When he delivered the first talk at Rainy Hall, the wood-paneled and stained-glass heart of New College, on April 24, his topic was not social revolution but "Man as a Problem to Himself." He emphasized less the problems inherent in unfair economic structures or vying social classes, and placed the real onus of human misery somewhere in the human soul. The time had come, Niebuhr told his audience, as reported in the *Scotsman,* for "a fresh study of human nature." A deeper look was important "because modern culture, whose great [scientific] achievement was the understanding of nature, could not be said to have been equally successful in its understanding of man."[4]

The monstrous dissolution of war was forcing liberal clergy, once confident in structural reform, to grasp for something deeper. Having swung about as far to the secular side as a religious minister could go, Niebuhr found a "road back to faith."[5] After two decades of telling American Christians—in sermons, journals, and speeches—that the living God was found not so much in theology as in reform of social structures, Niebuhr's religious comeback made his outspoken ways more controversial than ever.

His Giffords would look at society and history, but they primarily focused on the biblical analysis of human nature, with its story about the mystery of sin. In the biblical view, conflict arose from sin, not social forces alone. War was the sin of individual pride amplified on the stage of history. Although sociology had often interpreted human actions by a "simple determinism" that emphasized social forces and downplayed human free will, Niebuhr said the Bible did "not allow this interpretation."[6]

With its story of free will and sin, the Bible in a time of war had a more relevant, if often tragic, message: the conflicted nature of human aspiration will always produce both freedom and evil, progress and self-destruction.

Niebuhr lived three decades beyond his Gifford Lectures, and he continued to stand by most of the arguments he made in Edinburgh. His contribution to social ethics, both in the Giffords and afterward, would be called "Christian realism." It was a realism about the flaws in human nature. It held that social and political power must not be trusted but must in turn be checked by other political powers, all in the cause of curtailing the sinful propensity of humankind.

Arriving at such a stance involved both a political and a theological shift in what Niebuhr would call his "intellectual pilgrimage." If before he had endorsed radical reform and even social revolution, he now believed that "remorse and repentance" over sin, followed by "responsible action," were probably the better steps toward improving society and promoting social equality and justice.[7] Theologically, he had moved from a kind of secular view of gospel ethics to the new Protestant mood of neo-orthodoxy, with its emphasis on sin, a transcendent God, and distrust of utopias.

Both of these changes—political and theological—had their roots in an important era of modern sociology, the era in which Niebuhr had entered the ministry. Sociology reached its "classical period" from 1890 to 1930, a time of ripening theories and hardening research methods. It held out an alluring promise for any young minister who wanted to change the world. The science of society "was designed to overcome blind drift, fate, or the unforeseen and unintended consequences of man's action," according to two later sociologists.[8] Knowledge like this, which predicted social behaviors and consequences, also gave the power to control those forces in society.

To gain a perspective on the rise of social science, we can use the life of Adam Gifford as a road marker. Gifford was just eighteen in 1838 when August Comte, a teacher at the French Polytechnic, coined the term *sociology*. Comte was a disciple of the French utopians who spoke of "social physics." According to them, sociology was the highest science, flowering atop a historical progression that allowed physics to explain the workings of matter, and biology the workings of organisms.

After 1838, it was probably some years before Gifford heard of this science of "positivism" emanating from Comte's Parisian circles. Positivism was a cousin of naturalism and materialism, but as a scientific view of life it emphasized that human beings could work only with the "positive," or evident, facts of the material world, not its unknown underpinnings. Sociology was born, therefore, in the cradle of positivism, becoming a child interested only in facts about obvious objects.

After Comte and his period, the next two "founders" of sociology were Gifford's exact contemporaries. One was the English engineer and *Economist* editor Herbert Spencer; the other, the German revolutionary Karl Marx. The two Victorians accepted Comte's materialism but viewed social forces differently. Spencer, the preeminent apostle of evolution, coined the phrase "survival of the fittest" that Darwin borrowed. Spencer started out with his *Social Statics* in 1850. By 1876 he was publishing his three-volume *Principles of Sociology*. For the mechanistic-minded Spencer, life and society developed as matter differentiated into higher complexity. Both the brain and a free industrial society were results. Suiting well the temper of British economic thinking at the time, Spencer advocated individualism and laissez-faire capitalism.

Marx also propounded historical stages driven by matter in motion. But his was a "structuralist" science typical of thinking on the Continent, where social life had been constrained for generations by land, ancient custom, and layers of class, rank, and status. For history to change this kind of order, in Marx's sociology, a final class conflict was destined to play out. The working class had to overthrow the owners of capitalist production. A vanguard that understood these social laws would begin the process, which would finally expand into an ideal society under the "dictatorship of the proletariat."

All three—Comte, Spencer, and Marx—would make appearances in Niebuhr's Gifford talks. The lectures hinted that Marxian ideas lent themselves better to social justice whereas the ideas of Comte and Spencer were built on a delusional "dogma of progress," with all the dangers that entailed. Modern ideology had borrowed generously from Comte and Spencer, he said, and that would included even nazism, which was also bent on a dream of crowning human progress, by way of tanks, with the Third Reich.

In his Giffords, Niebuhr had another point to make about Comte, and he made it in his talk "The Easy Conscience of Man." By "easy" conscience, he meant the modern rejection of human sin. Comte had argued that in the evolution of society the family could teach "a sufficient measure of social love." This would produce citizens for utopia. Niebuhr rejected such optimism in the name of Christian realism. In a world of sin, the family was not so innocent as supposed by utopians and romantics. Families often united to destroy other families, oppress populations, or obtain glory and privilege, as the history of royalty and fierce clan loyalties had shown. Like any political body that can be corrupted, Niebuhr argued, "The family is also the source of that 'alteregoism' which is a more potent source of injustice than the egotism of the individual."[9]

Despite these realities, Niebuhr summarized, the modern conscience was easygoing because all its modern philosophies, from the materialism of Marx to the idealism of Hegel, held that the march of history would extinguish the problem of human sin. Such an easy conscience about human goodness had been justified by "the most varied and even contradictory metaphysical theories and social philosophies," Niebuhr said. "The idealist Hegel and the materialist Marx agree in their fundamental confidence in human virtue."[10] They differed only in their predictions of the historical time and social circumstances in which human goodness would come to fruition.

The United States would produce a few luminaries in sociology's classical period from 1890 to 1930, but European thought still had pride of place in the sciences, putting the theories of the French sociologist Emile Durkheim and the German legal and economic theorist Max Weber on the highest pedestals. For Durkheim, the son of a rabbi, the central scientific problem was to establish that society was an actual entity. Thomas Hobbes, the British philosopher of materialism and individualism, for example, had said that society was nothing but individuals joined to protect one from the other. Or as his British heir John Stuart

Mill would argue, "Men are not, when brought together, converted into another kind of substance," for people are made by "the laws of the nature of the individual man."[11]

Durkheim viewed matters differently. Society preceded individuals, he believed, drawing tenuously on the French Enlightenment philosopher Jean Jacques Rousseau, who asserted that nature expressed a "general will" in human collectives. Whereas Spencer's staff had collected long lists of social factors, Durkheim seriously used statistics. His chief curiosity was the rate of suicide in a society, which in the past had been tied to climate. In contrast, Durkheim said that every social group has a "collective inclination" for suicide "quite its own." The inclination increased, he said in his book *Suicide,* when social change produced "anomie," or hopelessness. The greater the change, the more suicides there were. Marriage and birth rates similarly revealed "a certain state of the group mind" in society. All these "social facts," his *Rules of Sociological Method* explained, fall in the domain of sociology.[12]

Durkheim's structuralist view of society was reinforced by the work of Weber, the German legal economist. Although Weber took society to be made up primarily of individuals, he also presumed laws behind "social action." Sociology sought a "causal explanation" for how society worked.[13] And he generalized "ideal types," or roles, as a way to analyze societies. When Niebuhr tied philosophy to social class, he was echoing Weber's argument: social *and* religious beliefs legitimize one political structure or another.

The European theories—evolutionism, the soft structure of ideal types, and the rigid structuralism of class struggle—shaped American approaches. Yet Europeans would eventually call sociology the American science because of its rapid spread in education. As early as 1915, courses in sociology were being taught not only in most American colleges, universities, and normal schools, but also in many theological seminaries.

The intersection of sociology and organized religion was logical enough. At the turn of the twentieth century, this knowledge of social action, facts, and forces would enamor America's liberal Protestants. Having rejected many supernatural creeds, they invested their optimism in human potential. If there was sin, it was in the distorted structures of society, not the brotherhood of the kingdom of God. "The fundamental laws of society are identical with the fundamental laws laid down by the social teaching of Jesus," said Josiah Strong, a Social Gospel leader. In this idealistic view, social problems could finally be solved when citizens loved their neighbors, distributed wealth more evenly, and through charitable or governmental action reformed those who had slipped into indolence, drink, or poverty. It supported government interventions in economic life, and for the churches' own part, thirty-three Protestant denominations with the Social Gospel impulse formed their own socially minded bureaucracy, the Federal Council of Churches. At its first national meeting in 1908, the council adopted a report, "The Church and Modern Industry," that urged churches to build a "new social order." The only thing the teachings of Jesus lacked, Social Gospelers believed, was the analysis provided by sociology. Therefore "religious organizations not only studied the social gospel, they invited social scientists to explain it to them."[14]

Liberal Protestantism would be disabused of this optimism, first by the Great War and then by the Depression. The call for social justice could not be quieted, however. Niebuhr heard that call even in a pessimistic era of war and unemployment. Reared in a German-American parsonage in Missouri and Illinois, Niebuhr attended denominational schools. When he entered Yale in 1913, its School of Religion had a course of studies with a sociological emphasis. It offered classes in systematic sociology, labor, immigration, American social conditions, and municipal problems. This was the program that Niebuhr entered as a third-year undergraduate fresh from his denominational college and seminary. Despite his interest in contemporary social problems, he must have found social science often very narrow and dry, for he overlooked many of the recommended courses in sociology.

The young Niebuhr's mind was on being a pastor and preacher, even an activist, and resolving his own religious questions as a pastor's son. He took a liking to biblical exposition and philosophical theology, and was particularly impressed by the new theological emphasis on "personality," not dogmas or miracles. In Germany in particular, liberal Protestantism had responded to the Enlightenment's scientific attacks on Bible literalism and supernatural events such as miracles by placing its confidence in the life and ethics of Jesus, a human personality who reflected the personality of God. This quality of personality—the human expression of character, creativity, love, wisdom, piety, and righteousness—was a biblical truth immune to scientific skepticism. At the end of his career, Niebuhr would argue that personality and history, both ultimately "ambiguous" to science, were the linchpins for Christian apologetics.[15]

The sources of Protestant liberalism reached back to a generation earlier than Comte's, a time in Germany when the pietist theologian Friedrich Schleiermacher repackaged Christianity for the bourgeois culture of avant-garde Berlin—whose citizens he called the "cultured despisers" of religion. He presented the essence of religion as feeling, the essence of Christian faith as "religious consciousness" and a sense of ultimate "dependence" on God. It was a cosmic scenario in which Christ redeemed humanity from ignorance rather than from sin. Upon these ideas, Christianity would adapt to its current culture, becoming positive to the sentiments of the romantic movement in literature, or appreciating the revival in the arts, and as a result being less despised by upper classes.

In his sometimes-merciless Gifford Lectures, Niebuhr added this naive side of Christian romanticism to his catalog of human errors. When mixed with sin and hubris, romanticism fed a primitive religious nationalism among modern Germans, summarized in the lecture outline as "Schleiermacher to Hitler." The German liberals and romantics were Christianizing Immanuel Kant, taking God on faith but putting their worldly energy into ethics. When Schleiermacher died in 1834, thirty years after Kant, he was succeeded in influence by Albrecht Ritschl. A contemporary of Lord Gifford, Ritschl put aside mystery and theory to seek practical, cultural Christianity. Faith rested on "value judgments," he said, not supposed Bible "facts." The supreme fact was Jesus, exemplar of moral perfection and ethical wisdom.

Across the ocean in America, that practical and liberal image of Jesus took on a life of its own in the Social Gospel movement. The popular Christian novel *In His Steps,* written in 1897 by Congregational minister Charles Weldon, showcased selfish businessmen and asked, "What would Jesus do?" Unbeknownst to American readers, such renditions of Jesus had been constructed on German scholarship, that of church historian

Adolf von Harnack in particular. A disciple of Ritschl, Harnack became the great figure in this liberal school, and indeed Niebuhr's pastor father reared his midwestern family on Harnack's ideas. Harnack's influence was spread most by his monumental history of dogma, which argued that much doctrine was either Hellenistic or political accretion. When these were stripped away, Christianity was a simple essence: "Firstly, the Kingdom of God and its coming. Secondly, God the Father and the infinite value of the human soul. Thirdly, the higher righteousness and the commandment of love."[16]

The work of Ritschl and Harnack was in step with the European ideas of social progress. During the Gilded Age in the United States, this credo captured even the vision of evangelicals. They recognized how industry tarnished urban life and became Social Gospel reformers. Merging the commandment of love with social science, they saw a kinship between socialism and the kingdom proclaimed by Jesus. Two events in 1912 made the kingdom seem closer than ever. The Baptist theologian Walter Rauschenbusch issued his *Christianizing the Social Order,* a list of capitalist wrongs to be righted, and Socialist party leader Eugene Debs received one million votes in the national presidential election.

This was a year before Niebuhr's father suddenly died, forcing Niebuhr to lead his German Evangelical Synod parish for five months before heading off to Yale. Ordained at age twenty-one, he gave sermons that from the start showed an interest in social issues. Before leaving for Yale, he preached on the church's mediating role between capital and labor. Another sermon was titled "Christianity and the Workingman." His studies at Yale helped him find personal answers to the great struggle among many college people of the time, and that was between belief and nonbelief in such a rapidly changing world. Like many idealistic liberals, Niebuhr found solace in William James's *Will to Believe,* which argued the human "right" to hold religious beliefs amid uncertainty, especially if they brought practical benefits.

For Niebuhr, one of the clear benefits was the call to live an ethical life and promote that vision in society. When he left Yale and returned to his home in the Midwest, he met his first practical ethical challenge: the First World War and the dilemma it created for German immigrants in the United States. The ethical thing to do, he decided, was to become unequivocal American patriots. As a pastor in the German Evangelical Synod, he worked to Americanize the German-American churches. The disaster of war was finally so great, however, that his ethics swung him from patriotism to pacifism, a Christian parallel to the Wilsonian call for a League of Nations to make the "world safe for democracy." As a Detroit pastor, he faced very local ethical problems as well, what he called the "typical" problems of a modern minister in an industrial city. Yet he was ultimately frustrated at middle-class Christianity's unwillingness to promote economic justice. The churches were unable to lift a finger as Detroit, with its mushrooming auto industry, became embroiled in racial and industrial unrest.

From his Detroit pulpit, Niebuhr's national star rose, boosted by his activism and his furious output of columns, editorials, and articles on "the social problem," especially in the liberal journal *Christian Century*. His first secular leadership role came when the mayor of Detroit appointed him to head the city's Interracial Commission. Although a critic of science in capitalist hands, he gave social science credit enough; it was a reformist tool, used increasingly by university or government specialists in the form of statistical analysis, social surveys, and case studies. For the racial work in Detroit, his commission employed "experts to make a real survey of race conditions," Niebuhr reported. "Then we are going to try to liberalize the mind of the community on race relations."[17]

During the heyday of the Social Gospel movement, Niebuhr had not paid too much attention to its thinkers and theorists. But through the 1920s he would join the Socialist party, echo the works of Rauschenbusch, and ridicule the lukewarm politics of church liberals. He also helped organize the Detroit chapter of the Fellowship for a Christian Social Order (and in 1930 cofounded the Fellowship of Christian Socialists). But this willowy, balding, almost frenetic bachelor minister was different from the early Social Gospel writers, set apart by the "stridency and breathlessness of his prose—as if verbal vigor alone could turn the tide against the culture of bourgeois contentment," said his biographer, Richard Fox.[18]

Niebuhr's radicalization came not over war and pacifism but because of the arrogance of capitalists. "The social realities of a rapidly expanding industrial community," he said, "forced me to reconsider the liberal and highly moralistic creed which I had accepted" as the Christian faith.[19] When labor leaders organized in Detroit, the board of commerce pressured clergy to cancel unionist speaking invitations. The churches were too lukewarm to fight Henry Ford, Niebuhr decided, so he attacked the powerful industrialist with his pen in *Christian Century.*

In 1928 Reinhold was recruited to New York City to edit *The World Tomorrow,* a Christian magazine for pacifists and socialists, and to teach social ethics at Union, a bastion of Protestant liberalism. Yet two decades into his public and academic ministry, Niebuhr had never hunkered down to make an in-depth theological study. The Giffords were his chance to be recognized as a thinker as well as a social prophet armed with a journalist's pen. When published as *The Nature and Destiny of Man* in 1941 and 1943, they did in fact become his greatest work, even if he had used the term *natural theology* only once.

Standing in the Lutheran tradition, Niebuhr was averse to moving from nature to God's existence or attributes. That Reformation insight, resurrected in the furious neo-orthodox revival, was about a "wholly other" God breaking into the world, and nothing else. But in another sense, Niebuhr followed the logic of natural theologians; he began with nature. In this sense, his stance was naturalistic, the opposite of supernatural. "His ultimate appeal in both politics and theology was always to the observed facts of human experience," biographer Fox explained.[20] So naturalistic was Niebuhr that he attracted "atheists for Niebuhr." For much of his intellectual life, his analysis of human affairs was not too different from the thinking of the sociologists.

Niebuhr was not the only churchman, philosopher, or social critic traveling on the path of socialism. From the United States and England would come two other Gifford lecturers who had notably put their faith in the socialist vision. They were John Dewey, the American philosopher at Columbia University, and William Temple, an Anglican bishop in England. Delivering his Giffords years earlier than Niebuhr, Dewey spoke on the topic *The Quest for Certainty* at Edinburgh in the springtimes of 1928 and 1929, just before the stock market crash plunged the world into economic depression. In 1932 William Temple, who in a decade would ascend to the post of archbishop of Canterbury, became the first high-ranking churchman to tackle the Giffords' question of God and natural theology.

All three of them—Dewey, Temple, and Niebuhr—had rallied around the vital hopes of socialism in the 1920s, believing that organized labor could tame the excesses of capitalism. A great difference in age divided them, however, as was evident in the year Dewey gave his Gifford lectures. He was sixty-eight and a world-renowned philosopher at the time. In that year, Temple was just forty-six and a socialist bishop in Manchester, where the "Satanic mills" of factories had so inspired both poets and communists. Niebuhr, by comparison, was a relatively young man at age thirty-five, an activist pastor in Detroit "hoping to 'debunk' the moral pretension of Henry Ford."[21] Despite the disparity in age, Niebuhr's career would intersect significantly with the lives of both Dewey and Temple.

Dewey was not only a liberal, he was a thoroughly science-minded one, a man who had jettisoned his former Hegelian philosophy and metaphysics for a new trademark: "instrumentalism," experimental science applied to human learning and social construction. Indeed, his Giffords, though coming late in his life, were heralded as a final blow to metaphysics *and* religion in favor of the only human certainty left: educating human nature with experimental science. In Dewey's social model, mass man—a conformist citizen shaped by the factory line and industrial age—needed a certain amount of liberating by the education of experts, many of them in the social sciences.

For this brave new future, Dewey said in his Giffords, both "outworn traditions and reliance on casual impulse" had to be replaced by scientific optimism. Optimism was due, for "nature and society include within themselves" all the necessary resources. On this premise, he endorsed the use of a new "tool" for educational advancement; in his experimental school at Chicago, for example, he taught children through cooking. The application of intelligence, he believed, would eradicate the dualistic world set up by idealists and religionists, a world that separated the physical and spiritual, ideal and real, essence and existence. The real and ideal could be one, Dewey said, and the use of experimental study and living could progress toward that unity. "An idealism of action that is devoted to creation of the future, instead of staking itself upon propositions about the past, is invincible," he told his Gifford audience.[22]

Such Deweyite naturalism and technocracy were alien to Temple and Niebuhr. But the social promise spoke to their youthful idealism. As a young cleric, Temple had "an 'in' with the Labour Movement." He joined the Socialist party but later in life admitted that socialism "is a vague term." Still, at the end of his life it was "not altogether uncommon to hear people say Temple was 'tinged with Communism,'" a Temple biographer notes.[23] Like Niebuhr, however, he went through stages. He declared "Socialism or Heresy" in his early ministry, moved to a free-market orientation (in light of Britain's bureaucratic inefficiency), and was a private-public pluralist in economics by the time of Niebuhr's Gifford Lectures. In that year, he said that a "perfectly good Christian" could hold either view, preferring either a socially managed economy or a free market; the Gospels did not propose one economic system or the other.

Temple opposed continental Europe's extreme neo-orthodoxy, which declared that God is known *only* in Scripture and Christ. His Gifford Lectures, *Nature, Man and God,* looked for revelation everywhere. Temple called his mingling of Christ and nature a "dialectical realism," consciously competing with the dialectical materialism of popular Marxism. We live in a "sacramental universe," not a brutally materialist one, he told his Glasgow audience, and the emergence of mind proved that. Mind knew God by revelation, to be sure, but also by a sacramental reality, a world of things, lives, and symbols that participated in a transcendent God. "Unless all existence is a medium of Revelation, no particular Revelation is possible," he said. "Only if God is revealed in the rising of the sun in the sky can He be revealed in the rising of a son of man from the dead."[24]

The bishop, whose stockings and great cassock bespoke Anglicanism, never met Dewey. But Niebuhr met them both and absorbed their influences. In 1928, for example, Niebuhr literally moved across the street from Dewey, who was then deemed America's dean of philosophy. Niebuhr had joined Union Seminary in upper Manhattan, catercorner to

Columbia, the turf of Dewey. They spoke on the same platforms, and they skirmished in journals. "Dewey and Niebuhr sometimes shared the same political causes," said a Dewey biographer.[25] Both joined the League for Industrial Democracy and the Socialist party, and both helped form the Liberal party. In later years, Niebuhr credited Temple with teaching him about an ecumenical spirit and moderation in social change. Niebuhr and Temple worked together in some depth at Oxford in 1937 while attending the World Conference on Church, Community, and State. The British churchman moved Niebuhr "from a rather tentative Marxist orientation to a position which is equally critical of both Marxist and liberal illusions."[26]

By contrast, Dewey became Niebuhr's nemesis. His instrumental gradualism threw cold water on Niebuhr's prophetic urgency. After nearly two decades, Reinhold's brother, Richard, reminded him how "you've fought the Deweyites etc. a long time."[27] Perhaps he should leave upper Manhattan and take a Harvard job offer. Dewey was equally exasperated by Niebuhr. Dewey was irked by Niebuhr's generalities and prophetic tones, but especially by the way he set up soft-headed liberal straw men to knock them down.

Before moving onto New York turf, Niebuhr had published his first book, *Does Civilization Need Religion?* His answer had been affirmative. Dewey's would have been negative, except in the sense of Schleiermacher's "dependence," a dependence on nature and experience, according to Dewey's Giffords. Niebuhr depended on social science in his book. He had gotten a copy of Weber's newly published *Protestant Ethic and the Spirit of Capitalism* in German, and his translation of parts of it into English was one of the earliest for American readers. Niebuhr introduced them to Weber's famed argument that Calvinism spawned capitalism. In Niebuhr's view, this Calvinist offspring was now bound for destruction, for capitalism had produced a smug and acquisitive Christian middle class blind to a roiling proletariat.

Between 1930 and 1933, the mercurial Niebuhr had two significant reckonings with his past Socialist loyalties, and the resulting change of mind was reflected in his Gifford Lectures. The first reckoning came when Niebuhr made his only Socialist pilgrimage to the Soviet Union. Although on his return he lauded the Russians' sacrifices for the welfare

of future generations, he was clearly distressed by the failings of state socialism. Then in 1930 and 1932 he made political bids for the New York
State assembly and then a U.S. congressional seat on the Socialist party
ticket. Both were failed efforts, and Socialist losses everywhere in 1932
devastated the party's morale.

This second reckoning pushed Niebuhr to a height of impatience
with electoral politics. Christian thinkers were being swayed by Deweyite
optimism and gradualism, and the cause of socialist revolution was losing
its glow. Niebuhr felt he had to take a stand before all was lost, and the
stand came a month after the 1932 electoral debacle when he published
his most controversial book, *Moral Man and Immoral Society.* It was an immediate sensation. The catchy phrase, in fact, had come from German
church historian Ernst Troeltsch, a sociologically minded theologian who
had shared a house with Weber and was eulogized by Harnack. *Moral Man*
would not achieve Niebuhr's full "Christian realism," for it lacked the theological reflection he took up later. But his pessimism about group morality (as seen in Comte's family) and naive social science was on full display.
"Social injustice cannot be resolved by moral and rational suasion alone,
as the educator and social scientist usually believes," he wrote. "Conflict is
inevitable, and in this conflict power must be challenged by power."
Dewey was cited "as a typical and convenient example" of the liberal illusion.[28]

Admitting himself to be of a "disillusioned generation," Niebuhr
agreed that he might come off as "unduly cynical." Still, he went on to
level his polemic at old liberal associates and contemporary church life in
general. "Modern religious idealists usually follow in the wake of social
scientists in advocating compromise and accommodation as the way to
social justice."[29]

For all his frowning on social science, Niebuhr had advocated a
Marxist sociology, the hard structuralism of class conflict. No wonder the
New York Times linked the "doctrine of Christ and Marx." Sympathetic
students at Union even hoisted a red communist flag on the seminary
pole. Union Seminary president William Sloan Coffin worried that
Niebuhr, his most famous professor, had "swung to the left in his justification of violence."[30] In 1956 Niebuhr would say that his *Moral Man* "was

not uncritically Marxist," but it did fail "to recognize the ultimate similarities . . . between liberal and Marxist utopianism."[31]

After he published the book, Niebuhr would experience another
change of heart, and it came through the influence of his brother, the
theologian H. Richard Niebuhr, at Yale. His brother opened Reinhold's
eyes again to the value of theology, indeed to the importance of religious
faith. H. Richard, who had used a Marxist sociology to analyze American
denominations, resigned from the Socialist party in the 1930s. Increasingly, he joined the neo-orthodox revival. As part of that shift, he criticized Reinhold's *Moral Man* for its romantic illusion about revolutions.
Reinhold, through his brother, came to see that God's grace targets the
individual sinner, placing human hope ultimately beyond history. *Moral
Man* was a catharsis. The next year Reinhold announced the "end of an
era." Capitalism would still decline and the working class would somehow ascend, but the crux was the trembling self before God. In 1936, to
expound this idea, Reinhold founded the journal *Radical Religion.*

By joining criticism of utopia, justification of power politics, and
espousal of existential faith, his "Christian realism" was developing its
final texture. As Niebuhr's biographer described the project, it was a
mission "to justify Biblical religion as the only adequate foundation for
self-understanding and political action in an age of lowered expectations
and inexpressibly horrible disasters."[32] The final contribution to that texture came in 1930, when Niebuhr met the Oxford graduate Ursula
Keppel-Compton, who came to study at Union for a year. In December
1931 she became his wife.

As early as 1923, Niebuhr's journal cited the old Methodist who wondered whether "I was so cantankerous in my spirit of criticism about
modern society because I am not married." With "about four children to
love," Niebuhr had mused, he might be milder; indeed, with the birth of
his first child, a son, in 1934, his prophetic voice started to calm.[33] This, of
course, was no apology to naturalists such as Comte shoehorning the
family into utopia. But it was another balance the latter-day Niebuhr
would strike between optimism and pessimism.

In the same year that Ursula had arrived at Union, so did John Baillie,
the up-and-coming Scottish theologian who took a professorship there.

It was 1930, two years into Niebuhr's tenure, and while older faculty remained confounded by his activism and Midwest accent, Baillie took "a
rapid liking" to him. Baillie returned home to fill a chair in theology at
New College in 1934. As he was the leading professor of divinity in Edinburgh, his nomination of Niebuhr for the Giffords was significant. It
came in late 1936. To play up Niebuhr's relative youth, Baillie had reminded newspaper readers how, at most Giffords, they had "grown accus-
tomed" to grandfatherly expositors. Baillie, who became moderator of the
Church of Scotland in 1943, was himself quite old when he was nominated to the 1961 lectures at Edinburgh, and he died a few months before
he was to give them.

Niebuhr was the fifth American nominated, preceded by James, Josiah
Royce, Dewey, and the idealist philosopher William Ernest Hocking, who
began his lectures at Glasgow in 1938. Of the five, Niebuhr had the least
in the way of academic credentials. He called the invitation a "doubtful
blessing." He was "driven like a demon" on the project. As a controver-
sialist, Niebuhr was accustomed to knocking off several articles a month,
taking stands, revising them, reversing, coming back. But the Giffords put
a man's ideas in stone. He knew he would be in "slavery to these lectures
for years to come."[34]

For the immediate experience, however, he was the lucky lecturer.
"Had a good audience all through," he wrote home. He didn't "know
much about how my stuff has been received."[35] But the punchy and pithy
newspaper headlines were compelling enough: "Mass Man," "Marxism vs.
Nazism," "Errors of Romanticism," "Inevitable Sin," "Hitler Explained,"
"Demonic German Nationalism," "St. Paul and Sin," "The Free Self," and
"Kingdom to Come."

The "slavery," of course, came with a published book. As busy as
Niebuhr was on his return to New York in November, the publisher
pressed for results. When the first volume of *The Nature and Destiny of
Man* appeared in 1941, it clearly enhanced Niebuhr's stature. Having resigned from the Socialist party in 1940, he was now the voice of American Friends of German Freedom and the Union for Democratic Action.
He founded an interventionist journal, *Christianity and Crisis,* elevating
democracy as the nearest approximation to God's earthly kingdom.

The sweeping topic of his Giffords allowed for a range of interpretations, some of them naturally self-serving. *Time* made it "religious book-of-the-year." Under the headline "Sin Rediscovered," the emphasis was laid on a medieval revival of sin, which overlooked Niebuhr's dialectical tendency always to speak of human beings as both sinners and children of God. Soon afterward, in declaring an "American Century," *Time* publisher Henry Luce tied the triumph of capitalism to Niebuhr's themes.

Years later, Niebuhr said that *Nature and Destiny* was a lightning rod for left and right. On the secular and Christian left, he took barbs for "my alleged overemphasis on the corruption of human nature." To that charge, "I must plead guilty," he said, but he made amends in volume two; there he offered the "cure" of God's grace and human beneficence.[36] The critics had trenchant accuracies as well. Niebuhr's presentation was not rigorous scholarship. "No cautious weigher of evidence here," said Robert Calhoun, the Yale historian of theology. This was "a preacher expounding the Word in line with his private revelation."[37]

If that was the plaint of liberals and pacifists like Calhoun, the strictly neo-orthodox felt that Niebuhr had given *too* much credit to human effort. To them Niebuhr had indeed allied with a kind of natural theology, which his critics, he said, might have called "a tortuous and ineffective return to the Christian faith." In their mind "there is no way of using experience to determine faith." Taking hits from both sides, Niebuhr defended a concept of "the circular relation between faith and experience," thus giving the nature side a foothold in the encounter with God.[38]

Once America joined the war in late 1941, Niebuhr felt free to turn to the second volume, released in January 1943. Subtitled "Human Destiny," it offered the hope absent from the sin-soaked first volume. The Hebraic, and thus biblical, worldview gave history a messianic purpose, he explained. That was inevitably corrupted by fanaticism and false messianisms, from Comte to Marx. By its very nature, human freedom produced both incredible creativity and enormous evil in history; it forced a yearning for eternal rest, which was thwarted by human finiteness. The frustration was resolved in Christ's incarnation—the infinite became finite. The incarnation pointed to hope beyond history, but left behind the law of love as the historical standard.

Niebuhr had ended his Giffords with a note of optimism that dissented from his *Moral Man and Immoral Society.* He acknowledged the potential for collective love in man, calling love the primary law of human nature. What was more, brotherhood was fundamental to social existence. In another reversal, he downgraded the worst sins of the Renaissance and Enlightenment, the great fountains of secular and Christian liberalism. At worst, they had merely exaggerated "that side of Christian doctrine which regards the *agape* of the Kingdom of God" as the main route to "a more perfect brotherhood in history."[39]

Ending on a note of hope invariably stripped Niebuhr's voice of its familiar excoriating tone. "No strident posturing, no dismissive oversimplification, but also no rumbling movement, no heated passion," his biographer said. The balance of volume two had drained off Niebuhrian drama: "It was respectable and decorous, a rhetorical design that sapped some of the vigor from his prose."[40] In 1952 Niebuhr suffered a stroke. His powers diminished considerably over the last nineteen years of his life. The kind of personal physical suffering he had seen as a parish pastor was now his own lot, a perspective quite different, and more interior, than the detached panorama of the social critic.

For better or worse, the great themes of sociology are rarely heard of now—the progress of whole societies, class struggle, or social forces that dictate rates of suicide, marriage, or births. When these themes are sounded, they come in the voice of technical surveys, census findings, or government data. Some sociologists yearn for the days of big-picture sociology, a yearning that sends them back to reading Weber for his sheer conceptual power. Although Comte and Spencer were the first "mainstays" of American sociology, it has now moved on "to tackle problems, small and smaller, of milieu, families, and small groups in general, until a sociologist was no longer just a sociologist, but a specialist in family sociology, public opinion, criminology, statistics, small groups, methodology and methods, race relations, and so on, *ad infinitum.*"[41]

In all of this, sociology's founding assault on God—whether in Comte's positivism, Spencer's "first principles" of agnosticism, or Marx's atheism—became ambivalent, part of the general secular noise behind modern life. "I have never asked myself, or heard anyone else ask: did

God set society up?" said the contemporary sociologist David Martin. Questions of human origins or design of the universe do not cross sociological paths, he said, though a "twofold consequence of theism" for sociology could be questions of Providence and human free will. What are social structures destined for, in other words, and how much do they control human choices? "For sociologists, human actions are both made possible and constrained by culture," Martin explained. Yet in a world of Niebuhrian ambiguity, even those social forces do not ultimately restrict the self and, by definition, God himself. In sum, "It is not that sociology can unveil beneath the social guise the lineaments of the invisible and eternal God," said Martin. "Nor can sociology invalidate His presence."[42]

As the conflict model of a Marx and the structuralist model of a Weber revealed their flaws or limitations, a new approach called the *symbolic model* was devised and grew in influence. The symbolists argued that social groups acted in response to "meanings," not just material factors. Others have argued that it was time for American sociology to jettison European structuralism. Not only had America's physical frontier decimated class early on, now the global village of world communication was finishing the job; rigid concepts of "capital and labor" were replaced by notions of multiple-group loyalty, sharing of interests, and the workingman's stake in the stock market. In the 1980s, with the rise of the conservative and libertarian movement in England and the United States, the argument was again made that there is no such thing as society, only individuals and families.

Whichever model or system was in favor, the attempt to find God in human affairs had certainly become more complicated. The great medieval theologian Thomas Aquinas, for example, had once argued that the world's hierarchical order mirrored angelic realms. That would no longer do. Still, the Christian belief in love seemed to require a certain amount of well-being and social equality in the entire human scheme. The question was whether socialism or capitalism produced a world closer to that kind of equality and prosperity. Sociology could keep a scorecard, but it could hardly suggest whether a god was pleased or enraged at the social outcomes.

Although Niebuhr had used the sociological tools available to him, he finally put them down in favor of emphasizing the dynamic role of the self in the midst of history. In this sense, his final analysis of how to "save the world" and address "the social problem" was to look into the human interior and then look beyond human history for the ultimate things of God. As one of Niebuhr's critics had said when *Nature and Destiny* was published, the lectures were not a truly rigorous work of history: "The real ground of the author's doctrine is not what he has read but what has happened to him as a struggling self."[43]

History and the self—these were the two great torrents of Christian thought into which Niebuhr had thrown himself. He had set up a major dilemma for the modern Christian believer as well. Having adopted such traditional beliefs as sin and the promise of God beyond history, Niebuhr would also embrace the non-traditional agenda of taking the Bible symbolically, not literally or supernaturally. The haunting question became how to talk about Christian faith now that a long list of doctrines—creation, Eden, Adam and Eve, the devil, the Fall, original sin, salvation, the Virgin Birth, the Resurrection, the Second Coming—were emptied of their supernatural and literal meaning.

For the idea of human sin, Niebuhr turned to two great Christian philosophers, Augustine of Hippo, who lived in the fifth century, and Søren Kierkegaard, a Danish thinker of the nineteenth century. While Augustine emphasized the corrupt desires produced by sin, Kierkegaard wrote again and again of the "anxiety" felt by modern man because of his alienation from God. Niebuhr had come upon this duo of angst-ridden Christian giants late in his life, and he read them deeply on the eve of his Giffords. When he addressed sin in his Giffords, in fact, Niebuhr quoted Kierkegaard almost as much as he quoted Saint Paul, Augustine, and Martin Luther—all exponents of the interior spiritual struggle for redemption. "Kierkegaard has interpreted the true meaning of human selfhood more accurately than any modern, and possibly than any previous, Christian theologian," Niebuhr said.[44]

Despite such efforts, however, critics of Niebuhr have given him a failing grade at adapting full-throated Christianity to modern times. According to American theologian Stanley Hauerwas, who gave his own

Giffords of 2000 at St. Andrews, Niebuhr had simply compromised the "truth-telling" mission of Christian language by obfuscating it in the ethical and sociological jargon of modern liberal society. "Niebuhr's work now represents the worst of two worlds: most secular people do not find his theological arguments convincing; yet his theology is not sufficient to provide the means for Christians to sustain their lives."[45] But as with any great thinker in Christian history, Niebuhr's poignancy spoke to a given time, and his ideas easily sounded dated to a next generation.

Niebuhr's affection for the socialist quest, and empathy with those Christians who gave it up, added one other important legacy to modern Christian thought. Traveling in Germany in 1933, Niebuhr met a young German theologian, a "religious socialist" who had tried to form a Christian socialist party but now was on the wrong side of the Nazi regime of Adolf Hitler. The young theologian's name was Paul Tillich, and Niebuhr pulled the necessary strings to have Tillich teach at Union Seminary. The invitation was ultimately the ticket for Tillich's flight from the gathering darkness in Germany to a long career in the English-speaking West. If Niebuhr had helped to save Tillich, Tillich also immeasurably helped Niebuhr: as a creative theologian, Tillich gave Niebuhr the kind of language he needed to talk about Christian doctrine and faith in a modern, secular world. The key, according to Tillich, was to phrase Christian doctrines as symbolic meanings for the individual self.

As will be seen in later Gifford Lectures, the philosophical emphasis on the "self" in European thought took on a variety of forms, both Christian and secular. All of them would finally come under the general heading of *existentialism,* a term coined by Kierkegaard, who struggled with the ultimate meaning of his individual existence before God. Tillich was by trade a historian of Christian theology, but seeing the trends, he had gathered up all the philosophical stances on the "self" and, putting aside his former political interests, developed a new vocabulary of faith for liberal Protestantism. In New York City, Tillich's brilliance on this front made up for his broken English, as he gave new psychological meanings to the litany of Christian doctrines. He used such terms as creation, fall, redemption, sin, faith, the demonic, grace, Christ, and resurrection. He was talking not about history but about each person's state of mind and spiritual encounter with God.

Niebuhr's Gifford Lectures barely mentioned the work of Paul Tillich, but the way he used Christian terms as symbols of personal faith and destiny was clearly borrowed from the young theologian he had rescued from Nazi Germany. In future years, however, while Niebuhr would stand by the main themes of his Gifford Lectures, he regretted his use of terms such as the fall of man and original sin, even as Tillichian symbols. Even as symbols, he wrote in 1963, the ideas "proved so offensive to the modern mind." He also regretted his symbolic portrayal of the end of history—a clash between Christ and anti-Christ—now that America was in a nuclear standoff with the Soviet Union. Indeed, he was "not so sure that the historic symbols will contribute much" to Cold War diplomacy or military strategy. No one can predict the future, the older Niebuhr seemed to say. Yet old men could at least hope "that the moving drama of history may validate a part of the truth they sought to discern."[46]

By their different emphases, the works of Niebuhr and Tillich would have different fates in the postwar era. Niebuhr's star had shown brightly in an age of ideology, but once that age passed, his works lost their luster. Niebuhr became known as a democratic pragmatist in power politics. After his death, American neoconservatives sainted him as a democratic warrior against communism. Around the 1950s, the American mood shifted from one of ideology to one of personal fulfillment, and this prompted the ascendance of Tillich's ideas. The American middle class was moving to the suburbs, where churches were built by the thousands and where the ideas of psychological meaning and fulfillment mixed with Christian belief. For the thinking Christian in this milieu, Tillich had far more to say than his older colleague Niebuhr could offer. Soon Tillich became America's most popular theologian. When he made the cover of *Time* magazine in 1959, it was titled "A Theology for Protestants."

Tillich gave the Gifford Lectures twelve years after Niebuhr, and his topic reflected the mood of that new era in the West. His Giffords, which began at St. Andrews in 1953, would make up the last two of his three-volume *Systematic Theology,* which was an attempt to apply all the new symbolic and psychological language to Christian history, doctrine, and experience. The starting point of Tillich's theology was the tendency of human beings to ask questions about their lives. For each modern question—ranging over topics of death, hope, guilt, vocation, love, loneliness—Christianity offered a corresponding answer. Hence, his theological method was called "correspondence," or finding a way to satisfy the modern human psyche by a symbolic use of classical Christian concepts such as the fall, the Garden of Eden, the devil, or the Christ—all of which, if interpreted in more personal and symbolic terms, had meanings for everyday life. While the demonic did not have to mean a literal devil, for example, it certainly could stand for pride, loneliness, or fear of death.

In the biographies of Niebuhr and Tillich there is one other common strand. As young men, and indeed as young pastors, their daily bread had been studying the life of Jesus as told in the Bible. Yet as they matured, becoming familiar with the history of Christianity and the new scientific methods of studying biblical texts, they realized how much myth and fact had been mixed together. They had to find ways to make the Bible meaningful even if they could no longer teach it as literal. When both Niebuhr and Tillich became deeply interested in history, they similarly had to find a way to locate Jesus' significance in a vast human drama covering thousands of years. Their solutions were somewhat different. While Niebuhr had put fulfillment in Christ beyond history, Tillich brought that fulfillment into the immediate encounter between Christ and the human psyche.

Both kinds of solutions would be adopted by modern believers confronted with the questions of history, questions now raised by a new group of scientists appearing at the start of the twentieth century—the professional historians. History had been written since ancient times, but its claim to be a kind of science had only now come upon the scene. As a science, moreover, it aimed not only at secular history but at sacred history as well, and in the West that meant the history of the Bible. As a young man, Tillich recalled how that avalanche of change arrived for many Bible believers, including himself.

The son of a minister, Tillich was reared on "biblical realism"—what the Bible said corresponded to what had truly happened in history. Tillich held to this belief until 1906, when he read *The Quest of the Historical Jesus,* by Albert Schweitzer. "I myself experienced a real crisis," he said. Until then the great questions about Christ had been theological, such as the "two natures" of the son of God, one human and one divine. Now, the perplexing question became one of distinguishing between the Christ of faith and the Jesus of history. One existed in eternity, so to speak, while the other lived and died in ancient Palestine. Until Schweitzer, Tillich said, his Bible teachers had resisted the methods of historical criticism, which carefully analyzed the sources and construction of documents that formed the Old and New Testaments of the Bible. "Whether one was more conservative or more radical in historical investigation of the biblical literature, the methods had to be accepted in the long run," Tillich recalled of those early times.[47]

In the age of science known by Dewey, Temple, Niebuhr, and Tillich, sociology had loomed large, and so had its practical application in a great age of enthusiasm for socialism. While the dreams of sociology and socialism obviously live on, the pondering of the "science of society" invariably raised new questions, whether about the angst of individuals in the social scheme or about the extension of societies across time—namely, the historical life of a human population. Science had advanced boldly as the twentieth century opened, and anthropology, psychology, physics, and sociology became its monuments. Science could not stop, of course, in the face of history, once seen as a poetic or liberal art. History too would become a science, with jarring implications for secular and sacred narratives, including the Bible, an object of increased "scientific" study in the West.

When Schweitzer wrote his autobiography, *Out of My Life and Thought*, not long before he left Africa to give his first Gifford Lectures in 1934, he also believed that the historical search for an authentic Jesus was inevitable. "Religion has," he said, "nothing to fear from confrontation with historical truth."[48]

SEVEN

HOW IT REALLY WAS

The Science of History

T HE OGOWE RIVER in Gabon, West Africa, flows more than
five hundred miles from the tropical interior to the Atlantic
Ocean. More than once it carried Albert Schweitzer back to Europe.
They were journeys past the famed shorelines of equatorial West
Africa—the Ivory Coast, Pepper Coast, Gold Coast, and Slave Coast—
namesakes of Western progress and guilt that Schweitzer, when he was
young and idealistic, had wanted to rectify.

Before becoming a medical doctor and opening his jungle hospital in
Lambarene in 1913, Schweitzer had succeeded in enough careers for several
men: musician, theologian, and historian. They all required that he return to
Europe periodically to lecture, publish, and give organ recitals to raise funds
for the hospital. Thanks to Schweitzer's eventual appointment to give the
Gifford Lectures in 1934, the Ogowe would be the most exotic vehicle to
ever deliver a speaker to the urban, rain-washed streets of Edinburgh.

The first accomplishment to secure Schweitzer's fame was his study
on the life of Jesus and how modern Europeans tried to grasp his world
of two thousand years ago. *The Quest of the Historical Jesus*, published in
1906, stimulated a new era of research on the ancient life of Christianity's

founder. For Schweitzer himself, that encounter with history stimulated an ultimate break with his calling as a minister and theology professor, for he moved to Africa and began a quest for a universal ethic for world civilization. When he arrived in Edinburgh to deliver the Giffords, he was introduced as "the man who sacrificed a brilliant future, not only as a professor but as a leader of European thought," to live in the jungle.[1]

Since the inception of the Gifford Lectures, there had been very few presentations that did not touch on history. The will of Lord Gifford did not cite "history" as a mode of natural theology. Yet history had become another tool of science, and depending on how it was used, could work toward undermining religious belief or explaining its nature even better. To be a science, history had to claim it was uncovering facts, an objective method that did not care about the myths that often surround great figures, nations, and events. Yet to speak of the religious search for God, or for a higher purpose, the science of history had to allow for some suggestive interpretation—an activity that secular historians shunned vigorously.

The Gifford Lectures imposed no such restraints, and while several leading historians gave the talks, none represents better the way that the science of history met religious belief than Schweitzer, who gave the Giffords in 1934 and 1935, on the eve of the Second World War, and the two English historians Arnold J. Toynbee and Herbert Butterfield, who gave their presentations in the years soon after the great world conflict. Although the science of history invariably produces a secular account of human affairs, human history includes religion, which in turn has tried to see Providence in historical events. That is a key theme in closing act two of the Gifford legacy: how the scientific discipline of history confronts the religious beliefs of the West. History, by all accounts, moves the human heart by telling "how it really was." Some historians, as the Giffords attest, want to go further. They ask about God's place in history, and whether it can be found among the pot shards, ancient documents, battle reports, and private letters that are the historian's quarry.

Although Schweitzer's topic, "Natural Theology and the Problem of Natural Ethics," was broader than Christian history, he drew on much that was historical to cover "the thought of all humanity." Toynbee had begun as an academic historian, applying the science of history to the classical past and then the Second World War. In his later career, however, he introduced ideas about God, religion, and providence into his work, making him a prime candidate for the Gifford talks, which he gave in 1952 and 1953 as "An Historian's Approach to Religion."

Herbert Butterfield, a brilliant historian at Cambridge, followed Toynbee's lead. His Christian background finally led him to apply his historian's skills to essays such as "God and History," in which he argued pointedly: "If God cannot play a part in life, that is to say, in history, then neither can human beings have very much concern about him." It was time, he said, to "recover the sense and consciousness of the Providence of God . . . as a living thing, operating in all the details."[2] Some years after the famous essay, in 1965, Butterfield used his Giffords to survey how secular and Christian historians had chronicled the details of history.

In the half century represented by the work of Schweitzer, Toynbee, and Butterfield, historical research had come on the world stage. Even though modern historians had more kinds of data and new "scientific" techniques to test the accuracy of artifacts and documents, their interest was not so different from that of the ancients. Herodotus, the Greek "father of history," said he wrote about the wars between the Greeks and barbarians "to give the cause of their fighting one another."[3] Beyond finding the cause of events, moreover, history was the quest to tell the truest possible story. In modern times, the German historian Leopold von Ranke, who founded his scientific school of historians in the 1830s, said their task was "simply to show how it really was," not to glorify a nation, hero, race, or institution.[4]

Ranke has traditionally been identified with the start of the modern and scientific school of history. Its quarry was documents and the attempt to separate myth, propaganda, and bias from objective facts and the authentic viewpoints of different parties involved in a historical event. The first application was to documents, whether the papers of Napoleon or the texts of the Bible. With an air of general suspicion, the new historian would begin by assuming that the earliest "layer" of an account was probably the most accurate, since it had fewer chances of being embellished by later copiers or interpreters. Next, the new historian cross-examined even this more reliable layer, first trying to match it with other accounts, but also looking for distortions based on the writer's vested interests—such as glorifying a political ruler or military general in order to curry favor. With this best effort, as Ranke said, the historian hoped to tell his reader "how it really was."

The next step was to draw conclusions about the various causes of events, which brought historians onto the precarious ground of interpretation. But without such humanizing and dramatizing, without inferences about why the Greeks and barbarians fought, why Napoleon lost to Wellington at the Battle of Waterloo, or why the early Christians recorded what they did about Jesus, history could be an intolerably dry mountain of facts alone. The scientific school rubbed off on a new era of Bible scholarship firing up in Germany in the 1830s. The Enlightenment of the previous century had been awash in the search for ancient sources, and by Ranke's time it had given rise to a movement called *historical criticism* of the Bible. Questions posed to secular texts were now posed to the most sacred book of the West, the Bible. Did ancient cultures influence its writing? What are its earliest layers and what are later embellishments? Why did Mark, the earliest Gospel, seem to influence the accounts of Matthew and Luke? And why was John, the latest Gospel, so different from the other three?

As young Albert Schweitzer came of age, growing up in the parsonage of his Lutheran pastor father, the sermons and teachings of the time emphasized the harmony of the Gospels, and most of all the ethical teachings of Jesus. Albert followed his father's vocation, studying church music and theology in France and Germany, being ordained a Lutheran minister in 1900, and eventually becoming dean of a theological school. At one point, he recalled, his students became entranced by the mysteries surrounding the historical Jesus. So he lectured two hours weekly on the topic. "The subject soon captivated me so completely that I devoted all my energy to its pursuit."[5] His interest took him back across 120 years of attempts to describe the life of the historical Jesus by using the new critical methods of research.

Researching secular documents was one thing, however, and probing sacred texts would become another. The historical critical method subjected all ancient texts to analysis for layers, earlier and later dates, plagiarism, and the whiff of self-interest. Yet this method put an additional onus on texts such as the Bible, which spoke of supernatural events. The new onus was scientific positivism. Although in secular history a date may be in question, there is no doubt that a date really existed. With the Bible, however, scientific positivism said that though a miracle may have been reported, a miracle is scientifically *impossible.* It could exist as a story, but not a fact, in history.

By the time Schweitzer was serving as a pastor and teaching, many of these doctrinal puzzles had been glossed over as matters of faith and tradition or as serving a symbolic purpose of the Gospel writers. As Schweitzer pursued such academic questions more deeply, he appreciated the new historical criticism. Yet as a pastor and believer, he doubted the usefulness of the most extreme skepticism, and set out to find a historical Jesus for both scholarship and faith. His first project was a "sketch" of Jesus' life, published in 1901 as *The Secret of the Messiahship and the Passion.*

From early on, Schweitzer had agreed that the key problem for the Gospels was, first, why the three "synoptic" Gospels, which seemed roughly alike, differed so greatly from the fourth, the Gospel of John. The other enigma was why Jesus had both hidden but also proclaimed his messianic role—the "messianic secret." Such questions dug deeply, especially for a figure considered the Christ, the second person of the godhead. Historical study nevertheless asked of this divine figure a very human question: what was Jesus' self-consciousness about his role and mission? In light of Gospel accounts, how did that reconcile with the way his disciples and the public saw him?

Schweitzer offered his solution to the two-part puzzle in his seminal 1906 book *The Quest of the Historical Jesus.* To arrive at his own conclusion, Schweitzer covered nearly seventy writers on the historical Jesus, focusing especially on a few trailblazers. Most of the renditions followed the liberal theology of the nineteenth century. They cast Jesus as a rational moralist, not an apocalyptic fanatic such as John the Baptist or a crazed personality sprung larger than life from the apocalyptic book of Daniel. Yet the failure of Jesus scholars to see the apocalyptic mood of Jesus' century, Schweitzer argued, was their downfall. The eschaton in ancient Jewish thought was the end of the world, and indeed Jesus seemed to be a very eschatological figure. The biblical account of Jesus, so full of contradictions, made sense only if Jesus was seen as a man proclaiming a supernatural end to the world.

To Schweitzer's pleasure, the earliest book about the life of Jesus had taken this exact stance. About the time of the American Revolution, fragments written by the German deist Hermann Reimarus were published with this apocalyptic outlook intact. Reimarus would be forgotten in the next century, glossed over by the liberal images of Jesus painted by rationalism, Hegelianism, and skepticism. The rationalist ones accepted some miracles, explaining most of them in practical terms. A "fully developed rationalism," as Schweitzer put it, peaked in 1828 with Heinrich Paulus's *The Life of Jesus as the Basis of a Purely Historical Account of Early Christianity.*

What Schweitzer called the "liberal lives" of Jesus began their pedigree with David Friedrich Strauss's *The Life of Jesus* in 1835. Instead of explaining miracles, Strauss deemed them literary myths—a theme adopted by writers thereafter. The liberal lives used all four Gospels to show a "development" of Jesus' ideas and ministry. Portrayed as the teacher of an ethical religion, Jesus was well suited to the "rational bourgeois religion" of nineteenth-century Protestantism, Schweitzer said.[6] Whether his resurrection was historical, psychological, or mythical, the main point was this: the belief had inaugurated early Christianity and retained its power for modern believers.

A turning point in scholarship came in 1838, when the German theologian Christian Weisse showed persuasively that Mark, the shortest Gospel, was the earliest. Mark seemed to have been the basis for composing Matthew. Soon came the "two document theory": the Gospels of Luke and Matthew were composed based on Mark and a second document, a list of Jesus' sayings, or *logia.* More controversial, however, was the growing consensus that John, the fourth Gospel, was an entirely separate composition, probably written by a church theologian long after Jesus' death.

By the 1880s, when Scotland built its great bridges and Lord Gifford conceived his will, historical accounts of Jesus had dramatically multiplied. Nearly all of the serious New Testament "lives" were coming out of Germany. The one great exception was the *Life of Jesus* (1863) by the Frenchman Ernest Renan, a former Catholic novitiate and now a famed Orientalist, a scholar of the language and terrain of ancient Palestine. By creating a novelesque Jesus, portrayed as an eloquent sage and veritable man of the people, Renan became *the* leading literary figure of nineteenth-century France. Still other lives of Jesus cropped up, liberal, skeptical, and increasingly speculative, portraying Jesus, for example, as born in the Essene sect.

After Strauss's *Life of Jesus* in 1835, a more radically skeptical group of historians had emerged. Bruno Bauer, writing New Testament "criticisms" between 1840 and 1850, was a master of applying scientific analysis to biblical texts. Schweitzer traced this skeptical line down to a contemporary of his own, Wilhelm Wrede, a theology professor who published *The Messianic Secret in the Gospels* in 1901. Indeed, the skeptical approach to the biblical texts could be so powerful, Schweitzer concluded, that only two kinds of views about the life of Jesus could possibly survive the work of historical criticism. One was the "thoroughgoing skepticism" of thinkers from Strauss to Bauer and Wrede, in which the Jesus read of in the Bible was at most a literary myth or fiction. The other view was more promising, Schweitzer believed, and he called that approach "thoroughgoing eschatology." The Bible told of a real, historical, eschatological messiah figure named Jesus. Either the Bible account as history fell apart under the harshest skepticism of historical criticism, or the Bible account held together in a remarkable new way when one viewed Jesus as an apocalyptic figure.

That was the conclusion of Schweitzer's most famous work, *The Quest of the Historical Jesus.* He summarized his own period as being marked by the revival of the apocalyptic Jesus by the work of the German scholar Johannes Weiss, a pioneer of "form criticism," a technique of Bible study that classifies biblical phrases as either oral or textual, authentic or borrowed. Weiss had boldly reasserted the thesis of Reimarus in the 1892 work *The Preaching of Jesus Concerning the Kingdom of God.* It was in reaction to this apocalyptic revival, according to Schweitzer, that Wilhelm Wrede offered his thoroughly skeptical *Messianic Secret* in 1901. Now it was Schweitzer's turn to weigh in, and his 1906 *Quest* sided with the apocalyptic messiah of Reimarus and Weiss. Schweitzer's summary of the historical trends was so persuasive that he became famous in New Testament studies almost overnight.

His own summary of the life and mission of Jesus has also endured. Whereas Weiss had summarized the eschatological message of Jesus' preaching, Schweitzer proposed an actual plot by which his ministry ran its historical course. Early in his ministry, Jesus had expected God's apocalyptic action to shake the world, and in anticipation of that he sent out the disciples to preach repentance. Yet nothing happened, so Jesus took a turn. He seized on the prophecy found in Jewish eschatological texts and beliefs, including the story of a suffering figure. Then Jesus set out for Jerusalem on the eve of Passover, expecting God's glorious action. By then, however, the disciples had spread the messianic "secret." Indeed, Judas had betrayed it to the religious authorities. When the ensuing controversy finally came before Pontius Pilate, the Roman ruler tried to use the case of Jesus to undermine the Jewish priests who were troubling him. The result, prompted by an unpredictable mob, was Jesus' execution.

This was not a story of Jesus that was comforting to nineteenth-century churchgoers, Schweitzer said. In addition to that, nineteenth-century historians had painted Jesus in their own genteel images. Since the early 1800s, the search had "set out in quest of the historical Jesus, believing that when it had found Him it could bring Him straight into our time as a Teacher and Savior," Schweitzer said. But Jesus always "passes by our time and returns to His own," although he remains a spiritual reality for every generation. Schweitzer concluded: "Jesus means something to our world because a mighty spiritual force streams forth from Him," a fact that "can neither be shaken nor confirmed by any historical discovery."[7]

Perhaps even Schweitzer did not realize the way in which his work would upset orthodox Christian belief, which prevailed in churches and missionary societies if not in the theological academies of the West. After finishing his medical degree, with the aim of working in Africa, he applied to work with the Paris Missionary Society in Gabon. But the board rejected him for his liberal view on the Gospel of John. Schweitzer insisted he "only wanted to be a doctor," not a minister. As the board evolved, good news finally came. He was approved on "the understanding that I would avoid everything that could cause offense to the missionaries and their converts in their belief."[8]

The year before he left for Africa, Schweitzer updated *The Quest,* responding to the latest trends in thoroughgoing skepticism. Some historical criticism of the Bible had made the case not only that early Christians
gave Jesus his divinity by borrowing from Greek ideas and mystery religions, but that behind these inventions, perhaps a historical figure like
Jesus was really a fiction, created almost out of thin air by ancient and su-
perstitious minds. Again, Schweitzer said these critics had gone too far
and overlooked the historical traits of the Gospel accounts. A chief proof
for a historical Jesus, Schweitzer repeated, was how well the earliest
Gospels of Mark and Matthew captured the world of Jewish apocalyptic
belief. "By his eschatology Jesus is so completely and firmly rooted in the
period," Schweitzer said. "He can only be represented as a personality
that really appeared in that period."[9]

While such a defense of the historical Jesus endeared Schweitzer's
work to many Christians, his use of historical criticism troubled other believers. He had his Christian response: "Truth is under all circumstances
more valuable than nontruth, and this must apply to truth in the realm of
history," he said. "Today Christianity finds itself in a situation where pur-
suing historical truth freely is difficult because it has been neglected again
and again in the past." Indeed, he said, "the early Christians published the
writings of the Apostles without being sure of their authenticity."[10]

Schweitzer wrote this at one point during his nearly fifty years of
work in Africa, where he had arrived in 1913, just seven years after *The
Quest* shook the world. Africa had grown large in his imagination since
his childhood, when he was deeply impressed by a bronze statue of a
heroic African he saw in a city park. As a theology student he read of
French missionaries in Gabon, a nation colonized by the French in West
Africa. While living in Europe, Schweitzer had displayed all the marks of
academic success, yet the conformist spirit of the age made him part of a
restless generation. One way to defy that spirit, he and his wife decided,
was to become medical missionaries in Africa. Moving back and forth
between his privileged status in Europe and the hardship of the African
jungles, Schweitzer was finally inspired to seek something higher than
even the Christian faith, and it was this search for an ethic for world civilization that became the topic of his Gifford Lectures.

The debate over the fate of Western civilization was just below the surface in Europe, which in 1900—the year of Schweitzer's ordination—was characterized by a mood of inevitable progress and the building up of proud nations and colonial empires. Beneath the veneer, however, were whispers about the dreadful price society might pay for its embrace of progress, rationalism, technology, and commerce. Not for the first time, German literature spoke of "cultural exhaustion." Some detected the "pervasive tragic spirit" so vividly told of in Greek literature.[11] And had not Greece and Rome fallen? Even the best civilization *must* decline. By Schweitzer's own account, his ear was attuned to some of these prophecies of exhaustion and doom.

For popularizing this mood, the most prominent voice of the early twentieth century was Oswald Spengler, a German high school teacher who became one of Europe's most widely read historians. Spengler began writing his *Decline of the West* before the first guns flared in August 1914, the start of the First World War. He did not hew to the dictates of the new scientific history writing, and that was part of his appeal. He drew on the images of organic life to show how empires sprout, grow, flower, and then wither away. He told this electrifying story often in the evocative imagery of the great architecture of one civilization or another, all of them ending mostly in rubble and dust, like temples overgrown with vines in some romantic painting. *Decline* was published in the summer of 1918, a few months before the end of the war, which left 15 million people dead. The book was a best-seller, and it offered a chilling prophecy of a "new caesarism" at a time when Adolf Hitler was virtually unknown.

As Spengler brooded over his prophetic history, Schweitzer was traveling up and down African rivers, doing the work of a doctor but also struggling over the fate of the world. When his vision of the future came to him, it was far more optimistic than Spengler's. Soon after arriving in Africa, Schweitzer was riding a small boat on the Ogowe River at sunset. At the moment they passed a group of hippopotami, he recalled, "there flashed upon my mind, unforeseen, and unsought, the phrase, 'Reverence for Life.' " This was the universal ethic he was looking for and would elaborate upon, applying it to Eastern and Western religion, humanitarian work, and respect for all sentient life. In this moment, Schweitzer saw a human ethic that denied selfishness, as the apocalyptic Jesus taught, yet affirmed the self. It was an ethic of "world-and-life affirmations" that could mend the technological West and tradition-bound East. On the Ogowe that day, "The iron door had yielded: the path in the thicket had become visible."[12]

Yet almost as soon as Schweitzer had embraced this outlook, the edifice of Europe crashed down, even in Africa, with the onset of the war. As Germans serving in a French colony, the Schweitzers were put under house arrest and herded into an internment camp. Imprisonment gave Schweitzer one advantage. He found time to write out his philosophy of civilization based on reverence for life. Liberated in a prisoner exchange, he completed the work in postwar Europe as lectures, for the African jottings had been lost. Like Spengler, Schweitzer traced the modern problem to a cultural subservience to machines, but Schweitzer went further: this subservience had also separated spiritual values from reason.

He first preached this message at his own church in Strasbourg, Germany, and then in Sweden in 1919, where he was invited by the scholarly Archbishop Nathan Söderblom, a fascinating ecclesiastic figure of that period. A pastor to the family of Alfred Nobel and now primate of Sweden, Söderblom won the Nobel Peace Prize in 1930 for his work promoting Christian unity among the Protestant, Anglican, and Orthodox churches. He was also invited to deliver the Giffords at Edinburgh. He presented the first series in the spring of 1931 but died a month later, unable to present the rest, although he settled on their title, *The Living God,* on his final day.

Starting with the Söderblom invitation to Sweden, Schweitzer expanded his European lecture tour, elaborating on his reverence-for-life ethic. The cumulative outcome was his two-volume *Philosophy of Civilization,* published in 1923. The following year he returned to Africa to rebuild his hospital. He spent nearly a half century in Africa and died there in 1965 at age ninety. Yet his forays back and forth to Europe produced the story of his "fourteen sojourns" to Africa. One of those trips home came in 1934, when Schweitzer was fifty-nine; his destination was Scotland, for the Giffords.

Although Schweitzer had kept a hand in New Testament studies, publishing *The Mysticism of Paul the Apostle* in the 1930s, his work on ethics was becoming foremost. For the Giffords, he apparently responded to the section of Lord Gifford's bequest that exalted "the Knowledge of the Nature and Foundation of Ethics or Morals, and of all Obligations and Duties thence arising." By natural theology, Schweitzer did not mean arguments for God's existence. He was looking for a universal ground for ethics amid "a wild confusion of uncompleted systems."[13] Schweitzer's pursuit of the universal ethic led him to a conclusion similar to that of Kant, who said proof of God was impossible but a moral sense in humanity could be argued for. A liberal Protestant, Schweitzer was primed for this kind of Kantian conclusion, witness his doctoral dissertation, *The Religious Philosophy of Kant*.

In espousing a universal ethic, Schweitzer made a primary theme of his Giffords the comparison of Eastern and Western religious beliefs, a contrast seen particularly in Hinduism and Christianity.[14] They reflected the two paths of world cultures across history, the Eastern suggesting a mystical ethic that denied the world and the West a rationalistic ethic that affirmed earthly enjoyment. Following this rationalist ethic, the modern West had ended up with an acute problem, which Schweitzer described as a stark separation of science and morals. Claiming all truth, rationalistic science relegated religion to the realm of emotion. Western religion seemed to agree. In such times, characterized by the apotheosis of machines, Schweitzer said, the ideals of the churches "had no power over the great majority."

The solution, he felt, was some kind of bridging of Eastern and Western ethics. It would be a matter of combining self-denial and detachment with life-affirming action. The new ethic would try to resolve the dichotomy between self-perfection (private) and devotion (public). Schweitzer proposed "reverence for life" as that kind of comprehensive ethic. It provided an absolute—namely, the sacredness of life—and yet allowed a person to decide "how much of life he may keep for himself and what he must sacrifice for others." It was a rational idea for modernity, but also a rational argument for mystery: "True ethics has the significance of mysticism in action." By his last two lectures in 1934, Schweitzer had argued his most controversial stance: ethics must include nonhuman life, for "every form of life represents value." As he said later, the world was a place in which all things had a "will to live," so these "wills" should co-operate, an outlook that might lead a human being to rescue "the worm that languishes on the pavement by placing it on the grass."

Schweitzer would return to Africa and then, in 1935, make the trip back to Scotland, up rivers, over the ocean, and then by train. In Britain, he was as busy as ever, his schedule including a recording session of Bach on the organ for Columbia Records. A universalist now, Schweitzer avoided use of Christian terminology in his final Gifford Lectures. Religion aimed for "spiritual union with the Infinite," a union "realized not alone in thought but in deed as well." When he touched on Buddhist and Hindu ethics, the headline read, "Self-Perfection and Helpful Service." Ethics had to come from within, Schweitzer said, but in a balanced way, for a "man need not torture himself [about doing good] every time with perfectly right thoughts, but whether he does it wholly."

Christianity had also produced two kinds of ethics, one of self-perfection and redemption and the other of devotion to the world, as seen in Jesus' Sermon on the Mount. After these two seasons of Gifford talks, short stretches in 1934 and 1935 when Schweitzer set out to cover "the thought of all humanity," he ended on themes of particular interest to his British audience: the idea of a personal God and eternal life. Again he came across as the Kantian he was, despite his new fusion of Eastern and Western thought. "The decisive thing is not whether we can imagine the Absolute and Infinite in a satisfying manner as a Personality, but whether we enter into personal relationship with It," said Schweitzer, whose every Gifford talk was reported in the *Glasgow Herald,* including this last. "Through ethics we surrender ourselves to the Infinite, which none of our words and none of our ideas will fit, and seek in It rest and peace and redemption from this world."[15]

If Schweitzer received his inspiration on the Ogowe River at sunset, the muses spoke to British historian Arnold J. Toynbee on a train, also at the end of the day. It was 1919 and the young Toynbee was riding the Orient Express across Greece. He had probably read Spengler's *Decline of the West*, and he was conceiving an even grander reading of history.[16]

Already an authority on Greek and Roman history, Toynbee began jotting down his outline for a world epic in 1921. His masterwork, *A Study of History*, was published in segments between 1934 and 1954. It totaled ten volumes, and Toynbee's name immediately eclipsed that of Spengler's. For a generation at least, he joined a list of mega-historians from Thucydides, Eusebius, and Augustine to Vico, Calvin, Voltaire, Hegel, and Marx.

Toynbee portrayed history in cycles, but cycles that human ingenuity and the interaction of cultures could direct. Whereas Spengler had been strictly deterministic, Toynbee was mostly volunteeristic. History was driven by "challenge and response," a dynamic in which societies strained to manage their physical and social ordeals. By 1939, when he published a second installment of volumes, Toynbee had increasingly inserted the idea of God into his work. The importance he placed on religion for the health of civilization would also increase. At first, he spoke of religion as simply an interesting by-product of the rise and fall of civilizations. By the 1950s, when he completed the last volumes of *A Study of History*, religion was *the* most important fruit in the life of a civilization, for it added new fiber to the moral backbone of the world.

As with his precise view of religion, Toynbee's ideas evolved. Once the ten-volume *Study* was done, Toynbee added a few more volumes with his second thoughts and revisions. "Toynbee changed his thought fundamentally more than once," said one historian. So it was "impossible to treat his thought as one system."[17] By the end of his life, Toynbee was remembered by some for his empirical skills and by others as a crank visionary with ideas about a future religion they criticized as "mish-mash" or "Toynbeeism." Kindly enough, his family had referred to his life's obsession, the *Study*, as the "nonsense book." Yet for all the buffeting, by the time he delivered the Giffords in Edinburgh in 1952 and 1953—titled "An Historian's Approach to Religion"—the essence of his grand vision was intact.

Toynbee's approach to life stemmed from both vision and tragedy. As a young Oxford student disqualified from military service, he had friends who perished in war. His uncle, also named Arnold Toynbee, was the famed economic historian who coined the term *industrial revolution* and died of a disease caught working in impoverished East London. His father was put into an asylum; at the time insanity was considered hereditary. Young Toynbee, a nominal Anglican, felt both afraid *and* chosen. Walking in London one day in 1919, he suddenly became "aware of the passage of History gently flowing through him in a mighty current, and of his own life welling like a wave in a flow of this vast tide."[18]

His obvious talent led to teaching classics and finally to a long career, from 1924 to 1955, at the Royal Institute of International Affairs in London. There, he produced the annual *Survey of International Affairs*, plus a multivolume record of the Second World War. In this, he was a consummate empirical historian. Toynbee was also distinguished from intellectuals of his day for opposing British accommodation of Hitler, an act of Whitehall that left him with a bitter memory. From 1930 onward, he began to craft his *Study*.

The ten volumes were not exactly historical narrative. The first six set out themes. Like a botanist, Toynbee put down a taxonomy of civilizations and subgroups across history, a set of categories that would change. He began with twenty-one "societies" that grew into civilizations. Five societies ended in "arrested civilizations," and 650 primitive societies stayed at that level. In his last work, Toynbee settled on fourteen "independent" civilizations and seventeen "satellites"—a total of thirty-one. They had nearly all died out, however. Five to seven survived into the present, and most of them were in decline, including the West.

Whatever the number, civilizations followed a process of genesis, growth, breakdown, and disintegration. The last phase was typified by the top-down oppression of a "universal state," followed by a "universal religion" rising up from the oppressed proletariat. Meanwhile, "barbarians," who smelled spoils ripening, harassed the borders. To conclude, Toynbee held out the law of freedom as the savior of history and looked at how historians have assessed human events. Although original enough, he had clearly borrowed from the ideas of his era, using Spengler's cycles here and Karl Marx's proletariat there. Toynbee acknowledged Henri Bergson's "vital force," and when it came to religious symbols, he drew on psychologist Carl Jung's study of common archetypes in every culture.

Although much of Toynbee's most influential history writing was done between 1937 and 1957, called his Christian period because of his Anglican devotions and church activity, two earlier spiritual experiences have been given credit for his introduction of religious topics into his monumental *Study*. The first came in 1929; while he was traveling abroad he felt that a presence liberated him from a dangerous temptation. But the second experience was far greater in benign power, Toynbee said. He felt the divine presence as he bent over his dying son, who committed suicide in 1939 because of spurned love. War, death, and apparition had apparently planted in Toynbee a desire to help save Western civilization. His work as a historian has been interpreted as a spiritual journey, one in which he felt obligated to put God back into history.

That journey finally led in 1952 and 1953 to the Gifford Lectures, in which Toynbee summarized his historian's view of religion. As usual, his lectures began on a grand theme: the two great blights that mark every time in history. First of all, every culture seemed to set up idols to man or nature, and second, the world worked in such a way that every living thing was naturally self-centered. Every group of people tended to view itself as the center of the world, and that was true not only for cultures, but for historians in their own professional setting. "One cannot be a historian of any school without trying to break out of one's own self-centeredness into a time that is not one's own life time," Toynbee told his Edinburgh audience.[19] He also started out his lectures defending the search for patterns in history, an idea scoffed at by "scientific" historians. The "complete renunciation of all patterns" in history was impossible for a historian, Toynbee argued, for even chaos was a kind of scheme for interpreting human events.[20]

Toynbee moved from culture to culture in his lectures, explaining one form of idolatry or another, and then in the last two talks revealed his provocative conclusion. Although he awarded the founder of Christianity pride of place in the history of religion, he gave nearly equal status to the bodhisattvas, or voluntary and sacrificial saviors, in Mahayana Buddhism. For Toynbee, all religions had proved their value by speaking to the human concerns about self, yearning, and suffering. But religion achieved its highest form when it offered redemption through the sacrificial love of a superhuman being, a process that Toynbee called the Christian-Mahayanian diagnosis. The highest saints also followed this path. They worked for human salvation "not by dissolving themselves, but by making God's will theirs."[21]

As history moved on, he said, the many different religions began to merge in a common religious impulse. The "annihilation of distance" between cultures allowed the merger, which in turn demanded a psychological shift of each culture around the globe. This movement of history foretold a time when the seven higher religions—Hinayana Buddhism, Mahayana Buddhism, post-Buddaic Hinduism, Judaism, Christianity, Islam, and Zoroastrianism—might "coalesce into a common heritage." Toynbee argued that their "essential truths and essential counsels" were about the same already. They all held that the universe is mysterious, man is not the greatest spirit, knowledge is a means to action, and life should center on an Absolute Reality.[22]

Around the essentials, however, every religion had developed nonessential accretions such as holy places, rituals, taboos, and social conventions. With a final provocative flair, Toynbee argued that for the sake of the future, religions must throw off these accretions, many of them cherished sacred myths. "The stuff of which myths are fashioned is mostly local and ephemeral," he said. Emotional attachment to locality is strong. Still, such prejudices should be given up in favor of "primordial experiences" shared by all humanity. Whereas communing with God is primordial, for example, the bread and wine of Christianity are local, rooted in Mediterranean wheat fields and vineyards. Finally, the hardest belief for religions to relinquish is that "we alone have received the revelation," Toynbee explained. Indian religion had weaned itself from this idea, but a "traditional Pharisaism" still colored Judaism, Christianity, and Islam.[23]

Still a classicist at heart, Toynbee ended by quoting the fourth-century Roman pagan orator Quintus Aurelius Symmachus. When Rome became Christian, Saint Ambrose demanded that all pagan gods be removed from public buildings, including the altar of Victory in the Senate House. Symmachus appealed for its return. "The heart of so great a mystery cannot ever be reached by following only one road," he pleaded to Ambrose, but to no avail. The many roads of Symmachus could be embraced "without being disloyal to Christianity," Toynbee argued. "For what Symmachus is preaching is Christian charity."[24]

At the time of Toynbee's Giffords, he was active in the Christian ecumenical movement. In 1954 he served on a panel for the World Council of Churches. Yet by using the Giffords to argue a common essence in higher religions, a topic he "left unfinished" in volume seven of *A Study,* Toynbee revealed his gradual transition from a focus on Christianity to a broader, more tolerant view of all religions, especially Hinduism and Buddhism, which he lauded for their tolerance. In his Giffords, Toynbee came very "close to abandoning his Christian philosophy of history," according to his biographer, but he retained Christianity at the pinnacle of higher religions.[25] In later years, however, the abandonment became more obvious, and it may have started earlier in life when his first wife left him as a convert to Roman Catholicism, which she then went on the public-speaking circuit to defend. In the end, Toynbee could not reconcile the diversity of revelation in religion with the exclusive claim of any one. His Giffords equated the essence of all higher religions, and in his last years he spoke of himself as a post-Christian.

Toynbee's career as a historian had contributed to research on ancient Greece, current events, global civilizations, and world religions. With the publication of the last volume of *A Study* in 1954, he made the cover of *Time* magazine. In other forums, however, he was subjected to an avalanche of pent-up academic criticism. Professional historians faulted him for using sources three times removed from original documents. They decried his wild speculations and his apparent claims (which he denied) to "discover" laws in history. He had pushed religion to the front of history and ignored economics and politics, a mushy-headed approach that was an embarrassment to historical science.

Yet it may come down to tastes. Others were inspired by Toynbee's visionary history, a stark contrast to a scissors-and-paste approach, or what one Gifford lecturer called "Dr. Dryasdust" history. Toynbee had also elevated the comparative method in history. As his Western peers railed, Third World historians applauded his respectful look upon other cultures, even his challenge to the self-serving linear views of the West. A modern symposium on Toynbee's legacy agreed that "no once-and-for-all dismissal of him is possible," and much of that staying power is in his vision of civilizations vying with and absorbing each other.[26] Years later, the American historian Samuel Huntington stirred this controversy anew by proposing an ultimate "clash of civilizations." Toynbee had ended with five to seven civilizations, all fragile. With the fall of the Soviet empire, Huntington counted eight cultural blocs as global rivals. The Chinese and Islamic cultures, Huntington predicted, are most likely to conflict with the West, opening a new era of "challenge and response."

As the work of Toynbee showed, trying to find patterns or laws in the ebb and flow of ancient civilizations, or predicting future clashes, was a risky business that often brought scorn from the academy. Each new generation of historians, especially after 1900, would consider their task a more or less scientific enterprise, a precise effort to grasp "how it really was." In this light, the historical quest to find God's activity or providence was obviously more scandalous than even the search for impersonal laws in history. Yet God and history were ideas the Gifford Lectures were ready to grapple with, especially as the West emerged from the Second World War.

In the two decades that followed, the West saw a small renaissance of Christian writing on the meaning of history, a wave of interest that included the likes of Toynbee and the earlier Gifford speaker Reinhold Niebuhr. It was a time in which "book after book and article after article" on the subject appeared. In 1951 Princeton historian E. Harris Harbison identified "a kind of Augustinian revival of interest in the meaning of history," referring to that great early work of Christian history, Augustine's *City of God*.[27] The rapid fall of the Roman Empire had prompted Augustine to interpret the grand ways of history, and the catastrophe of the Second World War spawned a similar search for the meaning of historical events.

Typical of the trend was the work by the Catholic historian Christopher Dawson, whose Gifford Lectures in the late 1940s were published as *Religion and Culture* and *Religion and the Rise of Western Culture*. A Welshman, Oxford graduate, and Catholic convert, Dawson wrote history that emphasized the restorative role of religion in crumbling cultures. He was unique among Catholic thinkers, according to some, for his appeal to "the religious experiences of men in various historical cultures" to illustrate how Christianity could be a force to avert cultural decline.[28] It was a pressing problem for the Christian West, and Dawson marshaled all the historical evidence he could to offer an antidote to despair.

One of the last historians highlighted by the postwar wave of Christian history writing was Herbert Butterfield, the Cambridge historian. Butterfield was never allured by the search for laws in history, but rather stood in the line of scientific historians from Ranke to John E. Acton, the famed Lord Acton who in 1900 was founding editor of the *Cambridge History of the World*. Acton hailed from an English family of Roman Catholic nobility, and in Germany, where he had studied, he learned the new historical methods. He applied them in the *Cambridge History,* a first great attempt in English objectively to tell the human story. Yet in Acton's vision, the objective facts were to illustrate human progress, for the very belief in progress was "the scientific hypothesis on which history is to be written," he famously said.[29]

Although Butterfield had no use for the ideology of progress, he too belonged in this "objectivist" school, even more so than Acton. While Acton tended to pass judgment on figures in history, Butterfield believed the historian must exercise a professional humility. Humility also applied to the facts, for rarely could a historian explain exact causes the way a chemist could write out a chemical reaction. The great events in history often turned on the most arbitrary occurrences; historians mused about such events as a stone cornice falling on a prime minister's head. In a dash, history has changed course. History was often detoured by the proverbial Cleopatra's nose, a small body part whose prettiness undermined the Roman Empire.

The application of these views—the search for objective fact, a humility about causes, and recognition of arbitrary events—decorated Butterfield's most famous early work, a brief book called *The Whig Interpretation of History* (1931). Like Toynbee, Butterfield was impatient with self-serving historians, and he used England's liberals, the Whigs, to illustrate his point. Whig historians, he showed, had written British history as if their party alone had crushed tyranny and produced representative government and religious liberty. Butterfield himself was from a working-class family in Yorkshire, and he was a nonconformist Methodist Whig as well. Yet he accused Whig historians of naked bias when they wrote about a straight-line march from Whig virtue to British liberty. On the contrary, religious liberty arrived "not by a line but by a labyrinthine piece of network" and involved agencies "that had little to do with either religion or liberty." By such candor, Butterfield brought one epoch of history writing "to an end" and "put historians into a state of self-analysis and scrupulosity," said Owen Chadwick, who followed him in the history chair at Cambridge.[30]

Butterfield's desire to get at the truth probably stemmed from the religious character of his family. Back in Yorkshire, his father would have become a Methodist minister had he had the financial means. The father's interest brushed off, and Herbert, his son, became a lay preacher. At Cambridge University, the history faculty turned Herbert Butterfield into a full-fledged historian, although he continued lay preaching and taught church history at Wesley House, which trained Methodist clergy. After the Second World War, the Cambridge divinity faculty asked Butterfield to lecture publicly on Christianity and history. Reluctant at first, he took up the task by giving seven Saturday morning lectures in 1948, drawing an audience of eight hundred every time. The BBC broadcast six of the talks a year later, and the corpus was published as *Christianity and History.*

By this time, Butterfield was at the pinnacle of his career as a secular historian. His technical works of history—from Napoleonic diplomacy to the origins of science—had firmly established his reputation, a seniority that also had led him to expound on the very origins of the historian's art itself. Yet with the 1948 talks, he had inadvertently begun a new career as a religious thinker as well. In that era, only Christopher Dawson would write more works on Christianity and history than Butterfield. Among philosophers and theologians, moreover, only Reinhold Niebuhr exceeded Butterfield in turning out volumes of material on the Christian view of history.

When he was invited to deliver the Gifford Lectures in 1965, Butterfield naturally spoke on what he knew best, the stories of secular and Christian history writing. "It was to be Christianity that helped to make the West so historically minded," he told his Glasgow audience.[31] The Christian writers of history took two kinds of approaches, he explained; the first showed clearly how God's providence helped believers, and the second showed that God's work was far more mysterious than that. In the first case, for example, Eusebius, the "father of church history," tried to demonstrate in the early fourth century that Emperor Constantine had won battles and lived long because he was a Christian. In the next century, however, Augustine argued that history was not so simple, especially as the Roman Empire crumbled around him. God worked through both the city of God and city of man, saints and sinners, for "these two cities are entangled together in this world, and intermixed until the last judgment separates them."[32]

The work of Butterfield has been called Augustinian, but that is untrue in one important respect: Augustine saw history as predestined; Butterfield held that human free will was an authentic force even in God's providence in history. *The City of God* also smacked of a sacramental body such as Catholicism, and with his Methodist calling, Butterfield was led to an individualist interpretation of the Christian faith. The two particular keynotes that Butterfield sounded in his writing on history and Christianity were his notion of "providence" and his assessment of the historical investigation of the New Testament, a field of study set afire by Schweitzer a half century earlier.

For Butterfield, the idea of Providence in history—the interpretation of great events and people as if God were playing a role—had three levels to be considered. First was the biography of great figures in history, a world of personalities, free will, and the clash of wills that Butterfield favored in his technical historical works. A second level of Providence might be seen in statistical and sociological laws such as population, food supply, distributions of wealth, and the limits of governments and armies. But for the truest reading of Providence, the third level was most important: that of faith. Providence is at heart a mystery "revealed only to those who take the Gospel home to themselves in that innermost region of the personality."[33]

To illustrate Providence, Butterfield turned to the Old Testament. It showed the justice of God, for evil nations met reckonings or received the accumulated grace of righteous groups in their midst. The Old Testament was a history of promise and judgment, a reflection of the world as it truly operates. Butterfield held the view that God finally brings purpose out of historical discord, much as the conductor turns a cacophony of instruments into a symphony. To surrender history to chaos was unacceptable for a believer: "It is never permitted to a Christian to despair of Providence," Butterfield said, adding on the nuance of human free will. "We must imagine Providence as doing the best that the wilfulness of men allows it to do."[34]

Turned pessimistic by the Second World War, Butterfield envisioned a more secular West and an increase in global religious conflict. His hope lay in the Christian faith being spread from one individual to another, and in this project he knew the veracity of the New Testament was an important tool. He approached the New Testament as a professional historian, and from this perspective there was much to disappoint. It was a "matter of regret that the followers of Jesus did not provide for us something more like an ordinary biography," Butterfield said in his Giffords. "I find it difficult to imagine the disciples failing to talk to one another and to other people about Jesus as an ordinary human being."[35]

Yet accurate biography writing was not the method of those ancient times. The early Christians, moreover, expected the end of the world, so collecting such a story was surely a low priority. Facts were lacking, he conceded, and "this lack may even make the Gospels less plausible" to nonbelievers. Butterfield's credentials as a historian nevertheless gave him a perspective that the ordinary person lacked. Even a biography of Winston Churchill, for example, could be riddled with hearsay and bad memory, yet still capture the man.

And so it could be with the biography of Jesus found in the Gospel accounts. From a historical critical viewpoint, Butterfield went on, the Gospel writers seemed to be authentic chroniclers. By ancient standards, the crucifixion was a model of disinterested reportage and was the New Testament's "most detailed, most consecutive, and most consistent piece of narrative." The Gospel writers, moreover, were remarkably self-effacing and showed no "personal ambitions or vested interest" that would motivate a "hoax." To explain how a rag-tag band mustered the "spiritual power" to found a religion, a historian might reasonably look for a dramatic event at the start of Christianity, Butterfield concluded, referring of course to the New Testament account of the resurrection.[36]

In the heyday of his work on the historical Jesus, Schweitzer had said that the truth about history was always preferred to falsehood. That had been his quest, and also the dream of Toynbee, Butterfield, and others, all of them hoping to meet the scientific requirements of the age and yet allow for a belief that God is present in history. Each in his own way knew that history could not be entirely scientific, and neither could faith be reduced to historical proofs. As Butterfield had said, Providence was best seen in the eyes of faith.

Behind the attempts to make history scientific, then, there always lingered doubt that this was the right approach. For certain, history *was* a science by the mere fact of working with fact, objectivity, and analysis. Yet the differences between pure natural science and the research and production of history remained striking. Natural science looks for generalized laws, while history focuses on unique events. History draws lessons from the *past;* natural science predicts the *future* according to physical laws. Finally, history was unavoidably a subjective study, bathed as it was in normal human feelings about nation, race, morals, and religion. Historians who wanted to know "how it really was" might have to invest their own personal feelings in the human dramas and events of the past.

While this idea of searching for the internal side of historical events bubbled up here and there, it was expressed most fully in the English-speaking world by the British scholar Robin G. Collingwood. An accomplished field historian of ancient Romans in England, Collingwood elaborated on the idea of the subjective consciousness of both the historian and the human actor in history. For the historian to write a true history, Collingwood said in a collection of essays published posthumously in 1946 as *The Idea of History,* the historian must "think himself into the action, to discern the thought of its agent."[37] The real story of history may at times be the human consciousness involved, not just the factual event, duration, or location, all of which might be inescapably imprecise in their reconstruction.

Bible scholars also heard this sort of analysis and, frustrated by the paucity of data on the historical Jesus, eagerly tried to think themselves into the action of the New Testament. What they found, even amid the historical critical method of studying the Bible, was a persistent consciousness in Jesus and his early followers—a proclamation of salvation in the New Testament that was called by its Greek name, the *kerygma.*

Before and after the Second World War, this idea of the kerygma took biblical studies by storm, and the search for the essential spiritual message of Jesus and the early Christians was the next great wave in Bible studies after Schweitzer inaugurated "the quest" for the historical Jesus.

The experienced historian will always find, however, that the truth never seems to lie at one extreme or the other, either in the facts of history themselves or in the consciousness of the actors in history. That conclusion descended upon many Bible scholars who had enthusiastically taken up the quest for the kerygma of the New Testament. Not everything in the Christian Bible could be reduced to a pure message, they decided; they also needed a reasonable amount of ordinary, mundane, and prosaic facts on which to construct the Christian story of salvation. In the 1950s, therefore, a group once devoted to the kerygma began a "New Quest" of the historical Jesus. Yet again, it was a project to try to assemble the historical reality in which Jesus and the early Christians lived.

Fortunately for the New Quest, the postwar period produced a mother lode of new textual and archaeological materials. Some of it was controversial and challenged traditional Christianity. Other finds were supportive of "biblical archaeology," which was a devout attempt to prove the Bible by unearthing its historic locations in Egypt, Palestine, Syria, Asia Minor, and elsewhere. Whether the postwar discoveries troubled or aided the traditional understanding of Christian origins, this was a cache of hard historical evidence like none seen in generations.

The unearthing had begun in Egypt in 1945, when a peasant found buried pots containing fourth-century Gnostic writings and "gospels" in the Coptic language—the booklike Nag Hammadi manuscripts. Two years later the Dead Sea Scrolls began to surface, a vast collection of texts from a Jewish eschatological sect, perhaps the Essene mentioned by the Roman historian Josephus. The scrolls were found in caves along the desert cliffs of the Dead Sea, which boasts the lowest point on earth and water so alkaline that travelers still thrill at floating on its surface.

Although the Coptic texts spoke of Jesus, the attribution was far more commonly to the Christ, the Savior, and the Lord. The Gnostic gospels not only portrayed Jesus and his teachings in the company of well-known New Testament characters, they added others as well. The authors of the Dead Sea Scrolls, on the other hand, referred only to a Teacher of Righteousness in their midst. The findings were explosive, of course. As ancient history went, Jesus was better documented than even the Caesars, but it was still sparse evidence indeed. Besides the New Testament, only five other historical sources referred to Jesus of Nazareth. Even though the historical facts were few, Christianity had successfully resurrected its founder with the help of faith and tradition. It was a potent combination, for the facts, faith, and tradition changed history and indeed built Western civilization.

The legacy of the Gifford Lectures was born out of that civilization, and in their first half century the lectures displayed some of the new scientific pillars of the West: anthropology, psychology, physics, sociology, and finally the scientific approach to history. They had become pillars because of the certainty that the sciences presumed to offer, certainty that they so persuasively provided in the realms of technology and medicine. Yet as the historian knows, every great pillar of society may meet its greater force, and for the West that earthquake was seen in the First and Second World Wars, which shook Western certainty to its foundations.

And thus a new search for certitude began—a search that was about subjectivity, not about the objectivity of the sciences. The sciences would endure and grow, but in the third act of the Gifford legacy, the advent of the new "subjectivism" is a force to be reckoned with. The sciences arrived bold and knowledgeable, but for many human souls, they simply had a limit: they could not provide meaning, stir the soul, or do what faith, art, emotion, and other liberating human expression could do, which was to bring warmth, juices, and life to a cool, dry, factual world that science wanted to put on a scale and measure.

Many decried the new interest in subjectivity as a descent into irrationalism, a desperate rejection of all that had been built up through careful research, study, and experiment. The new subjective impulse also pushed aside the older interest in natural theology, for it too seemed locked in the cage of reason, a kind of pretender to science. Still, even the subjective movement in the West could not dismiss the idea of God. Rather than jettison such a belief, it tried to find new ways to talk about the Creator, and some of that talk naturally reached the Gifford Lectures.

EIGHT

FROM BARTH *to* BEING

The Revolt Against Reason

O NLY A CATASTROPHE in history, such as a global war, seemed sufficient to shut down the Gifford Lectures. They took their only hiatus from 1941 to 1946 and then slowly made a return. The West was entering its postwar era, and as old philosophies and ideologies died, new ones emerged. Although the findings of science and knowledge of history may have challenged religious belief, they also had shepherded it into the modern world.

Owen Chadwick, the Cambridge historian, asked in his Gifford Lectures whether it was science or historical consciousness that had most secularized Europe since the late 1800s. He felt that science had dealt the sharpest blows, chiefly because of how it confounded the literal biblical precepts of many. "The historical revolution was not so upsetting for religion as the scientific revolution," he said. For its part, history could render religious beliefs as relative, products of a time or place instead of eternal verities. Still, said Chadwick, knowledge of history "often brought with it more understanding of the nature of religion" and its value for individuals and for societies.[1]

The case has also been made, however, that historical consciousness is what finally undermined "biblical civilization" in the West, a death knell that between 1880 and 1930 produced far more turmoil in the United States than in Great Britain.[2] But now, the impact of science and history would be overlain by two new movements before and after the Second World War. One revived a belief in God's almighty revelation, and the other, called existentialism, lionized personal freedom. Both offered certainty in turbulent times. Neither was friendly to natural theology.

Act three of the Gifford legacy had begun, and as often was the case, the characters, plots, and ideas had their origins in northern Europe. The Protestant revival in the theology of revelation declared that, fortunately for modern man, the Word of God had already broken into the world, much as if God had thrown a large stone at humankind. Once and for all, Christ and the Bible had been revealed. This "wholly other" God had little patience with ingenuous attempts to climb to heaven. The captain of this movement was the Swiss theologian Karl Barth. He would declare himself an "avowed opponent of all natural theology" but give the Gifford Lectures nevertheless.[3] Although Barth shunned labels, his movement received plenty. Some called it theology of the Word and dialectical theology, others neo-orthodoxy, biblical theology, or a theology of crisis—the crisis being each person's encounter with Almighty God. A Buddhist thinker, on hearing of God breaking into the world to foment crisis, called it a "theology of flying saucers."[4]

Existentialism, the second proposal, came from the heart of European philosophy. It presumed to offer the one kind of certainty every person could grasp: his own existence and choices of how to live. "The existentialists are nothing if not individualistic in outlook, precisely because they emphasize man's freedom," said the philosopher Frederick Copleston in 1948.[5] A vague movement, it flourished after the Second World War, adding its ethos to both philosophy and theology. A true founder is hard to determine, but the most popular candidate is Søren Kierkegaard in the nineteenth century, followed by philosopher Martin Heidegger, a German professor during the Second World War. Later came the French writer Jean Paul Sartre, an atheist and libertine.

With their separate emphases on revelation and individual choice, these two movements were born of the same litany of modern discontents, with totalitarianism at the very top of the list. From the 1920s through the 1960s, the two showered their influences across Europe and, later, North America. And they would be new impulses in the Gifford Lectures. Although one of them looked at God outside the world, and the other at people within, they shared a common fulcrum—the despair over personal salvation, whether it be in heaven or in an authentic life on earth.

In the rise of neo-orthodoxy, and Barth's eventual invitation to give the Gifford Lectures, one other European dynamic was at work. This was the centuries-long tension between Roman Catholicism and the churches of the Reformation. In general, Catholicism had been a natural home for natural theology, typified by the 1879 encyclical *Aeterni Patris,* issued by Pope Leo XIII in hopes of reviving Catholic philosophy on the model of Thomas Aquinas. At the heart of Protestantism, however, stood a rebuke to such human efforts. The alternative, as revived extremely in Barthian thought, was to put faith only in God's revelation, known exclusively in Christ and the Bible. They were the great stones that God had thrown into the fallen human world.

As a leading thinker in Europe, Barth was an intellectual celebrity, and that alone made him a worthy candidate for the Giffords. Yet in his forthright style, he had already waged war on natural theology in his pamphlet *Nein!*—a title that spoke for itself. In one of the remarkable episodes of the Giffords, he nevertheless gave the lectures at Aberdeen beginning in 1937. He warned his hosts of his stance and told his Aberdeen audience of his difficulties. Those who had given the lectures since 1888, he said, must have had to "rack their brains" to fulfill Lord Gifford's guidelines. But none had "so much trouble" as himself.[6] He declared himself a Reformed theologian, opposed to "all" natural theology, which included Catholic philosophy. In a small way, however, it was a Catholic turn in the Gifford legacy that may have gotten Barth the invitation.

The first Gifford invitation to a Catholic thinker came in 1922, a generation after the lectures began. For a bequest dedicated to natural theology, a Catholic seemed natural enough. Yet the barriers had their modern reasons. Not long before the Gifford Lectures began, Rome had declared papal infallibility. Two decades into the Giffords, in 1907, Pope Pius X condemned Modernism in his encyclical *Pascendi Dominici Gregis.* Modernism in France and Britain had been trying to reconcile science with the Bible, faith, and even the sacraments, with its obvious liberalizing effects. "Well, I think it is understandable in nineteenth-century Scotland," says a Gifford officer today. "It wouldn't be natural for people to look to Catholic thinkers."[7]

Obviously, Catholics inquired after truth as earnestly as anyone else. When St. Andrews began its Giffords, it advertised for candidates. One who wrote in was St. George Mivart, a widely known Catholic biologist in London whose book *On the Genesis of Species* was a major rebuttal of Darwinian materialism. "I beg leave, through you, to offer my services," Mivart wrote St. Andrews in 1888, though without success.[8] By the end of his life, Mivart would be censured by Britain's Catholic cardinal for heterodoxy, an incident that also suggests the delicacy of the Modernist debate for Gifford committees. The first invitation was finally issued in 1922 by Edinburgh to the popular Catholic writer Baron Friedrich von Hügel, an Austrian native long resident in England.

A scholar of independent means, Hügel was on the edges of Modernism. He was friends with both Alfred Loisy, the French priest who led the Modernist movement, and its exponent in Britain, George Tyrell, a convert who had become a Jesuit. Loisy and Tyrell were excommunicated soon after *Pascendi Dominici Gregis,* clashes that deeply saddened Hügel. Although Hügel supported "liberty of thought and action" in the church, "he emerged—almost alone—serene and secure and Catholic from the dramatic dark days of the Modernist conflict."[9] Yet the Gifford invitation came too late. The excitement, combined with his illness, plunged Hügel deeper into physical maladies, indeed a "nervous breakdown," preventing him from delivering the Edinburgh lectures.

That made the French historian Etienne Gilson the first Catholic on the scene. Crossing the English Channel in 1931, Gilson traveled to Scotland's northernmost great city, the port of Aberdeen, a town built of shiny silver granite. There in 1860, the ancient schools of Catholic and

Protestant origins—King's College and Marischal College, respectively—had merged into one institution of higher learning, now called the University of Aberdeen. A modern campus by the 1930s, it still centered around King's Chapel, with its original crown spire from 1500, a signature piece of architecture that survived only because the principal had defended it from Reformation zealots armed with axes and torches.

Gilson arrived at the old King's quadrangle to lecture on medieval philososphy (*The Spirit of Mediaeval Philosophy*). He did so well (though he spoke in French) that the work became a modern classic. Although he touched on the merits of natural theology, especially in the arguments of Saint Anselm and Thomas Aquinas, Gilson had a different point to make. Christian thought, he argued, did not simply cloak itself in Greek philosophical ideas. What it uniquely added to Western philosophy was the Hebrew Creator, with all its implications for a created order, the nature of being, and existence itself. "Neither Plato nor Aristotle knew anything of Genesis," Gilson said. "Had they done so the whole history of philosophy might have been different."[10]

Despite his injection of the biblical idea of a Creator, however, Gilson argued that good philosophy stands on its own merits as a system using evidence and reason. And that, he said, was what so many of the medieval Catholic thinkers had done, thereby giving basic tools to a list of modern philosophers: rationalists such as Descartes, Spinoza, Leibniz, and Hegel, and even the British empiricists. Without giving due credit, modern science had dismissed medieval thinkers as obscurantist and impractical "scholastics." From another side, the Reformation had denounced them as lacking faith in God and putting too much stock in reason. Indeed, their sin had been the idolatry of natural theology.

But Gilson had not arrived in Aberdeen to apologize in his concise set of twenty lectures in 1931 and 1932. What the Catholic medievalists gave to the world, he concluded, was a solid philosophical realism, the belief that God was manifest in external objects in the created order as well as in supernatural powers of the mind and the cosmos. Much of that was lost in the Reformation, he suggested, with Martin Luther's emphasis, for example, on faith alone. The Reformation was the seedbed of much speculative philosophy in Germany, a flight into the world of mysticism and supernaturalism. "It was natural that it should be Lutheranism that issued in idealism," Gilson said. "The true Lutheran is not drawn to seek God in nature."[11]

The Gilson series was a comprehensive piece of work. It was only logical, the story now goes, for someone at Aberdeen to think about summoning a Protestant version of the same depth and breadth. Who better to provide it than Barth, leader of the neo-Reformation movement? The Aberdeen senate invited him to give the next possible Gifford series. The intention of putting Gilson and Barth almost back-to-back, as if there were a grand tournament between two great belief systems, was never announced. Yet the idea of a war between natural theology and faith alone, or *sola fide,* was already in the air.

In 1934, a year before Barth was invited, the Swiss theologian shocked the English-speaking world with an attack on Emil Brunner, a fellow dialectical theologian and ally. Brunner had made the mistake of publishing, in full view of Barth, a forty-four-page booklet, *Nature and Grace,* which argued that even fallen man might use a little bit of natural theology, a modicum of "general revelation," to move in God's direction. Even John Calvin, the Reformed patron saint, had said as much. And it seemed apropos in an age of science. Barth fired back with his stentorian title *Nein!* He argued that natural theology gave rise to both religious syncretism and anti-Semitism.

Barth's public assault was more than a theological debate, as would be understood soon enough. His war on natural theology was being waged against the backdrop of the rise of National Socialism. Slowly but surely the Nazis were taking over German Christianity. Barth's belligerence in lashing out at heterodoxy in the theological ranks suggested a far more cosmic battle with new political idols being built in Europe. "Features of aggressiveness and totalitarianism are manifest in the Barthian tradition," one critic of Barth said, speaking in later Giffords. The real headline was this: "Barth confronts the unlimited power of fascism with the jealous power of God."[12] In that politically charged era Barth fought back with every theological tool in his possession. He argued that natural theology, in its quest to know God, fostered human hubris, and such hubris had its origins in the same sinfulness that gave fascism its vision of a godlike state. When the Nazi regime came to power, Barth felt that the liberal Christianity of his youth had made the demonic regime possible.

Born into a long line of Reformed Church pastors in Switzerland, Barth, the oldest of five, was grooming himself for scholarship even as a child. Once, when the family left for a hike, he was nowhere to be found. His brother found Karl in the barn, about ten years old and seated on a three-legged stool writing on the top of a barrel. What was he doing? "I am writing my 'Collected Works,' " he said.[13]

Barth wrote and spoke with a creative fury, penning volume after volume, changing his mind and drafting them again. He always used many words instead of a few. How many ways can a theologian say that God is God and that nothing else can be said about God? With imaginative power, Barth found hundreds, producing reams of pages, dense and repetitious but also inspiring. First by writing, and then by speaking, he launched a movement that rocked modern Protestantism.

Barth was nurtured in the liberal Protestantism represented by the church historian Adolf von Harnack, under whom he had studied. His crisis with liberalism came on the eve of the First World War. A large number of German intellectuals, including Harnack, signed a public letter supporting Kaiser Wilhelm II's military ambitions. Clearly, Barth decided, cultural Christianity had become a mistress to politics. He went back to the Book of Romans, much as Martin Luther had done. There he found Saint Paul's orations on how human sinfulness was revealed in the bright light of a holy and Almighty God, who alone gave salvation by grace.

Barth's immersion in Pauline theology was marked by the 1919 publication of his *Epistle to the Romans,* but this was not yet the Barthian revolution. Three years later, in 1922, he issued a second edition of *Epistle* that had been rewritten to emphasize his dialectical theology, and this was the book that sparked new ways of thinking in Europe. Translation of the second edition into English in 1933 announced the migration of Barth's neo-orthodoxy from northern Europe to the rest of the world.

Even before his work on Romans, Barth had been enamored of Kierkegaard's ideas about the individual human struggle. As Kierkegaard said repeatedly, human beings lived not only under sin, but an ultimate paradox. They lived in time, as if in a prison, while God lived in eternity. The twain would never meet, and that put humanity on the precipice, a status of "fear and trembling" and "sickness unto death," as Kierkegaard stated the case. "If I have a system," Barth said, "it is limited to a recognition of what Kierkegaard called the 'infinite qualitative distinction' between time and eternity."[14]

Kierkegaard represented the classic reaction of faith to rationality in the nineteenth century. It had been seen earlier in Blaise Pascal, a French Catholic enthusiast of the Augustinian notion of human sinfulness and divine grace, and the human "wager" that it was all true. In England the apocalyptic poet and artist William Blake rebuked that nation's scientific Enlightenment and its deistic attempts to figure out God. Blake's drawing of a white-bearded God measuring the universe with a two-pronged compass was hardly a paean to natural theology; it was an attack on its very presumptions. In Blake's poem *Jerusalem,* reason itself becomes the evil one: "I am God, O Sons of Men! I am your Rational Power! . . . Vain, foolish Man! wilt thou believe without Experiment . . . ?"

What was handed down by Kierkegaard, an apostate student of Hegel, was a version of the Hegelian dialectic, but a version that did not propose the Hegelian resolution of opposites. For Kierkegaard, and similarly for Barth, the dialectic was a synonym for paradox, a perpetual contrast and opposition between God and the world. As sinful, finite creatures faced such an eternal God, their lives were on an ocean of paradox, and the only way to cross it was a "leap of faith." Barth would take this further theologically. Kierkegaard had said that the incarnation of Christ was the greatest paradox of all, for Christ was the eternal God embodying time. For Barth, this idea diminished God. There was no paradox in Christ, Barth said, because God has the power to do what he will with time. God can simply swallow up time, so for him, where is the paradox?

Of the many labels given to the Barthian movement, the one that he seemed to like was "dialectical theology." Barth became a master of the dialectic, which explains why he wrote at such length on a single topic such as the otherness of God. In the dialectic, the theologian says something about God, but then quickly realizes that it falls short of the Supreme Being, so he must say something to contradict or correct the first statement. The correction too falls short, and something more must be said to correct its failing. Back and forth the process goes on, never finding rest, always a thesis, antithesis, and synthesis—the synthesis being knowledge that *nothing* can truly be said about God, except for the name of Christ and his word in the Bible.

Thanks to Barth, natural theology was pummeled by many dialectical blows. When Christopher Dawson, the English Catholic historian, gave his Gifford Lectures in 1947, he lamented that still "to-day Natural Theology is hard-pressed by the convergent attacks of Dialectical Theology and Dialectical Materialism."[15] The attack by atheistic Marxism was surely to be expected, but the attack by a theologian such as Barth needed a special explanation.

Barth had taught theology at German universities for thirteen years, ending in 1934. Late in that year he refused to open his class with homage to Hitler. Could anyone imagine, he asked the authorities, opening a commentary on the Sermon on the Mount with "Heil Hitler"? Although Barth's appeal was rejected in 1935, he escaped harassment or arrest. He wisely left for Switzerland, where he was offered a special professorship at the University of Basel. The year before, while still in Germany, Barth had courageously drafted and circulated the "Barmen Declaration," creating a movement of "confessing churches," confessing Christ as Lord, and no one else. In theological terms, Barmen opposed a state alliance being called the "German Christians." Barmen declared the churches' loyalty beyond this world, and above any regime. The stakes were high, not only in politics but in theology.

In the summer of 1935, Barth was nothing if not well known. The commentary on Romans had come out in English in 1933, marking the Barthian invasion outside the Continent. The next year had produced the Barmen Declaration and the little fusillade *Nein!* So the next summer (of 1935), the senate at Aberdeen sent him an invitation to deliver the Gifford Lectures in the spring of 1937 and 1938. As happened often enough, the Aberdeen letter spoke of "Natural Religion" instead of natural theology, a difference that most people overlooked. The letter included samples of two previous lecture outlines, Gilson's *The Spirit of Mediaeval Philosophy* and a more recent one by an Oxford professor.

Nearly two months later, Barth replied that he was "ready to take on the task that has been put to me." But as a "matter of loyalty" he wanted to speak with candor, and then let the "high senate" at Aberdeen take responsibility for the invitation. "If I am to take over these lectures I by far cannot think of holding them in the same sense as Lord Gifford in 1885 appears to have thought of them," he said. For it was "surely well known that I am an avowed opponent of all natural theology." At the same time, Barth was also impressed by Gilson's survey of medieval philosophy. Despite the issue of natural theology, he said, he hoped to "meet the purpose of the foundation at least partially, in that I have taken the liberty, following the example of Prof. Gilson, to suggest a mostly historical subject: the problem of the natural recognition of God in traditional Protestantism."[16]

Barth's reply was happily accepted at Aberdeen. Two months before the lectures, he wrote again, revising his topic. "You will find," he said in the January 1937 letter, "that I have approached the subject somewhat differently than I had planned at first." In recent years, Barth had been traveling in Europe to speak on the Reformation confessions of various countries, a topic on which he had long been an expert. He had spoken in Hungary, France, and Switzerland on their sets of Reformation beliefs. So why not do the same in Scotland?

For the Gifford lectures, Barth delivered a running commentary on the Scottish Confession of 1560, a document produced by the fiery John Knox, a former priest who studied at St. Andrews University. During the Reformation furor, Knox holed up with a group that murdered the reigning cardinal, and preached at St. Andrews castle until taken captive as a galley slave by French Catholic invaders. Once freed, he preached for the Crown in England, fled to Geneva to work with Calvin, and then repaired to Edinburgh, where he disputed theology with Mary Queen of Scots, a Catholic Frenchwoman, whose castle was a mile down Main Street from his pulpit. To adopt a Calvinist form of Christianity, the Scots Confession had two parts, the knowledge of God and service of God in the church. Barth organized his two series of lectures accordingly. To cover the first, the knowledge of God, he set off for Scotland at the beginning of March.

By elevating the old Scottish creed, Barth wanted to present a dramatic antithesis to natural theology. " 'Natural Theology' is thrown into relief by the dark *background* of a totally different theology," he explained in his opening talk, which like the rest was read in German, with an Englishman at his side to give a running translation. It was clear which theology Barth presumed to be the false pretender, for he put "natural theology" in skeptical quotation marks. "When 'Natural Theology' has this opponent no longer in view, it is notorious how soon it tends to become arid and listless," he said. "And when its conflict with this adversary no longer attracts attention, it is also notorious that interest too in 'Natural Theology' soon tends to flag." So here he was, being natural theology's "indispensable opponent," thus giving it life, as Gifford would have hoped.[17]

During that trip to Britain, Barth visited Edinburgh, St. Andrews, and London, where he met about thirty senior churchmen and spoke to German pastors and curates. He wrote letters and kept notes on his travels, observing that the British "have an admirably sound view of the world situation at home and abroad, an assurance in always finding the middle way which is quite unhysterical." Especially then, the British were banking on avoidance of war, right up to late 1938, when Prime Minister Neville Chamberlain returned from Munich to declare "peace in our time." Barth found British life wholesome, with well-stoked fireplaces and the joy of sitting, smoking, and talking by the hearth. They ate porridge in the morning ("ugh!") and drank whiskey at night. "It was the kind of atmosphere in which you cannot be cross with anyone, just as you cannot make anyone really cross unless you virtually commit atrocities."[18]

Still, Barth marveled at the naïveté of British Christians. They were open to "natural theology, pietism, 'historical criticism' in the style of the 1890s, the 'comprehensive church' (a favourite boast of the Anglicans), moral optimism and activism and so on." Indeed, 1937 was a very ecumenical year. Britain hosted major conferences at Oxford and Edinburgh, groundwork for the modern understanding of world Protestantism in a changing political and social situation. Barth contributed a paper on "revelation" but did not attend. "I have to some extent achieved my own 'ecumenical movement,' " he said. Ecumenism made good contacts but also forced compromises. When the British asked what the confessing church needed, Barth replied: "not expressions of sympathy or protest, but a solemn assent to Barmen."[19]

Barth was back in Scotland later in 1937 to receive an honorary doctorate of laws at St. Andrews, a university and town whose Christian traditions impressed Barth—he called the settlement a "corpus Christianum." But again he sensed that the British were missing the great theological revolution, if his chat with the Duke of Kent was any indication. "He didn't seem to be very clear about the significance of dialectical theology," Barth said. The Duke only asked if this was his first visit to St. Andrews. "I replied, 'No, it's the *third* time.'"[20]

In the spring of 1938, Barth returned to Aberdeen to deliver his second series of Giffords. And then the war threw a smothering blanket over most theological controversy.

After the war Barth mellowed, which has complicated assessment of his entire legacy. The touchstone of his softening was the essay "The Humanity of God," a surprise to many after two decades of God's total otherness. Of the stridency in his *Epistle to the Romans,* Barth once said, "Well roared, lion!" in parody of himself. Given the times, "it was necessary to speak in this way." On his only trip to the United States, in 1962, Barth was quoted in *Time* magazine as saying, "I had to show that the Bible dealt with an encounter between God and Man. I thought only of the apartness of God. What I had to learn after that was the togetherness of Man and God—a union of two totally different kinds of beings."[21]

Times had now changed, at least for a bygone National Socialism, if not for Soviet communism. As a young pastor, Barth had bridled at the economic inequality in Switzerland. He helped organize labor unions and was called "the Red Pastor." When Barth was perhaps the only prominent theologian who did not condemn the Soviet invasion of Hungary in 1956, many neo-orthodox leaders, such as Brunner, were deeply troubled. Barth urged pastors behind the Iron Curtain to mollify the regimes so they could continue their spiritual preaching. In the view of some, Barth put God so far outside the world that his natural theology amounted to atheism. In a more positive light, Barthian scholars have portrayed him finally as a contrarian. During the Cold War, he may have seen the United States as simply another kind of secular and militarist power in the world, now consumed in an anticommunist crusade. A wealthy nation like America, he believed, had propaganda enough without him.

The final road of the Swiss lion would be an emphasis on Christology, that all things were known and made real in the incarnation of Christ. Some would call this another kind of extreme, a Christianized form of Absolute Idealism, indeed a "Christomonism." If all things were essentially Christ, there was very little else to talk about, which for some theologians rendered their fields single-minded and dogmatic. By the time Barth visited America, his longtime foe Reinhold Niebuhr was active mostly in advocating American democracy against communism. For Niebuhr, Barth's theology had become "irrelevant" in the postwar era. Nevertheless, the Barthian tone and mood seemed to fill American sermons everywhere. Yet times change: when the Swiss lion entered the United States, Niebuhr spoke for many when he said he didn't read Barth anymore.

Still, this being the great Barth's first American visit, *Time* said that among Protestant clergy his "arrival has caused as much stir as would a visit by the Pope to a Jesuit convention." Barth's humanity and sense of humor were emphasized. He was the German professor from central casting, the ash of his pipe flicked about on his rumpled suit and his heavy glasses down on his nose. "Once, upon hearing that Pius XII had paid tribute to his work," *Time* reported, "Barth smiled and said, 'This proves the infallibility of the Pope.' "[22]

At Aberdeen in more recent years, theology professor John Webster, editor of *The Cambridge Companion to Karl Barth,* pulled from his bookshelf the slim volume of Barth's Giffords, printed in German and English. "The natural theology issue has been misunderstood, particularly by American and English interpreters," he said. "They thought, Why on earth did Barth make all this fuss." Barth felt he needed to side against all worldly religious projects, ambitions of reason, and hubris that he saw materialize in Germany. "He was being looked to by dissenting church groups in Germany as the leader of confessing theology. If he came to Aberdeen and suddenly turned on the charm of natural theology, they would have felt betrayed."[23]

Barth himself would turn out to be a topic in future Giffords, but how could he not have been? In one case, philosopher Brand Blanshard of Yale University, speaking at St. Andrews in 1952, pitted Barth against the use of reason to find goodness and faith. As reported in the *Scotsman*, Blanshard said that the dialectical theologians made religious belief irrational. "Barth and Brunner, like their master, Kierkegaard, revelled in paradoxes," Blanshard said. "They represented God as being so completely 'other' that He almost disappeared; we were supposed to believe things about Him that, by our standards, were self-contradictory, and ascribe actions to Him that our moral sense could only regard as evil." Indeed, Blanshard worried that crisis theology would prompt reasonable people to reject the Christian faith altogether. Having read the Edinburgh newspaper the next day, the Scottish theologian Thomas Torrance was "astonished" that Blanshard would fall for such a stereotype of the great Swiss theologian. First, he pointed out the changes represented by "The Humanity of God," and then he argued that Barth indeed used reason, but reason based on certain theological assumptions.[24]

At about the same time, Charles Raven, a retired professor of theology at Cambridge and a trained natural scientist, blamed the Barthian ascendance for the demise of a once burgeoning dialogue between science and religion. Theology had dropped that to look backward into the past, Raven said in his 1950–52 Giffords. "Christians clung convulsively to 'old paths in perilous times,'" as reflected not just in Barth's Calvinism but also in the Lutheranism of Scandinavian theologians, the Catholic reversion to Aquinas, and an Eastern Orthodox revival of Alexandrian Platonism. "Such revivals of the ideas of the third or thirteenth or sixteenth centuries were hardly relevant to the scale and character of the changes demanded in the twentieth," Raven argued.[25]

Raven, speaking elsewhere of politics and theology, had famously said that Barth could be so otherworldly because Switzerland stayed neutral in the two world wars. Barth had been a Swiss pastor during the first and a Swiss border sentry in the second, a stocky Bible theologian in a helmet and heavy coat, bearing a rifle, which was a picture that showed up in the *Time* cover story on Barth.

All was quiet for a few decades on the Barthian front, until Old Testament scholar James Barr took up the Barth phenomenon in his 1990 Gifford series, *Biblical Faith and Natural Theology*. For biblical scholarship, Barth's legacy had been significant, and Barr held the view that the Swiss theologian had essentially halted any serious historical study of the Bible, so focused did a generation become on proclaiming its paradoxes and absolutes. While agreeing with Barth that natural theology could not prove God's existence or attributes, Barr still contested his legacy in three major areas: historical, biblical, and finally ideological.

In fighting his battle against nazism, Barth in his own idiosyncratic way had equated the hubris of natural theology with the "German Christians" and the culturally established churches that sided with the fascist state. To deepen his argument, Barth had harked back to the original intent of the Reformation, which he claimed had jettisoned natural theology—because of its very tendency, as the future would prove, to produce human monstrosities such as anemic liberalism or ferocious nazism.

Barr called this line of argument utterly fallacious. Reformation leaders used natural theology quite openly, Barr said. First, Calvin in particular, and even Luther to a degree, recognized God through nature. The better-known Reformed creed, the Westminster Confession, had declared the same. Yet in his Gifford Lectures, Barth had used the far narrower Scots Confession, a brilliant maneuver to make his point. To cite Westminster would be to "recognize the very substantial part played by natural theology" in that Reformed document.[26]

On a second matter, Barr said, the great liberal German theologians—Schleiermacher, Ritschl, and Harnack—had all railed against natural theology and metaphysics, although Barth made it sound like they fawned over them. The fawners, in fact, were the evangelical and orthodox Presbyterians—part of Barth's family tree—who in England and America were famous for their biblical apologetics, using every piece of evidence possible, from science to history, to argue that the Bible was literally true and that the facts of the world seemed to say so. Of course, Barth was no friend of American fundamentalism. Its champions had decried Barth's allowance of historical criticism, while, in turn, he had spoken of their making the Bible "a paper Pope." Nothing had been stranger to plain-spoken American fundamentalists than Barthian crisis, dialectics, and paradox.

In his Giffords, Barr made a final point on natural theology and politics in Germany. If Catholics were the handmaidens of natural theology, a slippery slope to nazism, why were Catholics so opposed to Hitler? The Protestants made up the "German Christian" movement that so willingly said *Heil Hitler.* What was more, Barr argued, the Nazis were "absent" two fundamental features of all natural theology: rationality and universality. Hence, Barth's theological stance was more political than anything else. "It was thus the rise of German totalitarianism, whether rightly interpreted or not, that brought the issue of natural theology into an absolutely central position," Barr said. "Barth associated natural theology with the pro-Nazi position of the 'German Christian' movement."[27]

Barth's influence on biblical theology was also less than sincere, Barr went on to say. While claiming to approach the Bible just as it was, Barth actually picked what he wanted and threw out the rest. He was entirely christological. He handled "the Old Testament as if it was a portion of Jesus Christ," when in fact the Hebrew tradition did claim to have knowledge of God. Barth skipped over places in which God was revealed in nature, ranging from the Mosaic Law and the Psalms in the Old Testament to the claims of Saint Paul in the New. "There is substantial basis within the Bible for the acceptance of natural theology," Barr said. Yet Barth would not acknowledge it because his "theology was at bottom a dogmatic-philosophical system."[28]

And that was Barr's third area of criticism: Barth had claimed to be the naked soul of a theologian before God, and yet he was actually draped in an array of ideological systems. Any system is manmade, and thus a kind of natural theology. But the genius of Barth had been to keep his natural presumptions hidden, Barr said, or to browbeat questioners into silence. "A hidden natural theology accompanied by violent attacks on natural theology is a clear sign that something is wrong." It was hidden amid many borrowings from the philosophical milieu of Barth's own background: "Conceptions entirely modern, related to existentialism, to atheism, to Hegelianism, were cleverly compounded with biblical exegesis and Reformation formulae."[29]

If Barth needed a defender in the Gifford Lectures, then he got one of the best at the end of the twentieth century. In 2001 at St. Andrews, the theologian and ethicist Stanley Hauerwas came to the proverbial rescue. He was the first United Methodist to give the Giffords, trumpeted his Duke Divinity School and home denomination, although British Methodists had long done the honor. A Dallas native who spoke with a winning Texas twang, Hauerwas was at the same time pegged by *Time* as "America's best" theologian. He was "contemporary theology's foremost intellectual provocateur" and "a thorn in the side of what he takes to be Christian complacency," the magazine reported. Professor Webster, the Barth expert back at Aberdeen, said Hauerwas today is not too different from Barth in his own day. "There are temperamental similarities between the two," he said. "They are both vigorous and absolutely clear minded. They map the world on the basis of their own convictions."[30]

Speaking at St. Andrews, Hauerwas turned the whole Barth debate on its head by declaring *him* the true natural theologian. Barth "rightly understood that natural theology is impossible abstracted from a full doctrine of God." The villainy of twentieth-century Christian thought was seen not in Barth but in two other famous Gifford lecturers, William James and Reinhold Niebuhr. Hoping to make Christianity palatable, James and Niebuhr had gutted it of all supernatural meaning. "Barth, in contrast to James and Niebuhr, provides a robust theological description of existence."[31] For anything such as natural theology to work, Hauerwas said, God must be a specific God, and for Christianity that meant a trinitarian Creator who worked in the world by way of Christ and the Holy Spirit in the church.

Natural theology had become "unnatural," and in the face of this Barth helped "Christians recover a confidence in Christian speech." All talk about God was a missionary activity of the church, Hauerwas said, so even natural theology could not go off on its own, as if on speculative holiday. Even it must witness to Christ. "Barth shows us the way that theology must be done if the subject of theology, that is, the God of Jesus

Christ, is to be more than just another piece of metaphysical furniture in the universe."[32]

Still, Hauerwas spoke in a tragic voice. Barth had electrified a generation, but finally he did not reach all its members. Galvanizing the present generation, Hauerwas said, seemed ten times harder. It was hard to claim that God threw his Word, like a divine stone, at the world and expected the world to simply believe. "We live in times when many people believe that Barth is but the name that confirms the judgment that Christian theology is not capable of rational justification," Hauerwas said. In truth, Christian faith was justified only in its testimony to the Father, Son, and Holy Spirit, a trinitarian doctrine that defied public logic and scientific explanation. Although Barth gave a "stunning intellectual performance," his time had clearly passed. A new witness must carry on. Hauerwas, a Methodist with Anabaptist leanings, cited two such witnesses, and one was a leading theologian of the Anabaptist tradition. But the other was Pope John Paul II, an irrepressible modern voice for the trinitarian God.[33]

Although Barth had moved beyond Kierkegaard, he testified to the Danish prophet of paradox and existential dread as a rite of passage: every theologian had to encounter him at least once. When historians assess the opening of the twentieth century, they rank only Kierkegaard with Karl Marx and Friedrich Nietzsche for setting the tone of philosophy and religious belief.

A relative nobody in his own day, Kierkegaard became a celebrated thinker with the revival of his works after the First World War. He was a storyteller, an observer of what goes on deep inside the subjective self. A product of bourgeois privilege and comfort in Denmark, he criticized cultural Christianity and the state church; he struggled to deny himself worldly fulfillment, famously rejecting marriage to the woman he loved. This kind of subject matter, the interior life of a person versus the snares of conformity, was ripe for both religious and secular application.

And thus was born, in Kierkegaard's remarkable writings, modern existentialism. It became a system of thought by way of a succession of teachers and students, beginning in Germany. In the shadow of Kierkegaard came the German philosopher Wilhelm Dilthey (d. 1911), an heir to Schleiermacher's Protestant liberalism, with its emphasis on personal religious consciousness and subjective experience. Dilthey famously announced that human sciences were different from natural science. Human beings have interior subjectivity, whereas planets and atomic particles have nothing of the kind. Only human beings have "facts of consciousness." In the pursuit of the human sciences, the facts of consciousness are far more reliable than either scientific measure, which is mechanical, or history, which is transient. He said a true philosophy finds its "inner coherence not in the world but in man."[34]

Dilthey's seminars and writings influenced a young philosophy professor and psychologist, Edmund Husserl (d. 1938), who watched the Kierkegaard revival take place. When Husserl established himself at a major German university, he had followers too, most significantly his assistant, the former Jesuit novice Martin Heidegger. Dilthey, Husserl, Heidegger—the three names are almost unknown in the West. But they were the first tremors of the avalanche called modern existentialism. They were a passageway for the Kierkegaardian spirit into the present. Their common project was to observe human consciousness and, as Husserl would argue, watch how its "intentionality," or focus, is always filled with concrete and particular things. These things were the stuff of individual existence, the very things worth talking about.

Heidegger systematized this approach to life, inventing new philosophical categories to describe how individuals operate. During Hitler's reign, Heidegger retained his post as a university rector, although he eventually ran afoul of the regime—and after the war he denied complicity. The tarnish would not derail his philosophical influence, however. The topic of an individual life—who am I and what should I do in life?—was just too compelling for the coming generation. For Heidegger, the fundamental question was about being, an enigma he believed only the earliest Greek philosophers had had a grasp on. Ever since those Hellenic wise men, he said, philosophy had veered off track, crashing into the ditches of abstraction, first causes, and rational hierarchies.

Heidegger called his new categories *existentialia*. The most fundamental was the human propensity to care about existence. Next in priority was the propensity to care about the possibilities in a life. All possibilities, in turn, were hemmed in by facts. The facts were produced by the past and the present, and they put limits on every situation in life. With no control over these given facts, everyone experienced a "throwness" into the world, as if arriving disheveled and totally unprepared. What was more, life inexorably moved to the ultimate possibility, the abyss of death. Between birth and demise, however, life was an abundant opportunity to exist, to decide, to resolve, to take responsibility, to experience joy in the face of death. This was the "authentic" life, Heidegger said. The unauthentic tries to escape possibilities, usually in mindless conformity. Looking for the marks of an authentic life can be done with the "existential analytic," as Heidegger called the method. Life was as much a process of asking questions as finding answers. For all life's uncertainties, the asking was a far more guaranteed activity. The answers might not exist at all this side of the grave.

Before long, so malleable a methodology as the existential analytic was taken in both religious and secular directions. While some accuse Heidegger of espousing atheism—and Jean Paul Sartre praised him for seeming to do this—Heidegger's Christian defenders say he still professed God's existence. What he had focused on, however, was the human experience, which could be probed on its own merits, without a cosmic backdrop.

Existentialism was taken on its secular tangent by none other than Sartre, who as a young French student attended lectures by both Husserl and Heidegger in Germany. Sartre returned home and in 1943 published *Being and Nothingness,* a work that launched existentialism in France. Soon to follow were his novels, with titles such as *Nausea* and *No Exit,* which told stories of human angst between birth and death, vivid stories of disgust, embarrassment, and discomfort with other people. Yet a person could still take life on the chin nobly enough. For Sartre, human life was egoistical, godless, sexual, and narcissistic, just the kind of mood that suited many Europeans after a devastating war. His writing was also of Nobel Prize quality—a prize he rejected—and others called it fashionable, clever, cynical, and "intellectually pitiless."[35]

As every freshman philosophy student learns, Sartre turned Christian doctrine on its head by saying that human beings had no universal nature but were made by their choices. Hence, Sartre said: "existence precedes essence." Christian counterproposals would come, and some in the Giffords, but perhaps the best known secular counterproposal would emerge from Sartre's contemporary and fellow Nobelist in literature, Albert Camus, who protested Sartre's nihilism in his "philosophy of the absurd." Given the apparent absurdity of modern life, Camus said, the authentic individual "rebels," and by rebelling reveals a common sense of justice shared by human beings, and thus a common human nature. Christian theologians, in fact, found promise in Camus's stories and analysis of rebellion, both in his argument for a common human nature and in his plea for moderation (escaping the extremes of nihilism or Marxist tyranny, so fashionable in French thought at the time). In the pithy French tradition of René Descartes, Camus said, "I rebel, therefore *we* exist."

Averse to Sartre, Camus denied that he himself was an existentialist, yet the English translation of both their writings carried existentialism, with its exotic and playful appeal, to North America. As a result, when Americans spoke of Continental philosophy in the postwar era, they usually meant the ethos of Sartre and Camus. For all its dabbling in nihilism, existentialism nevertheless demanded that life be lived in action, responsibility, and decision, and despite its vague shape, the movement touched professional philosophers, artists, and many in the reading public. *Existentialism* became a "catch-all term for the cultural and artistic avant-garde and for radical critiques of universal principles and absolute values," said one history, much as people today use the nebulous term *postmodernism* when speaking of rebellions against conventional life.[36]

As Heidegger's teachings bore this fruit in French literature, he was also having an impact on Protestant biblical studies and theology in Germany. On the biblical front, the major figure was Rudolf Bultmann, the son of a Lutheran minister who had studied under Heidegger. He concluded that what counted for the early Christians, for the writers of the Bible texts, and for Christians today, is the existential encounter with God in Christ, not the historical stories, miracles, or proofs about God's work in a particular time or place. Hence, Bultmann urged a "demythologizing" of the New Testament, a scholarly and rigorous stripping away of myths borrowed and invented in ancient times. What was left was the proclamation, or kerygma, of God in Christ—and that was everything. "Hidden in the myth is a kerygma, a divine word addressed to men," Bultmann said.[37]

Bultmann argued that demythologizing had begun even in the New Testament, when the Gospel of John turned the eschatological end of the world into the individual's encounter with God, not a global conflagration. Paul turned the death and resurrection of Christ into spiritual dying and rising at baptism. For Bultmannian enthusiasts, a new meaning had been given to hackneyed concepts such as sin, faith, flesh, and spirit.

It is no surprise that Bultmann is called the father of a biblical form of existentialism. When he gave the Gifford Lectures at Edinburgh in 1955 under the title *History and Eschatology: The Presence of Eternity,* he emphasized the existential immediacy of God in history. God-events are not things of the past, but experiences taking place right now. "The meaning of history lies always in the present," he said. "And when the present is conceived as the eschatological present by Christian faith, the meaning of history is realized." The eschaton—the end of the world and arrival of the kingdom—takes place every moment. Bultmann called this "realized eschatology." Indeed, "In every moment slumbers the possibility of being the eschatological moment," said Bultmann, the Heidegger student. "You must awaken to it."[38]

More broadly, Christian thought was developing what Presbyterian-turned-Anglican theologian John Macquarrie, an expert on Heidegger and his first English translator, called "philosophies of personal being." Following Heidegger, existentialists all engaged in a "radical interrogation of human existence."[39] They all interrogated skeptically the rationalist approaches, whether in Christian theology or in natural science, and harked back to *Being* itself, and its corollary, the *being* of all things. Why is there being rather than nonbeing? For arguments about God, the focus on Being was quite helpful, for modernist theologians felt the public had given up on God as a heavenly sovereign with a white beard.

No one took Being to its theological conclusion more completely than the Protestant existentialist Paul Tillich, who began this creative project after fleeing Hitler's Germany for New York City in the 1930s.

While teaching at the University of Marburg, Germany, Tillich had been impressed by fellow professor Heidegger's "profound influence on his students," said one historian. Tillich traced existentialism to the German romantic idealist Friedrich W. J. Schelling, a man on whom he wrote two dissertations. But Heidegger was the one "who furnished [Tillich] with a large part of the philosophical vocabulary" for which he became famous.[40]

A brilliant synthesizer of Christian thought after he arrived in the United States, Tillich would expand on three of the cardinal principles of existentialism: being, human care, and asking questions. Most significantly, he proposed that God was "the ground of being." Christ was the "new being." Changing the biblical God into the *ground* of being, as if by the flick of a magic wand, furrowed some brows. But Tillich pleaded little choice in his new terminology. "I would prefer to say 'being itself,'" he said. "But I know this term is even more disliked."[41]

Also in the Tillich lexicon, the idea of existential care was expanded into a new definition for religion. Religion was "ultimate concern," which prompted some agnostics and atheists to jibe Tillich for turning them into believers when they insisted they were not. Finally, when Tillich gave his Giffords in 1953 and 1954, the series title characterized the third daily feature of an existential life: "Existential *Questions* and Theological Answers" (emphasis added). For such questions, the answers came from the Bible and Christian tradition. But now the biblical and doctrinal answers were presented in symbolic language. Being is always beyond being, an ultimate reality that can never be captured in mere human words. So literal words such as Adam and Eve, the resurrection, the devil, miracles, and all the rest had to be probed for new contemporary meanings. Symbols participated in ultimate reality. But they were always a dark mirror, and could become idols when believed in more strongly than Being itself.

Christian existentialism was not just a Protestant phenomenon. Despite papal displeasure with the trend—stated in the 1950 encyclical *Humani Generis*—the kind of Catholic thinkers who were candidates for the Gifford Lectures had taken it on squarely. As a part of the neo-Thomist revival, or study of Thomas Aquinas, Gilson argued that the great doctor *was* the first existentialist, having first expounded on Being. Aquinas had argued that God's very nature (essence) was his fact of Being (existence), which is "to say that in God essence and existence are identical."[42] Aquinas then extended that to humanity, for a God-given human nature (essence) preceded the creation of human form (existence). Sartre, of course, had turned the great doctor of the church on his head by declaring that "existence precedes essence." So Gilson, just as shrewdly, borrowed the parlance and style of existentialism to take the traditional order of being, which Sartre sought to topple, and set it aright.

Another leading Thomist who tackled existentialism was Jacques Maritain. His plan to deliver the Giffords at Aberdeen between 1940 and 1942 was cut short by the Nazi invasion of France, and then he became the French ambassador to the Vatican after the war. Next came Gabriel Marcel, a French teacher, writer, and philosopher received into the Catholic church in 1929. He shaped existentialism into its most personal Christian outlook, describing and analyzing personal relationships, love and charity, and living an unselfish life. Marcel had no previous knowledge of Kierkegaard but ended up in the same place. He added another existential category: hope in the metaphysical world of God. His Giffords at Aberdeen in 1949 and 1950 were titled *The Mystery of Being*.

From such works, Marcel popularized the distinction between mystery and "problems." Mystery was to be lived with in awe and wonder, whereas only problems were to be solved. Unhappiness in an age of materialism, he counseled existentially, arises from too much "having" and not enough being. Marcel's Christian existentialism mitigated pure individualism. It proposed community, a process he called "person–engagement–community–reality." The goodness in human experience was fidelity. As Copleston explained Marcel, "Fidelity to another enables me to transcend the flux of becoming," and it is an intimation of immortality, the ultimate life with God.[43]

When Emil Brunner gave his Gifford Lectures at St. Andrews beginning in 1947, he did not even mention Karl Barth. The two old dialectical warhorses had agreed to disagree. Brunner was not in postwar Scotland to talk about a wholly other God but about Christian civilization. Amid the ashes of war, he was fairly pessimistic. His neo-orthodoxy did not waver, however, for while Christianity may shape culture and make it truly humane, culture does not save souls. Systems that try to make the new man, he said, invariably produce a Nazi Germany or a Soviet Union.

What Christian belief had done, Brunner believed, was limit evil, for a godly utopia was not of this world. The dam against evil waters, however, was weakening. "The history of civilization during the last hundred years has made clear beyond any doubt that the progressive decline of Christian influence has caused a progressive decay of civilization," Brunner said. Modern problems were made more complex by a secular outlook. "Because people do not believe in eternal life any more," he said, "they are seized by a kind of time-panic." New ideologies try to "make this world a paradise" but turn it into a state "more akin to hell than to heaven." What believers should aim for is a "cultural soundness." They can contribute to this by teaching values, living by virtue, supporting both freedom and community, and embodying the commandment of love. Still, Brunner said, "I do not prophesy an epoch of general return to Christianity, any more than I accept the myth of the Christian culture of the past."[44]

After the war, the social leadership of the English-speaking world was also mulling the sources of values for the future. God in general and Christianity in particular seemed still potent enough. The British electorate voted Churchill and the conservatives out of power after the war, and Parliament created Britain's first welfare state. An early portent of that was the 1944 Butler Education Act, which required schooling for everyone, and which included religious education. Yet the war had seriously subverted family patterns of religious observance, especially among older children. Secularization came on rapidly. A 1990 survey found that although the vast majority of British citizens had been "brought up religiously" before the war, a steady decline followed, down to a third of everyone born after the mid-1960s.[45]

In the United States the story was different for a while. Church attendance rose, fueled by the cultural prominence of religious institutions, economic prosperity, and the baby boom—and a mood of common cause galvanized by the Cold War. The word *God* was put on the dollar bill and in the Pledge of Allegiance. By the 1960s, however, the "myth of a Christian culture," spoken of by Brunner, was looking true enough on American shores as well, highlighted by consumerism, secularization, and new urban problems.

For all its emphasis on revelation and personal experience, the movement for "subjectivism" in the West produced its own kind of reaction. In the Gifford legacy, if thinkers such as Barth represented act three, with its enduring influence, the story had to move on. Barth, the existentialists, and many others had rejected the very idea that human reason could understand God, or that God was finally a rational type of experience to be shared by an ever-expanding group of rational people. That was the great vision of natural theology at one time, a means to nourish a common belief, perhaps, in times of religious war and ecclesiastical rivalry when one revealed doctrine was pitted against another and proving their veracity became matters for heresy trials or the rack. But a new kind of modernity, as if the fourth and final act, had arrived. While religious thought was as liberal as ever before, its cooperation with reason was making a comeback, and this to the advantage of natural theology.

Some kind of threshold had been reached for both the British and the Americans on the question of God and the modern world, and that may explain why in 1963 a little book about God caused such a gigantic stir. In March of that year, the Student Christian Movement (SCM) in Britain published *Honest to God,* a pithy title written by the Anglican bishop of Woolwich, John A. T. Robinson, a former dean at Cambridge. What Robinson had provided verged on old news, for it mainly summarized the ideas of Bultmann and Tillich, among a few others. The overall effect was stunning, however. A sitting bishop seemed to suggest that the Bible should be demythologized and that God is the ground of being, not a God "up there" or "out there." Institutional faith might be drifting into obsolescence, he said, amid proposals for a "religionless Christianity."

A week before publication, the London *Observer* asked the bishop to write a preview essay. He had no control over the headline, which read, "Our Idea of God Must Go." When the little book came out it was an international sensation. In a few months the book's printing shot up to

350,000 copies. The book had blown the lid off theological publishing, said the SCM editor, David L. Edwards, having "sold more quickly than any new book of serious theology in the history of the world."[46]

The bishop received a thousand letters in the first few months, and commented that people still needed a simpler treatment than *Honest to God* to understand these new frontiers in Christian belief. Readers who had struggled with religion nevertheless felt a great catharsis upon reading his book, especially since it was by a bishop. In the end, four thousand readers sent letters to Robinson, and a later study of them showed among the Christian public "a motley assortment of beliefs that were sitting side by side in the same church on Sunday."[47] Traditional believers, however, viewed Robinson as having slid into doubt. The philosopher Alasdair MacIntyre, soon to be a Catholic thinker and eventual Gifford lecturer in 1987, said the bishop had espoused atheism, plain and simple. The Christian writer C. S. Lewis pointed out that few people really believed God was up in the sky, despite the good bishop's incredulity.

With all the flurry over *Honest to God,* some connoisseurs of the Gifford tradition felt a measure of satisfaction. The book had raised the very kinds of topics Gifford appointees had wrestled with for seventy years. Robinson cited one Gifford lecture in particular, but more important was the general overlap with the Gifford legacy, which offered up both liberal and traditional answers to the God question.

That was the stance taken by a young Methodist scholar, Bernard Jones, whose 1966 doctoral dissertation on the history of the Giffords was the first of such breadth in Britain. In 1970 Jones published an anthology of lecture excerpts back to 1888. "The publication of *Honest to God* in 1963 led to a far wider interest in natural theology," he wrote. If the 1960s debate was over the "problems of language and of the possibility of God-talk," Jones said, it was hardly novel for the Giffords. At the turn of the century R. B. Haldane was already asking "how must we conceive and speak of him?" And in the thirties the historian of religious symbols Edwyn R. Bevan, was giving Giffords on what Jones called the "the problem of the '*up-there-ness*' of God."[48]

Such times could well provide an opening for one particular type of God-talk, namely natural theology, which had been declared "the sick man of Europe" in the 1920s and 1930s, thanks in great measure to Barth. Noting this, Jones said that secular times may prove an advantage for such reasoned approaches to the Creator, once eclipsed by faith and tradition. "In a theological situation where it is almost a virtue to doubt, natural theology can only flourish," he said. "There will be a place for the natural theologian, the man who is prepared to examine by the impartial light of reason the claims of religion whether they be based on so-called miraculous revelation or human speculation."[49]

The publication success of *Honest to God* became a defining moment for historic Protestantism, and perhaps even among some Catholics, whose "updating" Second Vatican Council was in its second year when *Honest to God* was hot off the presses. The God question had never had so much publicity, especially in the popular press. Yet like God in the sky, a book or council can be more a symbol than anything else. "How influential is a popular book?" asked the historian Chadwick, looking at examples in the nineteenth and twentieth centuries, including *Honest to God*. "Does it follow that the book changed so many people's minds?" He was skeptical, especially regarding *Honest to God*. "Perhaps a popular book, upon a theme among the most profound and difficult known to man, is rather a symbol, the focus of inarticulate longings of a decade."[50]

A little book such as *Honest to God* was not the only symbol of change for the West, with its urbanization and loss of colonies, or for the Gifford project and all it historic trappings. The look of the ancient university cities and the settings for Gifford Lectures were changing as well. About a decade after Gifford's death, more than a hundred acres of his rural Granton neighborhood was turned into a gas works, a giant industrial yard with three tall holding towers to light up modern Edinburgh. Overlooking the Firth of Forth, his antiquarian home, which originally sat amid thirty acres and was ringed by willowy trees, survived halfway into the twentieth century. Then on January 2, 1954, Granton House caught fire and burned to the ground.

Lord Gifford never paid attention to fine architecture, so tireless was his "appreciation for beauty in the abstract."[51] But he would have surely noticed Britain's new urban look, with its tall blockish buildings. Most of it represented "the 1960s public architecture at its worst with their bold use of concrete and glass," according to a history of the University of Glasgow, which bucked the useful-but-ugly trend with its attractively towering Boyd Orr Building, put up in 1972.[52] In Edinburgh, back in 1914, the university had begun to buy land in Old Town, especially around grassy George Square, an eighteenth-century neighborhood of terraced houses. The development was slowed by war and preservationists, but finally in 1963 the thirteen-story David Hume Tower became a centerpiece of the high-rise transformation—and another monument to Hume, the local boy who made good. The eye of the campus, a daily storm of students coming and going, shifted to George Square, with its new buildings and Hume lecture theaters.

One Gifford lecturer who visited the University of Aberdeen praised its city as having "nothing but a few fishing villages and oil platforms between us and the North Pole."[53] Yet the North Sea oil boom, which began in mid-1960s, allowed it to build as well. At the university, the Fraser Noble Building opened, a complex topped by an expansive white dome. It was a space-age look for a campus centered on the medieval King's Chapel, an ancient quadrangle whose facade would soon hide a high-tech conference center as well. St. Andrews had a medieval village ambience to protect. So instead of growing upward, it expanded outward. The famed golf courses, which plunged into the sea, had been discovered by an age of jet airplanes.

The changes bespoke technology, a multiplying force of the age. Technology built the new architecture, but also built weapons of mass destruction. It made the world seem more machinelike, more anonymous, perhaps more closed off from God. Yet it did more than that. It provided tools to probe the largest and smallest scales of nature, revealing an unknown architecture of the universe.

NINE

A DESIGNER UNIVERSE

God and the New Sciences

T HE MANHATTAN PROJECT, which developed the atomic bomb, was the beginning of government science in the English-speaking world. Yet the era of truly *big* science was still on its way, and that began with the faint sound of a small satellite, the Soviets' *Sputnik I*, orbiting the Earth in 1957.

Soon enough, the "space race" between the United States and the Soviet Union expanded into a wider science race, and this in turn raised two commanding issues. The first was the future use of atomic weapons and the second "scientism," an ideology that put science foremost in human affairs. "Faith in science plays the role of the dominating religion of our time," one physicist said in the Gifford Lectures a year after America sent its own rival satellite aloft. With the scientific watershed of the 1950s, the Giffords began to address the split between values and technology, the "ambiguity of the effects of science." Science was a "two-edged sword" with its unintended consequences, its ability to both heal and destroy, its power to undergird religious belief or unravel its very fabric.[1]

As science uncovered more of the universe's architecture, it declared the cosmos to be not only cold and vast but utterly meaningless. At the same time, this bleak stuff of the universe produced order and complexity and, almost miraculously, life itself. This closer look at the nature of things came in biology, computers, and cosmology. Traditionally, a mechanistic world seemed less in need of a "God hypothesis." When nature seemed more open or inexplicably complex, there was more room for the God-seeking imagination to work. One physicist who gave the Giffords at Aberdeen in the 1970s took that imaginative leap. "The more I examine the universe and the details of its architecture," he said, "the more evidence I find that the universe in some sense must have known we were coming."[2]

The God hypothesis had one more opening in the age of scientism, and that was the limits to what science, a project of finite human tools, could see and measure. On its humble side, science had always acknowledged such boundaries. Yet there was also a proud and bold side to science, a philosophical side that said human reason could in the end figure out just about everything in the physical universe, a project that in some circles came to be known as a "final theory." As the twentieth century unfolded, the promise of that vision seemed proved with every passing year, especially by the explosion of new technology: electron microscopes, atom smashers, X-ray crystallography, space-based telescopes, and gene-hunting computers.

Even as it was acknowledged that this ultimate power of technology would always be in question—in other words, no one can travel faster than light to do scientific research—science came up against a new debate that it was probably not expecting. In the 1950s, the question was asked, Could scientists *themselves,* even with the best technology, be free of human bias? Might not scientists find what they want to find, since who can say what meaning the universe itself gives the "facts" it offers up to science? The problem was age-old, as when the Aristotelian scientists of the Catholic church, quite happy with their own earth-centered philosophy and their comfortable careers, rejected the sun-centered theories of Nicholas Copernicus and Galileo Galilei. Later in history, in about 1900, Lord Kelvin at Glasgow University was saying that physics had discovered everything, and others clung tenaciously to their belief in ether, a universal substance, even as Einstein's theory of relativity proved it to be a fantasy.

By the 1950s, scientists of the West were as singularly confident in their current knowledge as any of their forebears. Yet it was also a time, according to many Gifford speakers, when the philosophical agenda of scientists became a matter for intense debate. Were scientists really more rational than other people, or did their scientific cultures also reflect deeply held and even irrational beliefs, like everyone else's? It was an era, for example, when scientists said the idea of the universe having a beginning offended their atheism; an eternal universe was more pleasing. Advances in computers and genetics prompted scientists to declare that, just as they had suspected, human beings were biological machines. Some mathematicians, human as they were, declared that only "beautiful" equations could be true. The human side of science showed, in other words, that the scientific "facts" were usually combined with human interpretation. If it was fair to draw atheistic conclusions, why not theological and metaphysical ones as well?

This argument would show up in an increasing number of Gifford Lectures, raised by presenters who were well aware that despite the remarkable advances in anthropology, psychology, physics, sociology, and historical research, science *did* have limits. Having broken into a growing number of specialties, science no longer stood for a comprehensive knowledge of the universe, at least by a single scientist once called the "natural philosopher." What is more, science had proved that its advances were morally neutral. Technology could be used for both good and evil, as the war years had dramatically shown. Perhaps the first assessment of this kind, which inaugurated a new attempt to revive natural theology, a kind of act four in the Gifford legacy, was by Charles Raven, a Cambridge don who was a veteran of the early science–religion dialogue in Britain.

When he delivered his Gifford Lectures beginning in 1950, Raven believed that the new findings of science, its acknowledged limits, and the desire to make science moral had reinvigorated its great interchange with theology. Science and theology had struggled to reconcile earlier in the century, Raven believed, but that momentum had been lost in the years between the world wars. An old man now, Raven had nevertheless lived to see the discovery of DNA in 1953 and the beginning of the space race in 1957, and for him, these developments only expanded the opportunity for a dramatic comeback of discussions about God and nature.

Ordained in 1909, Raven was a theologian trained well enough in science to write popular books on natural history. When he gave his Giffords, *Natural Religion and Christian Theology,* he was ending his active duty as regius professor of divinity and master of Christ's College at Cambridge. Early in the century, he said, the search for God's immanence had been going full bore. The period gave birth to a range of creative works in metaphysics, notably by people such as Hans Driesch, Samuel Alexander, and Arthur Eddington. Although Whitehead's process ideas were still not widely known, Raven predicted that they would "probably have more influence upon the future than that of any recent philosopher." All of these trailblazers took advantage of "the breakdown of the doctrine of the closed universe," one of those humanly held scientific beliefs that said reality was a self-running machine that needed no God. Similarly, Raven predicted another "breakdown of the old mechanistic and compartmentalized concept of science."[3]

Between the wars, a philosophical shift toward a transcendent God had made science passé for much of Western theology. But after the war, for both theological and moral reasons, religion and science were reaching for each other "again with vigour." On the theological side, Raven said, Christians are "committed to the belief that the universe like the Incarnation reveals the nature of God." An equally urgent reason to bridge faith and science was moral, for modernity had produced a shocking "irony," the fact that two Christian democracies—he was suggesting that Great Britain had complicity with America in this—had used the first atomic weapons, and had done so on civilian populations.[4]

Values and technology had been split apart. More was at stake than intellectual curiosity in reconciling religion and science, Raven argued. "Without such understanding and cooperation there can be little prospect either of progress or of peace," he said.[5] Not a few Gifford lecturers to follow, especially those with experience in government science, would sound a similar warning about values and technology. When the sciences grow overly confident, or are never questioned, they produce the ideology of *scientism,* a term used often in the Giffords. Science and technology become a dogmatic belief system. Operating in this way, science can take on the trappings of a "false religion," German physicist Carl Friedrich von Weizsäcker said in his 1959 lectures at Glasgow, lectures that declared the "ambiguity" of science as to whether it always finds the truth and always produces something good. "The role of priest does not become the scientist and good scientists know that."[6]

Weizsäcker was a student of Heisenberg and an associate of many of the great prewar German physicists, including Erwin Schrödinger. Once inclined to become a minister, Weizsäcker went into science instead. But he traced his worldly wisdom and caution to living under a German "dictatorship" for thirteen years. His German colleague Schrödinger turned philosophical as well, taking an interest in Eastern thought. Having won a Nobel for his quantum wave mechanics calculations, Schrödinger began to probe the essence of biological life, with its vitality, consciousness, and apparent free will. Though inconclusive in his theory, Schrödinger put it all down in 1944 in his little masterpiece of a book, *What Is Life?*

Whatever life was, he said, the frontier of discovery would be the functioning of the mysterious gene, the tiny molecule that seemed to perpetuate life. In 1946 the physicist Francis Crick read *What Is Life?* and changed his career, "leaving physics and developing an interest in biology," according to his research colleague, James Watson.[7] Crick desired to show that life was indeed mechanistic. He worked at one of the most famous addresses in science, the Cavendish Laboratory at Free School Lane in Cambridge. When Watson, a young American researcher, arrived in 1951 to work with Crick, the race was on to discover the architecture of DNA. Two years later, they announced that the "secret of life" lay in a double-helix molecule.

In his storied self-confidence, Crick wrote in 1957 that "biology is getting nearer and nearer to the molecular level." And there, in the realm of molecular mechanisms, chemical codes, and heredity, lay "the decisive controls of life." Thirty years later, he had not budged in his convictions. "What is found in biology is *mechanisms,* mechanisms built with chemical components," he said. "Biologists must keep in mind that what they see was not designed, but rather evolved."[8]

Across the way at Cambridge, the theologian Raven was a science enthusiast of the first order, but he nevertheless saw a less decisively materialist trend at work in scientific thinking. Whereas Crick and Watson wanted to reduce the universe to a mechanism, others in science viewed nature as open to other kinds of laws and forces, Raven believed. If they were not supernatural, they were not purely electrical and chemical either. In modern biology, Raven said, "the problems of the scientist and the theologian are not, in fact, dissimilar." What causes creative variety? Raven asked. What has "survival value" in organisms, and how do they adapt to environments? How did altruistic and sacrificial behavior arise in a Darwinian world of ruthless self-preservation?[9]

In his Giffords, Raven revived the argument about the power and the limits of science, but raised it on the eve of a whole new movement called the sociology of knowledge, which challenged the objectivity of much scientific argument. From this sociological point of view, scientists had a whole range of vested interests—professional, ideological, and cultural—that influenced the kinds of research projects they conducted, the kinds of questions they asked, and very often the kinds of results their experiments produced. It was, indeed, a suspicious view of the objectivity of much scientific knowledge. For his part, Raven was far more generous toward the sincere quest of scientific researchers, but he could not ignore the testimony of scientists around him. He called it a "change in the scientific outlook." The thinking of scientists, Raven said, was being transformed from "an almost arrogant confidence to an almost despairing hesitation about the possibility of reaching real knowledge."[10]

The next year at Aberdeen, a Hungarian-born chemist gave Western culture permission to talk about the problems of subjectivity in "objective" science. Michael Polanyi, whose Giffords became well known as *Personal Knowledge,* was a scientist-turned-philosopher, a medical doctor and brilliant chemist born to a Jewish family but a convert to Christianity. He had fled to England to escape both Nazi and Soviet threats and had become head of a research laboratory at the University of Manchester. In time, however, he crossed over to philosophy, doing his advanced studies at Oxford. Although baptized a Catholic at his marriage, Polanyi joined the intellectual circles of the Anglican church and took an interest in the symbolic theology of Paul Tillich. Unlike existentialists or the neo-orthodox, who would not touch science with a ten-foot pole, Polanyi got up close and questioned its precepts.

Although his Giffords were concluded by 1952, *Personal Knowl-edge* was not published until six years later. This was the first of three kinds of contributions that Polanyi made to the modern debate on science and belief. In the Giffords, he argued that scientific discovery was prompted as much by personal insight as by objective procedures. "We know more than we can tell," Polanyi said.[11] His idea of insight, or "tacit knowledge," supported a mystical view of the scientist, waiting on revelation as much as on data to figure out nature. Like Weizsäcker, Polanyi warned against the ideology of scientism, or even excessive government science. He espoused an independent "republic of science," free of the political and military agendas of regimes, even democratic ones.

Raven, Weizsäcker, and Polanyi all represented the new mood of inevitability and caution about science in the 1950s. They realized that values must attend the growth of technology, and that politics should be separated from research as much as possible. They also believed that for religious belief to remain relevant for the modern public, it had to engage science and even allow for reform of certain theological doctrines in light of new findings about the physical universe. Where Polanyi departed from the others—his third contribution—was in addressing a specific area of science, namely chemistry and biology, a level of nature where he found an openness to higher principles and even to God.

As a chemist, Polanyi watched as the discovery of the DNA helix led to wider research on how the molecule functioned, producing proteins, the building blocks of living things, and then biological life itself. For Polanyi, this was something new, more than a mere chemical pattern, which only explained the repetitious production of crystalline structures. Those were "redundant" structures, not creative and living ones. Although physics and chemistry could explain crystals, Polanyi was doubtful that they could explain nonrepetitious, creative, and growing life. So he wrote articles such as "Life Transcending Physics and Chemistry." He asserted that "irreducible higher principles are *additional* to the laws of physics and chemistry."[12]

The outspoken and brilliant Crick, who was headed for a Nobel

Prize in 1962, dismissed Polanyi's higher principles as mere vitalism, the ghostly force that Crick hoped to exorcise from all natural science. The Scottish theologian Thomas Torrance, who was fascinated by the problem of reconciling faith and science, was won over by Polanyi, however, and became executor of his papers when he died in 1976.

In the nineteenth century, materialists had spoken of human consciousness as nothing more than a steam whistle on a train or the product of a gland, much as bile is secreted from the liver. To challenge that, a whole generation of Gifford lecturers, especially in the biological sciences, argued instead for "emergent" qualities, either by evolution or from a vitalistic force in the universe. The strength of these arguments had been to avoid the quicksand of a starkly dualistic universe, a cosmos of two entirely different kinds of substance, one of souls, angels, and demons and another of the physical objects upon which they acted. In emergence, there was only one real substance; it was a monistic whole and yet with a "dual aspect," one a material quality and the other a mental quality. For some critics, of course, this was only a kind of mushy dualism, which materialist science rejected vehemently. In any case, the argument for classic dualism, a two-substance universe of spirit and body, mind and matter, was not altogether dead in the sciences, as one Nobel laureate would show in his Gifford lectures.

The Nobel Prize, with three categories in science, was first awarded in 1901, thirteen years after the Gifford Lectures began. Over the years, if any Nobel recipients in science were inclined to talk about natural theology, it was not obvious. These laureates were far more likely to espouse a quiet naturalism, if not a vociferous secular humanism, as some would end up doing. If there was any theism among the Nobel scientists, what more likely magnet than the Giffords to draw them out, as would be the case with Sir Charles Sherrington, the famous Oxford neuroscientist. Of the eight Nobelists who gave the lectures—either before or after receiving the Nobel—Sherrington was the fourth, delivering them in the spring of 1937 and the next year as well.

Sherrington won the Nobel in physiology or medicine in 1932 for identifying the brain's nerve cells, which he called neurons. He also discovered what he called the synapse, or the point at which one cell transmits its impulse to another. Two other Nobel laureates in physiology would give Giffords. John Eccles, a Nobelist in 1963, was a student of Sherrington's and a theist, while Sydney Brenner, who won the prize in 2002, was a materialist and close colleague of Crick's. This double honor so to speak, the Nobel-Gifford honor, had no particular order, as illustrated by physicist Niels Bohr, who won the Nobel in 1922 and gave the Giffords nearly three decades later. Henri Bergson gave the Giffords in 1914 and won the Nobel in literature in 1927. The three others were a mixture. Söderblom and Schweitzer, Nobelists in peace, gave their Giffords first. Heisenberg, who won in physics, gave his Giffords later.

After his retirement, Sherrington began to espouse a dualistic view of the physical brain in relation to the mind. One realm of the brain was expressed as physical energy, while the other, mental, realm of the brain, though overlapping the physical, did not show up in the electrical activity of brain matter. When Sherrington gave the Giffords, it was not his first time to hint at dualism, but it was perhaps the most prominent. He kept hinting that the maligned and dismissed "ghost in the machine" was still a concrete scientific mystery. In 1950, two years before his death, Sherrington appeared on a BBC broadcast about the mind, where he conspicuously chuckled at the assertion that it was a mere "secretion" of the brain. "Aristotle, 2,000 years ago, was asking how the mind attached to the body," he said. "We are asking that question still."[13]

Dualism has always had theological implications, which is one reason why science disapproves. Sherrington never spoke of the "soul" but preferred the term *recognizable mind.* Besides his high reputation in science, Sherrington brought to his lectures at Edinburgh perhaps the most articulate exposition yet of a biological topic. After the talks, though, he expressed regret that they had "so little 'comfort' to offer" listeners, especially clergy, who might have been waiting to hear how the soul was proved by neuroscience.[14] What scientist, indeed, could prove such a thing as the immortal soul?

The furthest Sherrington went was the existence of mind and altruism, two human factors hard to explain by energy, chemistry, or the nervous system. These two qualities seemed to arise from the "I," or person, which was so elusive to scientific investigation, for who can verify what only the "I" can know? This "self" seemed entirely voluntary in action, not determined by matter. "We have, it seems to me, to admit that energy and mind are phenomena of two categories," Sherrington said in his Giffords, published as *Man on His Nature*.[15] The picture of dualism was hardly cut-and-dried, however, for some mental events were tightly knit with physical events in the brain, whereas others were mysteriously independent. For example, speech and hearing arose in certain bundles of nerves around the brain, and yet the thoughts behind words or the linkage of sounds with emotions or memories could not be isolated in any physical location. Despite this difficulty for pure materialism, modern science deemed dualism an entirely unacceptable, even medieval, view of nature. But Sherrington said the evidence left the question open. "That our being should consist of two fundamental elements offers I suppose no greater inherent improbability than that it should rest on only one," he wrote elsewhere.[16]

This was the sort of question that could only galvanize the ebullient Francis Crick. Once he had discovered the "secret of life," there was only one greater challenge left. "I decided that my main *long-term* interest was in the problem of consciousness," Crick recalled. He was not going to be looking for dualism, however, but for mechanism, the machinery of mind itself. Reared in a Congregational (or nonconformist) church, Crick was familiar with beliefs in the soul, and his French wife, reared a Catholic, also reminded him that *soul* was still a perfectly good word, even if it had become a fantasy to them and others in a scientific age. When Crick began to pursue the mystery of the mind, the field of brain science had already become materialist, working on the assumption that *all* mental events were brain events. It was a grand old tradition. David Hume had said the mind was nothing but a "bundle of perceptions." Crick hypothesized that the "I" of every person is "no more than the behavior of a vast assembly of nerve cells and their associated molecules," as he wrote in his book *The Astonishing Hypothesis*. For his readers, he imagined how Alice of *Alice's Adventures in Wonderland* would have stated the case, probably as, "You're nothing but a pack of neurons."[17]

Yet not every brilliant scientist agreed with Alice, and Crick was mindful of one great figure to be reckoned with. "Not all neuroscientists believe that the idea of the soul is a myth—Sir John Eccles is the most notable exception—but certainly the majority do," he said. "It is not that they can yet prove the idea to be false. Rather, as things stand at the moment, they see no need for the hypothesis."[18]

Eccles would be invited to give the Gifford Lectures. When he did, it was not the first time a Gifford performance polarized opposing views, forcing people to take sides on great scientific disputes. Schrödinger was a side-taker of sorts, though he never gave the Giffords. His 1958 book *Mind and Matter* praised the Sherrington lectures. Later in his career, Crick acknowledged that Schrödinger's little book, *What Is Life?*, may have gotten him started on the path to scientific discovery, for it "certainly made it seem as if great things were just around the corner" in molecular biology. Now, however, he wanted to emphasize Schrödinger's "limitations—like many physicists, he knew nothing of chemistry."[19]

Eccles was not exactly a chemist either, but he had figured out the chemical and electric signals of the brain sufficiently to win the Nobel Prize. He was the seventh laureate to give the Gifford Lectures, and a happy revivalist of the mind-brain controversy. Delivering his second series in Edinburgh in the spring of 1979, Eccles wanted to describe how the mind and brain might interact. So he laid out a scientific argument for the soul, a capstone to his career, delivered to a large audience in an enthusiastic—so he claims—atmosphere. Eccles called his astonishing hypothesis "dualist-interactionism," the interacting of two "worlds," the material and the mental. He hypothesized that the mind interacted with the three-pound brain's smallest receptors at the quantum level of physics, where the fuzzy world of quantum uncertainty allowed an openness in brain matter.

Picking up the theme of many nonmaterialists in science, Eccles spoke of mind as a kind of information, not the typical process of energy transfer seen in purely physical systems. Yet even an influx of information, if it produced an effect in the physics and biology of a system, had to be postulated under the law of conservation: energy may not be created or destroyed in a closed system. As Eccles explained it, "the flow of information into the modules could be effected by a balanced increase and decrease of energy at different but adjacent micro-sites, so that there was no net energy change in the brain."[20]

Eccles believed his theory had advantages. The materialist solution, with the brain's "closedness," faced "the impossible task of deriving a mental world out of neuronal circuits," Eccles said in his second series, *The Human Psyche.* "By contrast dualist-interactionism postulates the existence of both a mental world and a material world, and is concerned essentially with their interaction." The British neuroscientist Donald C. McKay, who gave the Giffords in 1986 at Glasgow on "What Brain Science Has to Say About Human Nature," disagreed with Eccles on dualism. McKay was also looking for a "spiritual" mind over matter, though in a nondualistic way; he saw the mind as a sort of head office "supervising" brain events. Eccles sympathized with McKay's search for a mind that emerged from matter, but said that McKay's construct ultimately failed to show how mind operates on neurons: "What is the neural status of the postulated 'supervisory system,' and where is it located?"[21]

Two influential philosophers of the era, meanwhile, also took on the topic of the soul and immortality. They were Richard Swinburne, a leading Christian philosopher at Oxford, who gave his lectures from 1982 to 1984 at Aberdeen University, and Anthony Flew, considered at the time the most famous atheist philosopher in the West, who spoke two years later (1986) at St. Andrews. Flew's Giffords, "The Logic of Mortality," argued that finitude in death, not eternal life, was more valid scientifically and philosophically. Already, Flew had famously argued for "the presumption of atheism," that the burden of proof was always on theism. Then came a surprise in 2004. In that year, Flew declared that the findings of modern science—from life's origin to a fine-tuned universe—had persuaded him that God existed, at least a God of power and intelligence. "It could be a person in the sense of a being that has intelligence and a purpose, I suppose," Flew said.[22] Yet he still rejected the afterlife.

Just as Flew concluded that God was more likely than no God, Swinburne, a self-professed "natural theologian," used his Giffords, *The Evolution of the Soul,* to argue that a soul and an afterlife were more likely than not. In effect, Swinburne held that dualism of soul and body could be arrived at on purely philosophical grounds, since so much of mental experience revealed its independence from states of the physical brain. For example, pain, visual images, and sounds may be caused by brain events, but can also operate independently—for example, one can feel dread, hear noises "in the head," or smell roses and taste honey when they are not present. Still more independent of brain events, it seemed to him, are thoughts, purposes, desires, and beliefs. Just because science cannot explain how this works, Swinburne said, one does not have to reject the hypothesis of dualism. A decade after his Giffords, he renewed this dualist argument, saying that although brain science was finding an increasingly tight fit between brain events and the mind, it was never completely tight.

In the end, Swinburne proposed a scheme in which Darwinian evolution of the physical body *and* a nonmaterial "self" lived in apparent harmony, a plan that gave its due to both scientific and philosophical observations. By asserting the reality of a soul, he argued for "soft materialism," which recognized that mental properties can operate in matter, and a "soft dualism," which did not presume to describe the exact nature of the soul. "All we can say," Swinburne went on, "is that under normal mundane conditions the functioning of the soul requires the functioning of the body." The soul is like the light bulb (mind) that needs an electric socket (brain); the "light will shine" when the current is turned on, but a damaged socket will turn it dark—and so it is with full human consciousness. All in all, Swinburne concluded, the immortal soul is as plausible as any other description of human identity.[23]

As brain science advanced in leaps and bounds, so did the revolution in genetics. One important effect was that genetics, combined with Darwin's theory of natural selection, finally gave the modern theory of biological evolution its full breadth of argument. Biological evolution, of course, was both a challenge and a tool for natural theology. So now, genetics would also become a topic in the metaphysical debate that characterized act four—the final act—of the Gifford legacy.

The genetic code was understood in stages. Although the double-helix structure of DNA was not figured out until 1953, the 1930s had seen advances in Mendelian genetics, or the basic laws of heredity. In that decade, Mendelian heredity was being combined with Darwin's theory of evolution by natural selection, which stated that the environment "selects" which organisms survive. By this combination, a new evolutionary theory called the "modern synthesis" was born. The mystery that had stumped Darwin—the source of variation in offspring—was explained by the way parental genes mixed and mutated, sometimes producing new and "better" traits in progeny. Then Darwin's mechanism of natural selection went into action: the environment selected the "fittest" organisms, which survived and multiplied. Over millions of years, this two-part dynamic of mutation and selection, beginning with the simplest forms of life, produced every organism on the earth.

The great advocate of the modern synthesis was Julian Huxley, a grandson of Thomas H. Huxley, the ally of Charles Darwin when they both battled the orthodox creationists in nineteenth-century British science. Decades later now, Julian Huxley had also trained in the sciences. He became a tutor in zoology at Oxford in about 1920. One of his most promising students, a close colleague for years, was Alister Hardy, who went on to make a name for himself not only in evolutionary studies but also in marine biology. Hardy would in time give the Giffords, but as a very changed man from his days with Huxley.

In 1942 Huxley wrote *Evolution: The Modern Synthesis,* a magnum opus that declared the triumphal advent of "neo-Darwinism," with its combination of genetic change and natural selection. Hardy was well on his own career path by then as regius professor of natural history at the University of Aberdeen, and in 1945 he attained the prestigious Linacre Chair of zoology at Oxford. Although Huxley stayed with strict materialism, even as he wrote books such as *Religion Without Revelation,* or the preface to the 1959 English edition of Pierre Teilhard de Chardin's spiritual evolutionism, *The Phenomenon of Man,* his prize student Hardy took a different direction. And that was why he ended up giving the Gifford Lectures at Aberdeen in 1963 and 1965.

As a student at Oxford, Hardy chafed under the dominant zoological method called morphology, which studied the innards of creatures, comparing organs to decipher evolutionary branchings of the animals. Hardy became "fed up" with such laboratory reductionism and left for Naples to study marine plankton in the Mediterranean. Then he joined the

British Ministry of Fisheries to study North Sea fish, having his first of many adventures as chief zoologist on an RRS *Discovery* expedition to Antarctic seas. Marine ecology became his passion. "He became a firm advocate of zoology outside the laboratory," said a biographer.[24] Hardy believed the ecology of animals—their relations to a diverse natural surrounding—was key to evolutionary theory.

His love of observing ocean ecology was manifest in his book *The Open Sea: Its Natural History,* released in 1956. It focused on the history of the oceans, their currents, and the story of plankton, and he produced watercolors of life trawled up from the ocean depths. A second volume looked at the world's fisheries just as colorfully. "He is devotedly obsessed by, and interested in, animals; he is eternally curious about the nature of their adaptations and lives, brilliantly critical in the examination of their mysteries," said his Boston editors.[25]

Yet through all this scientific activity, Hardy had kept to himself one of his own mysteries, his interest in religion. Back in 1925, when his science career was just beginning, Hardy had read a newspaper account of someone's religious experience. He hired a news clipping service to look for more such accounts, but it was a relatively dry well. Still, he developed a field scientist's interest and worked toward the day when he could gather them "like a collection of naturalist's specimens." Here was "an exercise in human ecology," he would say, a quarry that was neither fish nor fowl, but "man the religious animal."[26]

As was clear to most biologists, Darwinism espoused an ultimate materialism. Even the modern synthesis, adding the variable of genes to natural selection, viewed chemical mutations as being shaped by chance forces in the environment. For this reason, Hardy said, he remained agnostic on spiritual matters until late in his life, when he finally resolved that Darwinism and a nonmaterial spirit could dwell in the same world. "I eventually became convinced that the two could be combined." The solution, which was hardly novel since the days of Henri Bergson, was to accept that a creative force or divine spirit—"call it what you will," Hardy said—shaped not only "the direction of mutations" in genes but also the selective environment.[27]

Accordingly, when Hardy opened his Giffords he declared himself "a biological heretic." By the end of his life he was the author of popular books such as *The Biology of God,* which viewed evolution as culminating in human religious experience, giving the psychology of religion, Hardy said, an important role in the new natural theology. His Giffords, which were his first major treatise as a heretic, were a twofold plea. He urged dogmatic theology to surrender its obstinacy and then urged an equally dogmatic biology to welcome a "truer biology" that recognized a "para-physical" reality, the kind of mental world pointed to by Sherrington or Eccles. "The living world is as closely linked with theology as it is with physics and chemistry," Hardy said in his first Giffords, titled "The Living Stream."[28]

In his second, "The Divine Flame," he argued that the "spirit" of Christianity, not its dogmas, had produced the good fruits of modern democracy, including freedom of thought. A progressive theology, he said, was needed to retain that spirit in the face of so much modern un-belief, tyranny, and alienation. The Western foundations for the future were the spirit of Christianity, religious experience, a sense of the sacred, and the gospel teachings. There was no work left for dogma, he argued, thereby tying it to the medieval past. Admittedly, as a philosopher, Hardy was second tier. He likened himself to a Unitarian in thought but one who still loved to pray in Anglican chapels. He quoted nouveau theologians such as Tillich and Bishop Robinson (of *Honest to God*) and said Whitehead's process metaphysics, as a natural theology, harmonized with neo-Darwinism, at least if "biologists themselves are not dogmatic in dismissing consciousness as a mere epiphenomenon of the physico-chemical brain."[29]

Hardy's strength, however, was as a researcher, and he pledged in the Giffords to make natural theology a science of accumulated evidence. A few years later, in 1969, he opened at Manchester College, Oxford, the Religious Experience Research Unit.

As Hardy would put the case, his generation knew more about the religious experience of South Seas natives than of its own Western contemporaries. To remedy that, his research team first asked religious journals to alert readers to the need for religious testimonies, but this produced only two hundred, mostly from elderly women. So Hardy turned to the popular media. When the *Guardian* ran an article on "Sir Alister's 'Trip' to Heaven," interest snowballed. The *Observer* followed, and the *Times* of London covered a Hardy lecture with the banner headline, "A Scientist Looks at Religion." Soon came coverage by the *Daily Mail,* the BBC, and some American outlets. The research team also distributed a pamphlet, "Research into Religious Experience—How You Can Take Part." By the mid-1970s Hardy reported, "We have now received close on 4,000 personal records."[30]

Hardy sorted the different experiences by categories, holding them out as real scientific findings, and certainly equal to human experiences such as the beauty of art and the thrill of knowledge. "Spiritual experiences, from all accounts, appear to be just as real to those who feel them," Hardy said, although admittedly they had to come from subjective testimonies and recollections. It being the 1960s, with the Freudian overhang on biology and culture, Hardy also spoke of the "unappeased and frustrated desires" of modern man, prescribing not sex but a spiritual philosophy in harmony with the scientific outlook.[31]

In compiling his work, Hardy credited William James and James's pupil Edwin Starbuck as the pioneers. James had used such testimonies in his *Varieties of Religious Experience.* He had also drawn heavily on Starbuck's surveys around Boston, material that was published as *The Psychology of Religion* in 1892. With James and Starbuck, Hardy said, "the biology of God really began."[32] Their project had been forced off the road, however, by the Freudian bandwagon, which was negative toward religion, and the Barthian phenomenon, which distrusted the paranormal. Hardy wanted to revive the sort of "scientific" project James had begun by showing, now in the 1960s, that there was both a universality and a great variety in concrete human religious experience. For his commitment to this well past the Giffords, Hardy, in 1985, was the first scientist to receive the Templeton Prize for Progress in Religion, which gave a roughly $1 million award, slightly above the value of the Nobel Prize—and was founded in 1972 to rival the Nobel in reputation.

Before long, the world that Hardy knew, a world of large organisms and their environment, was overshadowed by a revolution at the microscopic biological scale—the revolution in genetic research. The promise of genetics guaranteed that it rapidly became a part of big science in the West, with massive amounts of research and a great number of professional laboratories. DNA itself became a focus of Gifford talks. Nobel laureate Brenner, a South African scientist who worked closely with Crick at Cambridge, had made the remarkable discovery of RNA, or the messenger molecule that assisted the reproduction of DNA. When he gave his Giffords, "The New Biology," at Glasgow in 1979, Brenner presented the purely materialist case for the remarkable power of the molecular world. There was obviously a moral and hopeful side to the mechanistic view of biology, Brenner and others would argue, for the mastery of genetic mechanisms could well vanquish disease and the great suffering that accompanied its ravages.

Yet to put God into this equation would require "something more" than the mechanisms of the new biology, for if God was relevant, the mechanisms could not be everything. If God was to be an active presence, then something more than individual genes had to be producing the miracle of life, or nature had to have an "openness" that allowed for creative guidance by a nonmaterial and mental property in the universe. Perhaps the best example, although not the only one, to offer that alternative was the 1992–93 Gifford series presented at St. Andrews by former Oxford biochemist and Anglican priest Arthur Peacocke on the pursuit of a modern theology, "Christian Theology for a Scientific Age."

Peacocke's main theme was that, despite the advances in the mechanics of genetics, the "reductionist" method in biology—reducing explanation down to the smallest, machinelike parts—still could not explain the phenomenon of life. Every time biology reduced "life" to the next smallest part, it seemed to become more complex and elusive than before. Peacocke was among those modernist theologians who said that, given the findings of science, Christianity was morally obligated to reconceive many of its medieval doctrines, painful as that might be for simple believers. Yet its claim on God was as resilient as ever, Peacocke said, even in the face of the genetic revolution. When the human genome was finally mapped in 2001, and the project revealed that it took fewer genes to accomplish a greater range of things than once believed, Peacocke was not surprised: divine miracle again had exceeded what natural science could know. For this way of thinking, Peacocke was not only a Gifford candidate in this fourth act of the Gifford legacy but also a recipient of the Templeton Prize, that other great badge of honor in modern-day natural theology. It was a double honor that, by the twenty-first century, seven Gifford lecturers would receive in their lifetimes.

One of those seven, Holmes Rolston III, was also a devotee of the God and genetics topic. A Presbyterian minister who became a professor of philosophy in the United States, Rolston was an expert in natural science, theology, and ecology. When he gave his Giffords at Edinburgh in 1997, the topic could not have better reflected the times: "Genes, Genesis and God." Like Peacocke, he was an antireductionist: genetic mechanisms were necessary, but not sufficient, to explain the depth of values in both nature and human culture. Something more than matter in motion, he believed, was needed to bring about a world presented so pictorially—and so ecologically—in the Book of Genesis.

For the God argument, the rejection of reductionist science—in which life and soul were "nothing but" mechanisms—became the primary rampart, a wall of defense that had been accumulating more layers over the years. Sherrington, in his Giffords, said that "mental phenomena on examination do not seem amenable to understanding under physics and chemistry."[33] Schrödinger alluded to the same in *What Is Life?* Thinkers ranging from Polanyi to Hardy seconded the motion. Hardy even questioned whether a brilliant man such as Crick could tell the public he had found "the decisive controls of life" in DNA. "But is it really true?" Hardy asked.[34]

By the power of such antireductionist arguments, nature had stayed open. But finally, under the Gifford aegis, the question became, Open to what? Open to what kind of deity? The scientists, philosophers, and theologians brought many views of God to bear, but when it came to finding God in nature, there were two primary options: a process God or a traditional omniscient monarch; a God who grew with the universe or a king who imposed his will on nature. A third kind of deity, which was more

Platonic, was a borderline case, but was generally viewed as transcendent, as far outside of nature, whether as an Idea, mathematical form, the ground of being, or the "wholly other."

For a biologist such as Hardy, process philosophy was a natural choice to bridge religion and science, for it did not propose an interventionist God "up there" or "out there." But the greatest theological proponent of process metaphysics—as recognized by even Hardy—was the theologian Charles Hartshorne. In the opinion of many, he should have been appointed a Gifford lecturer long before his death in 2000—at the age of 103. Hartshorne, a former assistant of Whitehead, would propose a more personal and active God, and it was he who coined the term *panentheism,* an alternative to pantheism by arguing that God is also outside nature. In panentheism, God is equally inside and outside of nature, although fundamentally outside and therefore its Creator.

Some of the scientists enamored of Whitehead's legacy had the pleasure of meeting Hartshorne. The son of an Episcopal minister, Hartshorne now divided his professorial career between Chicago and Austin, Texas. One physicist who met the process theologian was Princeton theorist Freeman Dyson, who gave the Giffords at Aberdeen in the easly 1980s.

Dyson, a scientific humanist like H. G. Wells, was "loosely attached to Christian beliefs by birth and habit but not committed to any particular dogma." Until he encountered Hartshorne, he said, he had "not concerned myself seriously with theology." When the two of them met at a Minnesota conference, they set their minds to putting a label on Dyson's inarticulate beliefs. "After we talked for a while he informed me that my theological standpoint is Socinian." Socinus was a sixteenth-century Italian heretic who held that "God is neither omniscient nor omnipotent," Dyson remembers being told. "He learns and grows as the universe unfolds." Dyson said he found that viewpoint "congenial, and consistent with scientific common sense. I do not make a clear distinction between mind and God. God is what mind becomes when it has passed beyond the scale of our comprehension."[35]

For an earnest inquirer such as Dyson, a cosmic Mind was enough, and a fruitful universe of diversity, and even a great variety of beliefs held in science and religion, was the proof of a basic cosmic goodness worthy of human curiosity and interest. Indeed, he titled his Giffords "In Praise of Diversity." Dyson had criticized dogmatism, both in science and in religion, as enemies to appreciation of a diverse universe. To give his praise of diversity a still more cosmic dimension, he titled his published talks *Infinite in All Directions.* Dyson's humanism, like that of others before him, viewed science and religion as international projects, brakes on the warlike nationalisms so prevalent in the twentieth century, and also forums of the new ecological consciousness.

This ecological concern had been the theme of the German theologian Jürgen Moltmann, whose 1984 Giffords at Edinburgh were published as *God in Creation: A New Theology of Creation and the Spirit of God*—considered a major theological treatise as environmentalism began sweeping Europe. For a final merging of science, process thought, and ecological ethics, however, the benchmark Giffords were those presented by the American physicist Ian Barbour, an early American reconnoiterer of the science-religion discussion. In the legacy of Whitehead's process philosophy, Whitehead had been a mathematician and Hartshorne a theologian, but Barbour had been an actual laboratory physicist and professor of the science. When he had first entered the field, he worked with the Nobelist Enrico Fermi in Chicago on the early development of the atomic bomb. By the 1960s, however, Barbour had turned to theological studies and famously proposed four kinds of relations between science and religion: conflict, separation, dialogue, and integration.

Barbour proposed that integration was the higher goal, and that process philosophy was the means to that goal. Although Whitehead had not lived to see the advances in particle physics, Barbour had been an eyewitness, and he felt that the physical data reconciled with Whitehead's metaphysical scheme. Barbour thus produced a "theology of nature," which argued that out of reverence for God's immanence (the physical "pole" of God in Whitehead's description), humanity needed a new environmental ethic toward technology and natural resources. Barbour began his Giffords in 1989 at Aberdeen, one of eight physicists who would receive the appointment between 1959 and 2000. His first series looked at reconciling theology and scientific findings, with an emphasis on the promise of process thinking. His second set looked at the wide array of ecological issues.

For Barbour's part, the question of "God's action in the world" remained a relevant puzzle for modern believers. What process thinking saw in the relationship of God to nature was something like the contact of individual beings, not machine parts. A "causality within interpersonal relationships, rather than mechanical force," he said, "seems to provide the basic analogy for God's relation to the world."[36] Just as people cannot coerce others to love them, neither can God.

A process view sets up no conflict between God and material laws, for God is seen neither as a physical cause nor as an intervening force. As Barbour puts it, God "is not *before* all creation but *with* all creation." God is integrated into reality, "present in all events in a role different from that of natural causes." Finding God by science is therefore not the question. By saying that "God is the source of order and novelty," process thinkers present God as a metaphysical answer to how the world works, not a scientific answer. Yet human language must have its way in describing this divine influence in all things, and people want to talk about the actions of God, Barbour seemed to say, holding out a consolation: "We can speak of God acting, but God always acts with and through other entities rather than by acting alone as a substitute for their actions."[37]

As a result of these realities, Barbour concluded, natural science and metaphysics will see cosmic history in different lights. Natural science sees a torrent of trial-and-error, messy and unpredictable, yet remarkably fruitful. In the metaphysical eyes of the process faith, "God works patiently, gently, unobtrusively." God is alluring both matter and mind to higher states, working most actively in the realm of consciousness: divine influence "is more effective at *higher levels* where creativity and purposeful goals are more prominent." Such a gentle God, so opposite to the monarch, is not simply an absentee landlord, however. "The process view does allow for *particular divine initiatives*," Barbour explained. "If God supplies distinctive initial aims to each new entity, no event is wholly an act of God, but every event is an act of God to some extent."[38]

When a hurricane or lightning destroys human property, it is legally called "an act of God," to put it in a nether world beyond human responsibility. But in the 1950s at Princeton, when the earliest and most powerful electronic computers were being built by John von Neumann, the idea was more like playing God. It was believed that computers could master the tumultuous laws of the weather. As consumers today realize, even that worthwhile science met severe limits. The reason is that the weather, unlike a clock, follows a nonlinear pattern, a pattern that came to be called a complex system, a nonlinear dynamic, or simply an orderly type of chaos.

This is what the MIT meteorologist Edward Lorenz found in his work with computer-driven "toy weather" in about 1960. By chance one day, he punched in a small set of equations to be repeated at high computational rates. He could not believe his eyes: on the printout, the loose figure-eight system that emerged showed that every looping of the lines was along an entirely different set of points, as if an eternal thread never repeated the same exact figure-eight loop. The chaos had stayed intact. Lorenz's 1963 report, "Deterministic Nonperiodic Flow," was ignored in the small meteorological journal in which it appeared.[39] But by the 1970s it was perhaps the most photocopied and circulated paper on weather in the history of science. He had documented a stable chaotic system.

The rise of chaos theory began with "a handful of scattered scientists, mostly unknown to one another," says its historian James Gleick.[40] It became formalized first among mathematicians, then moved into biology, ecology, physics, and finally medicine, all fields where dynamic systems were too sensitive and chaotic to be studied like clocks and missile trajectories, both of which followed fixed Newtonian laws.

As early as Lorenz, however, some of chaos theory's main features emerged, in particular the idea that a small and simple variation can generate a large outcome in a dynamic system. This was called the butterfly effect, for the flap of a wing could, hypothetically, alter a giant storm pattern a continent away. In the past, such small perturbations were believed to average out on a larger scale. But now, in chaos systems run on computers, the slightest deviation, by multiplying its effects, transformed entire systems. Technically, it was called "sensitive dependence on initial conditions."[41]

Two other features of chaos soon emerged, both of them with perhaps metaphysical implications for the theological imagination. The first was called the *attractor*. What held Lorenz's toy-weather system together, even though it never repeated itself, was *something* attracting it to maintain an overall order and shape. That something, still not understood, spurred a scientific search for a new kind of law. Second, when applied to biochemistry, evolutionary biology, and neuroscience, the field of complexity and chaos argued for the reality of "self-organization" in nature. An early example proffered was the giant red spot on the planet Jupiter, which seemed to maintain its existence—but neither as topography nor as weather—as the gaseous planet revolved. The answer, science believes, is that nonlinear dynamics created a self-organized spot. "It is stable chaos," said Gleick.[42] Invariably, natural scientists asked whether the origin of life was such a self-organized miracle, or whether the consciousness exuding from the physical brain might also be self-organizing.

For some chaos theorists God *did* play dice, although he loaded them to win in the long run, just as nature finally keeps order amid the surface of chaos. As with all the surreal findings of physics, whether quantum theory or relativity, chaos theory can seemingly be custom-fitted to either an atheistic or a theistic scenario, depending on one's outlook. Attractors and self-organizations could be their own gods, for example, born of the materialist laws of nature. On the other hand, cloudlike chaos could also suggest an open and "sensitive" realm in matter where the mind of God, making the slightest contact, could produce larger dramatic effects, unbeknownst to human measurement.

When chaos theory began to multiply its experimental examples in the mid-1970s, physicist Dyson said it came like an electric shock; science was jolted out of its exclusive search for laws that were *not* chaotic. Whatever the aftermath, Dyson was a happy Socinian. For other process thinkers as well, the complexity of evolving nature just made more splendid the emergence of a higher mind, or a God who did not intervene. But chaos had other implications, much as quantum physics did in its early years. For a traditional God, chaos was an opening. That was a cautious argument made by another prominent Gifford physicist, the Cambridge don and Anglican priest John Polkinghorne.

Polkinghorne, who spoke at Edinburgh in the early 1990s, was not called "the C. S. Lewis of science-religion" for nothing. Like Lewis, he was an apologist for classic Christian doctrines, most prominently the

Creation, the Incarnation, the Resurrection, immortality, and the End of the World.

Friendly enough to the Socinian and process scientists around him, Polkinghorne represented the highest level of reconciling scientific details and theories with revealed doctrines. "I do not find that a trinitarian and incarnational theology needs to be abandoned in favor of a toned-down theology of a Cosmic Mind and an inspired teacher, alleged to be more accessible to the modern mind."[43] To make the point, Polkinghorne formatted his Giffords, *The Faith of a Physicist,* according to the Nicene Creed, the basic faith statement of orthodox Christianity. For example, under the heading "We believe," the first words of the creed, he delved into current views of nature.

As a particle physicist, Polkinghorne felt that quantum theory still was fraught with ambiguity. He was reluctant to talk about God's activity amid so much "cloudy fitfulness at the subatomic roots." Microscopic uncertainties tended to average out at higher levels of biological life, and so would make no difference in that realm. Quantum fields were still essentially mechanical, and thus not truly open. "My instinct is to doubt whether the role of quantum theory is more than a small part of the story," he said.[44]

More significant would be the "exquisitely sensitive dynamic systems" being talked about in chaos theory. "Even in a Newtonian world," he said, "there would be more clouds than clocks." And the cloudlike system of nature, which combined both order and disorder, might be plausible territory for God's contact with the physical world. Dynamics in chaotic systems acted on "information-input," which suggested a "role for the 'mental' (information) in the determination of the material." Although the reductionist view was still central to scientific research, he said, for entire systems a holistic view was required. Thus, Polkinghorne endorsed the emergentist notion of "dual-aspect monism," spirit and matter joined in nature, a way for God to interact continuously with nature and yet for nature, including human beings, to have physical and moral freedom.[45]

For the natural theology of a traditional God, there was nothing in modern science so poignant as the big-bang theory. It was never the consuming topic of a Gifford series but was invariably touched upon, either as a mystery or as a plausible Creation event. The Gifford type of speaker was perhaps reluctant to go too far with the topic because of its scientific uncertainty, or perhaps because of famous faux pas, such as that made by Pope Pius XII when he addressed the Pontifical Academy of Science in 1951. To the embarrassment of even some Catholic astronomers, he said the big bang showed that "creation took place in time. Therefore, there is a Creator. Therefore, God exists!"[46] This was fourteen years before anyone had produced the first real evidence of an idea that was only a plausible mathematical theory.

Yet after Einstein proposed that the universe had a curved structure of space-time, astronomers began to wonder what that implied for the origin of the universe. Put simply, Einstein's theory of relativity had two parts, the special theory (1905) and the general theory (1915). The general theory applied the special theory's relativity of motion, unchanging speed of light, and unity of matter and energy to the real world—which meant the universe on its largest scales, where particles accelerated to the speed of light and the masses of stars and galaxies were enormous. At these scales, according to relativity, what looked like "gravity" was actually the curvature of space-time. When gravity was prevailing, the universe curved in on itself, but when energetic matter was strongest, the universe spiraled out expansively.

Einstein saw this problem of instability but preferred the long-held belief in a stable and eternal universe. He therefore created a "cosmological constant," or mathematical equation, to explain how an equilibrium had been achieved between gravity (pulling in) and energetic matter (pushing out). The constant worked, but only until the 1920s, when stargazers at the world's largest telescope, set on a southern California mountain, documented an eerie phenomenon called red shift. Discovered elsewhere in 1913, red shift appeared when starlight moved into the red spectrum as it receded from the viewer. Even Einstein predicted red shift in dense stars as a result of general relativity. But in the 1920s, astronomers at Mount Wilson found that entire galaxies were speeding away from each other, leaving the telltale red shift behind. The universe appeared to be expanding. Einstein did not like the idea, but in 1930 he conceded that the red shifts "in distant nebulae make it appear likely that the general structure of the Universe is not static."[47]

Now, an expanding universe was added to what Raven, in his 1950s Giffords, had called a new "vigour" in natural theology. The idea that the universe had an origin—that it was expanding from a single point—would turn out to be an easy concept for the general public to understand, and thus for stirring a new controversy over God and science. Opponents of the apparent "big bang" that started the universe—a term invented in derision—argued that the expansion was illusory, simply the continuous creation of new atoms in an otherwise static and eternal cosmos. Amid such speculation, one Russian cosmologist theorized in 1948 what a big bang might mean for scientific measurement today. If that primordial explosion really happened, it would leave a precise and persistent afterglow across the universe, much as the heat of an oven, opened rapidly in one corner of a kitchen, would in time spread evenly to every corner.

In 1964 two American researchers were baffled enough by a persistent buzz in their large New Jersey radio antenna to investigate further. They were trying to fix the antenna's reception, not find the first evidence of the big bang, but that was the outcome. They were the first to publish the news that the omnipresent static they heard might be the elusive "background radiation." Now the theory had concrete evidence, and the idea of a beginning point in the universe took wings. Although still contested by some, the big bang has become a conventional theory, strengthened in 1990 by the first outer space probe, which successfully found the same background radiation, confirmed in greater detail.

Along with the big bang came the idea of the atomic, molecular, chemical, stellar, planetary, and biological evolution of the universe over some eight to fourteen billion years. Everything that happened was based on a small number of "constants," or unchanging amounts of force, weight, quantity, and ratio, that apparently came with the universe in the beginning and apparently never change. If they had changed, even in the slightest degree, the universe seen today would not have evolved; if they changed slightly today, that same universe would dissolve into a chaotic soup. Hence, as the philosophers would argue, there are two choices. Either the universe was designed to produce intelligent biological life or else human beings, winning the cosmic lottery, are remarkably lucky to be here.

Here is where Polkinghorne was willing to draw attention to the so-called anthropic principle, which Dyson described as the view that the universe, given its architecture, "in some sense must have known we were coming." Even in Gifford's day, scientists such as Clerk Maxwell were commenting on how the laws and forces—namely the "constants"—governing atoms and molecules could not be explained by Darwinian evolution, which claimed to explain the origin of everything. The constants, however, just came with the cosmic architecture. They underwrote an evolutionary scheme that produced conscious life. For Polkinghorne, there was no great puzzle here. Both the constants—which had to be taken on faith—and an evolutionary process that could be studied by science seemed to work together as if the very biblical picture of creation. "Belief in creation *ex nihilo* will always be a metaphysical belief, rooted in the theologically perceived necessity that God is the sole ground of all else that is," Polkinghorne said, referring to what must be taken on faith. Taken on science, however, belief in *creatio continua,* or God's continuous creation, is better understood in a universe of evolving complexity, a material process with the potency to produce biological life.[48]

When in 1970 the Anglican cleric and professor of historical theology Eric Mascall gave the Giffords, he was perhaps unique in surveying "natural theology today," the subtitle of his series. Mascall was a neo-Thomist, believing that classic arguments for God's existence still had relevance, as did the realistic metaphysics of Aquinas. For several decades, Protestant thought had walled off science or addressed it only in generalization, but in the 1940s Mascall defied that trend by trying to relate specific theological doctrines to scientific details, the claims of creation, the soul, death, human origins, and the universe to the latest findings of physics, anthropology, biology, and astronomy. Although his early writings—such as *He Who Is* (1943) and *Existence and Analogy* (1949)—mostly laid out logical proofs for God, the attempt to jibe theology and scientific detail came to fruition in 1957 with his *Christian Theology and Natural*

Science. Mascall recognized that scientific theory is always changing, one historian said, but he saw this as "no reason not to ask how Thomistic theology connects with science at any given time."[49]

Mascall was a historian of theology who argued that the partnership of realism, reason, and Christian theism, as characterized finally by Thomism, had produced science in the Christian West. Lacking this formula, the Greek world and ancient Indian culture had not produced a progressing science.

This was a contentious theme in the history of science, yet it was one that another Gifford lecturer, the Benedictine priest and physicist Stanley Jaki, took up with great relish in his Edinburgh talks of the mid-1970s. In his published lectures, *The Road of Science and the Ways to God,* Jaki declared that the "cosmological argument" about the contingency of the universe (its existence depends on something greater) was not only proved, but has been the "realist" outlook necessary for successful science. By implication, only Christian culture could have produced a science that understood the natural world. Having taken this cultural argument into several settings, Jaki learned what a turbulent debate it had become. Once while lecturing at Oxford on the subject, he was pelted with insults by Muslims who felt that science had originated in their sacred lands.

A member of the Pontifical Academy of Sciences, Jaki argued that there was something wrong with the idea, believed by so many people, that "science progresses." He suggested that this was more of an ideological argument made by scientism, with its confidence in explaining the entire material universe, than a true reading of how science works, especially if one assumes that God exists. "In a very deep sense science is not progressive," Jaki said, with no little controversy attached. "It is anchored in a few basic philosophical, nay metaphysical, propositions about the mind and the universe."[50] So science pretty much stays the same as it looks for evidence to reinforce its metaphysical view of the world. What can progress, Jaki argued, is natural theology—the ways in which recognition of God can be harmonized with science.

While the excesses of scientism were not attractive to traditional Christian thinkers such as Jaki and Mascall, they also had a deep aversion to the popular process philosophy of Alfred North Whitehead. In effect, Whitehead proposed a God that changed with the universe, which for both science and theology, Jaki and Mascall agreed, was a catastrophe. If all was process, how could human knowing be constant, especially in the sciences to which process philosophers had turned? Many theologians had been "eager to be 'scientific' by espousing Whitehead's process philosophy," Jaki said in his Giffords.[51] But by doing so they had undermined the reliability of science itself. Science was not in process but was a constant kind of knowledge: once the scientific method and identity was established, its "being" has stayed the same, despite new generations and new technologies.

Although Jaki and Mascall differed on how much science can talk about God (with Jaki being more conservative), they agreed, as friends of Thomistic thought, that natural theology's key task is to present an unchanging Creator, not a Being in flux. Mascall dedicated some of his lectures to this argument, namely that "the main tradition of Christian theology has asserted both the timelessness and the changelessness of God." A *changeless* God gives permanence to a world that changes. What is more, Mascall said, a *timeless* God offers what a believer might expect of an almighty Being. For a God who knows every instant, past, present, and future, he said, "has a vastly greater scope for his compassion and his power than one would have who could attend to only one moment at a time," as suggested in process thought.[52]

In this opening scene of the fourth act in the Gifford legacy, the limits of science and its diverse new findings gave lecturers significant handles for talking about God's immanence in the world. Openings for God were found in science's need for a moral compass, the human subjectivity of scientific discovery, and the universality of religious experience. As a mechanical matter, the apparent duality of the brain, the massive creativity of genes, the big-bang universe and its fine-tuning all suggested an architecture and thus, for the believing mind, an Architect. In this process, scientists were rejected as high priests of the age. But in turn, they were revered as professional, respected, and awestruck guides into the marvels of the universe, and indeed as people who also could draw metaphysical conclusions.

Natural theology, meanwhile, found a new support on the back of these discoveries in biology, astronomy, and computer science. They gave new resilience to projects in reviving old and new theologies, whether Platonism and Thomism or theological constructs such as the process theology born of Alfred North Whitehead's world. Not to be forgotten, moreover, was the traditional Christian belief in a God who made the universe from "nothing" and continues to sustain the world and create through its laws. On occasion as well, the cumulative evidence of science would promote a drastic change of thinking in even the most atheistic of philosophers, men such as Anthony Flew, who in 2004 conceded the existence of some kind of Creator.

When Mascall gave his Gifford Lectures, he summed up one other significant convergence of science and theology. In the past, God's revelation had often been taken as a reality separate from nature. Now, nature—in all its scientific detail—was taken as a revealer of God. As Mascall explained, "Many modern thinkers find the old barrier between natural and revealed religion difficult or impossible to maintain."[53]

For years, Mascall had taught at the University of London, a city that stood for the new cosmopolitanism of the West. When he traveled to speak in Edinburgh, similar changes were under way. Since the 1970s a small Muslim community had begun to save money to build a mosque. It grew in numbers but was never able to muster the financial wherewithal—at least until the oil boom in Saudi Arabia. Thanks to Saudi largess, the Edinburgh Mosque, with its brown sandstone edifice, opened in 1998 a block from Hume Tower. With minaret aloft, it was designed to mingle Arabic lines with the castle architecture of old Scotland.

In 2001 Edinburgh appointed Mohammed Arkoun, a retired professor of Islamic studies at Sorbonne University in Paris, to speak. His lecture, "Inaugurating a Critique of Islamic Reason," was a call to modernize and democratize Islamic thought. By now, Christian, Muslim, Jew, Hindu, adherents of other Asian faiths, and advocates of none lived cheek-by-jowl in the West. Such pluralism strained the very idea of earnestly inquiring after "truth."

TEN

RELIGIOUS PLURALISM

The Limits of Knowledge

T HE SUN FINALLY SET on the British Empire in 1962. After a fourteen-year experiment in open immigration, Britain closed its doors almost entirely to old colonial domains, where millions of former "subjects" adhered to other faiths. At about the same time, in 1965, the United States inaugurated a lenient immigration policy, allowing as never before the entry of Asians, Africans, and residents of the Middle East.

Whether doors were opened or closed, the age of modern pluralism had begun. It was religious pluralism at first, but soon to follow was the question of truth itself. If there are so many truth claims, now vying in the close quarters of a single culture, how is one to decide? The monotheistic West had always relied on the one true God to provide a single truth. But now a plurality of truth claims haunted the land, and it seemed just the topic for the modern-day Gifford Lectures.

Immigration to the West was only the final stage of Christian contact with world faiths. It had begun when missionaries went abroad, reached a second phase when they brought back their bewildering reports, and crested at a third stage: the substantial migration of non-Christian religions into the West. The Western response also went through a kind of threefold experience, beginning with the *exclusive* claims of Christian, and even Western, truth against all other rivals in the world. As the world opened up, such exclusivism became harder to maintain, and two additional approaches, the *inclusive* and the *pluralistic* outlooks, emerged as other ways to deal with the conflicting truth claims of various religions.

Quite naturally, the inclusive and pluralistic stances were typical of the Gifford Lectures from the start. Even Lord Gifford, according to an acquaintance, despite his Calvinist culture, "managed to rise above the brimstone smoke, into the clear cold region of comparative religion."[1] Yet the interest in world faiths at the end of the Victorian age was a quite different affair than it would be after the 1960s, when immigration dramatically jostled the West and the mission fields were changed forever. The world became smaller, and pluralism more rampant.

The legacy of the Gifford Lectures had begun as a battle between the old speculations of idealistic philosophy and the new hubris of scientific materialism. On that stage, of course, the great disciples of science appeared: anthropology, psychology, physics, sociology, and the science of history. Each in its own way, they challenged religious belief, and touched on one aspect of it or another as held by Western believers. Yet all these disciplines claimed the mantle of science, joined together in the drama of science versus the untested theological claims of the age. Now, however, it was not science that challenged religious belief, but a whole circus of other religions and philosophies. As some viewed it, this was a contest more confounding to Western belief than the acids of materialist science. For that story, Edinburgh, in the summer of 1910, sets the scene for this great intellectual storm of the century.

In the great Assembly Hall of the Scottish church high atop the craggy cliffs by the ancient castle in Old Town, the first World Missionary Conference made history that summer by gathering twelve hundred delegates from around the world. "About the biggest thing that ever struck Scotland," a local churchman said at the time. Though mostly a British and American event, participants arrived from India, China, Japan, the Philippines, South Africa, and Australia. Mission workers shared seats with public figures, including the Archbishop of Canterbury, statesman William Jennings Bryan, and the president of Christian Doshisha University in Japan. The archbishop reckoned the assembly had "no parallel" in history.[2]

Religious pluralism, although unexceptional today, was an exotic challenge in 1910. Missions had naturally aligned themselves with colonial expansion and belief in progress. Few at the Edinburgh event quibbled over which culture had the truth, for it was clearly deposited in the Christian West. The mission before them was to spread the gospel, regardless of its cultural garb. Yet the difficulties involved were already on the table. First of all, how should Christians respond in faraway countries when people rejected the truth? Sooner or later that question would evolve. It would become, When other truth claims arrive in the West, how will the very *idea* of truth be preserved?

The 1910 event reflected the middle stage of the Christian encounter with these issues. Secure in their homeland, Christian leaders in Edinburgh sat back and listened to emissaries from abroad. These messengers were Christian workers, missionaries, and converts who traveled West from the heartland of Hindu, Buddhist, Taoist, Muslim, Sikh, and animist cultures. In one session, a "Chinaman in picturesque, flowing native garb" made his appeal.[3]

One of several key topics in 1910 was "The Missionary Message in Relation to Non-Christian Religions." For an entire day, mission workers gave their opinions, which ranged from fire-and-brimstone evangelism to a willing accommodation to native customs and language. But in the end, self-confidence was most important for this grand assembly. A closing speaker called for "the frankest comparison of Christianity with other religions," said a report in *Christian Century* magazine. "Such a comparison can result only in the enhancement of the glory of our holy faith."[4]

Edinburgh in 1910 was the start of regular world meetings. Mission workers convened in Jerusalem in 1928 and then Tambaram, Madras (South India), in 1938. Each faced a new world challenge. At Edinburgh, the experts urged missionaries to sympathize "as far as possible" with other faiths, seeking their "nobler elements" while knowing only Christianity could satisfy the yearning soul.[5] In Jerusalem, the new challenges were secularism and syncretism, or the mixing of Christianity with native beliefs. But a new tone was set at Madras. By then, the world faiths had begun to assert themselves in the face of the Western and Communist presence.

Many of the world faiths, in fact, were borrowing Christian ideas about organization and proselytizing, and marshaled indigenous doctrines that closely matched Christian creeds. As one report said in the 1930s, the passive Hinduism, Islam, and Buddhism of the colonial era had changed. Now they were "stirred, and no doubt stung, by the presence of an aggressively critical Christian movement," the report said. "They have responded not merely by defensive argument or counter criticism, but by various movements of internal reform."[6] Clearly, the encounter of world faiths was becoming a two-way proposition, and a far more complex theoretical matter for Western belief.

In 1960, *Christian Century* magazine featured another report, this one expressing a viewpoint far different from the zeal of the Edinburgh gathering fifty years earlier: Christians were "theologically unequipped for living in the twentieth century, with its pluralistic mankind."[7] The Christian West was still uncomfortable with diversity, which was the growing characteristic of the global village. Now that so few Christians believed that adherents of other faiths simply went to hell, the article said, Christianity needed an alternative view. In just half a century, the pluralist challenge had become clear enough. A mission leader from London, for example, was inclined to say that evangelism abroad had failed, unlike the fabled victory of Joshua at Jericho. "We have sounded the trumpets," he said in the 1950s. "And the walls have not collapsed."

With world religions arriving in the West, agreed others, the clash of Christianity and agnostic science would seem like "child's play" compared to the gathering storm of pluralism.[8] Social scientists tended to agree, for after secularization, the next great force changing society was an increased diversity of beliefs. "Modernity has plunged religion into a very specific crisis, characterized by secularity, to be sure, but characterized more importantly by pluralism," sociologist Peter Berger, a modernist Lutheran, said after his studies of the developing world in the 1970s.[9] Everywhere it seemed, pluralism was undermining the authority of even the greatest religious traditions.

When Lord Gifford had looked at this topic through his Victorian spectacles, it seemed far less troublesome. The Christian West was solid enough. It was in a position to be generous and ameliorative, and maybe even paternal, with the other religions. One spring evening in 1881, he presented that case to the Granton Literary Society in a lecture on Hinduism. Gifford was an early inclusivist, opening his talk with a proverb: "Look not, every man on his own things, but every man also on the things of others."

When Gifford had read up on Hinduism, he had obviously looked for what it had in common with Christianity. To spice up the talk, he chose one tantalizing aspect of Hindu mythology, the story of the ten avatars of Vishnu. The avatars, or divine "manifestations," were somewhat mysterious to churchgoing Scots in Granton, of course. But before Gifford could explain them, he had to lay out Hindu cosmology, and he did well enough as a lay student of the subject. Absolute reality for Hinduism was Brahma, an eternal and impersonal source of all things. Then came three principal deities: Brahma the creator, Vishnu the preserver, and Shiva the destroyer. Ancient Indian thought was far more complex than this, Gifford said. Still, the "highest essence" of Hinduism was a kind of theism. "Pure Brahmanism knows only one God," he said, "in whom all things consist and exist, apparently for ever."[10]

What was more, the three chief deities look very much like a trinity, Gifford was delighted to say. "The doctrine of a Trinity, though not the earliest, is one of the most prominent doctrines of Hinduism," he said. "Three essences in one God!"[11] Before the talk was over, he would build a bridge between his listeners and Hinduism, the chief religion of India. The reality of Brahma, a name for both the cosmic essence *and* the creator, was beyond mere mortals. So Hindus worshiped active deities such as Vishnu and Shiva, and particularly the first. Indeed, the ten avatars of Hindu cosmology were incarnations of Vishnu, who took on human or animal form. Nine avatars had come already, culminating in the Buddha (also recognized in Hinduism) himself. But a tenth avatar was expected. This avatar would inaugurate a new age.

At this point, Gifford was trying to strike a note with everybody's Christian background. In Hinduism, he explained, "the descending god

must take some form" on earth. In other words, the "*avatar* has come to
have very nearly the sense of the Christian word *incarnation*." Avatars are
quite a different breed from the Christ, but why not emphasize similarities? Gifford seemed to say. "God's revelations are not over," he said. "We
look for His *coming* still, and hymns like the Christians' are also sung in
Hindustan." At the world mission conferences, Gifford's analysis might
have been condemned as syncretism. But the missionaries *might* have
warmed to the final allusion in his address to the Granton Literary Society. By talking about Hinduism and Christianity this way, Gifford might
well persuade a Hindu to cross over. For Christ was the "most magnificent Avatar;—God manifest in the Flesh!"[12]

Long before such modern Christian attempts at comparative religion,
there was the original exclusivist view: Christianity is the only way. For
centuries it fought heresy and confusion to form its truest doctrines,
handed down once and for all by the saints. By the time the Christian
West sent missionaries to truly unknown lands, it was assumed that God
had been absent from the world after the human Fall, and only present in
the world since Christ, who left behind his church. The old religions
were therefore falsehoods, while the religions formed after Christ were
either heresies or false prophets. Yet, so exclusive a view, although logical
enough, did not square entirely with the biblical accounts.

When Paul traveled through Asia Minor and Greece he suggested
that God had worked in the world before Christ, and not just in preparing the Hebrew people, of whom Paul was one. In Lystra, Asia Minor, the
local polytheists thought Paul was sent by Zeus and Hermes. To clarify
the matter, he told them that while God had "allowed all the nations to
walk in their own ways" before Christ, God was not entirely silent in the
past ages. "He did not leave himself without witness," Paul said, suggesting that God had revealed himself through the benign workings of
nature. Then at Athens, as Paul stood before a pagan altar "to an unknown
god," he said the true Creator had given every nation a history and
boundary "that they should seek God."[13] Here was a seed for the inclusive
view of truth, and modern-day Catholic theologians in particular would
cultivate it.

By having the earliest missions, Catholics first developed an inclusive

"Logos" theology, a belief that Christ as the Logos, or Word, was active in history even before his earthly incarnation. This view presumed "seeds of the divine Logos in the heathen environment of the Old Testament; in Indian, Persian, Egyptian, and above all in Greek, 'philosophy,'" said one theologian.[14] Like the exclusivist stance, however, Logos theology tended to deny that any authentic religions arose after Christ, still a hard claim to uphold in the face of the real world and real consequences. As one mission expert asked, "Is the preaching of the Gospel directed to the total annihilation of all other religions?"[15] Or was the better part of wisdom the other extreme: could Christ be present in all religions, saving just about everyone?

In the United States, one man was feeling he had spent considerable time on that problem. His name was William Ernest Hocking. As the newly named International Missionary Council was convening its 1938 meeting in Tambaram, Hocking was beginning his Gifford Lectures at Glasgow on the topic "Fact and Destiny." His 1938-39 talks were never published, as it turned out, perhaps because of an entirely different reputation he was developing in the 1930s—for the topic of world missions and world religions. The Cleveland native had been the young disciple of Harvard's idealist philosopher Josiah Royce. Hocking went on to teach at Andover Seminary and Yale, and then joined the Harvard philosophy faculty to expanded on Royce's work by making Absolute Idealism more individualistic and pragmatic.

For Christian thought, Hocking came up with the novelty of negative pragmatism. "That which does not work is not true," he said. In a similar vein he argued for the existence of God by asking people "to try to get along without God and see what happens."[16] Known on both sides of the Atlantic since 1912 for his first work of philosophy, *The Meaning of God in Human Experience,* Hocking would eventually argue, using both his idealistic and pragmatic views, that Christianity could transform world civilization. This would not be achieved by converting others, but by example and the spread of democracy, freedom, and setting the stage for a possible world faith, a pluralism aimed toward the one God.

For Hocking and others, North American Protestantism needed to put its missionary house in order. With funding by John D. Rockefeller, a

Baptist layman, Hocking chaired the first-ever review of missions abroad. While the International Missionary Council was trying to boost the evangelical reach, Hocking and company tried to take an accounting of the real effectiveness of mission work. Organized in 1930, the fifteen-member commission traveled across Asia and produced *Re-Thinking Missions,* a report that rattled church attitudes when it came out in 1932. By 1940 Hocking was expanding on the theme of the "scandal of pluralism," delivering Hibbert Lectures at Oxford titled *Living Religions and a World Faith.* His outlook hardly diverged from the mission study, which said: "The relation between religions must take increasingly hereafter the form of a common search for truth."[17]

The *Re-Thinking Missions* project was a mirror-opposite of Edinburgh in 1910, when missionaries came in from the field. Now, a commission of independent lay leaders set sail for points in Asia, determined to obtain an unfiltered picture. For Hocking, the experience was life-changing. A somewhat isolated academic, he was for the first time allowed to see "domestic religion in various parts of the world—the religion of villagers, farmers, artisans—not always in evidence in the books."[18]

Over the course of the two-year project, the commission first sent fact-finders to view Protestant missions in Asia and then began its own nine-month excursion, arriving in Bombay after weeks on the open sea. It spent three months in India and Burma before moving on to Sri Lanka and Hong Kong. During ten weeks in China, it reached Shanghai in the interior, and then points west. At the end of March the commission headed for Japan, meeting up at the ancient temple city of Nara after some members traveled by way of Korea and Manchuria.

Finally, after sifting through a mountain of detailed inquiries, gathered in rural and industrial districts near and remote, the commissioners—who held evangelical and liberal views—agreed that missions had "in some measure finished their work," which was to plant seeds. Local cultures were best equipped to nurture the growth. The days of transient proselytizers and amateurs were over, the commission said, envisioning a more permanent cadre of high-caliber religious diplomats taking their place. Putting a single face on the Christian West and pooling funds seemed best. So the commission recommended a "single organization for Christian service abroad" for cooperating Protestant churches.[19]

The entire project stood as a monument to the inclusive outlook. Although not definitive on the problem of rival truth claims, *Re-Thinking Missions* urged religions to ally against an age of unbelief and secular tyranny. "Christianity finds itself in point of fact aligned in this world-wide issue with the non-Christian faiths of Asia," said the report. Religion itself seemed to be at stake in every culture. "It is no longer, Which prophet? or Which book? It is whether any prophet, book, revelation, rite, church, is to be trusted."[20] Naturally, the globe-trotting research project was conspicuous for its avoidance of the great Western centers of Christianity, such as Edinburgh and Rome. But even in those two cities, which stood for the intransigence of the institutional church, a quiet revolution was brewing.

For a few centuries, the cities of Edinburgh and Rome had very little in common, except that both were built on seven hills. The Reformation and Catholic heritages had made them different worlds, at least until the world became smaller, especially in the 1960s. The American report on *Re-Thinking Missions* three decades earlier had said that secularism was a common foe of all world faiths. For their parts, Edinburgh and Rome seemed to be saying that the pluralism of religion demanded more Christian openness to other religious traditions.

The Gifford legacy, as part of act four in the twentieth century, would abet that openness. Well before such projects as *Re-Thinking,* in fact, the Gifford vision had been one of providing an antidote to secularism by finding the broadest religious vision possible, that being a kind of rational understand of a universal God and a universal ethic. Locked in his Victorian times, Lord Gifford had been able to see only so far, but as time passed, the horizon would widen—thanks in part to the lecture series.

An early voice in Scotland was the theologian John Baillie, a former moderator of the Church of Scotland and the senior theologian at its main school, New College, whose two towers looked down on the ravine between Old and New Towns. Baillie had spent a lifetime trying to bridge European and North American thought, but his last great work put an emphasis on pluralism. Chosen to give the Giffords in 1961, Baillie died just months before. His manuscript for *The Sense of the Presence of God* was so complete that it went directly to press.

In the Giffords, Baillie reflected the growing concern of many when he dedicated an entire lecture to the topic "Faith and Faiths." More was at stake than simply meeting different ethnic religious groups around the world. What if they represented, in some mysterious way, Baillie asked, different forms of God's revelation? While defending Christian unique- ness, Baillie was being inclusive. Typically, Christians made this concession by acknowledging "natural religion," the idea that "heathens," while mired in superstitions, had at least attempted to respond to a vague flicker of God in nature. Yet in his lectures, Baillie felt it was time to move toward a doctrine of revelation in other faiths, especially, he said, the recognition of God's general revelation to non-Christians.[21] In natural re- ligion, human beings strained to find meaning in natural mysteries; in general revelation God put a foot forward to inform the human mind. In this case, all religions would have apprehended aspects of God in nature, in the human conscience, or in mystical experience.

As Baillie began writing his Giffords, the first signs of a new approach by the Roman Catholic Church began percolating at the Vatican. In 1958 the new pope, John XXIII, announced his intention of holding a Second Vatican Council to throw open the windows to fresh winds and update the church, what the church fathers later called reading the "signs of the times." When the three years of council proceedings ended in 1965, one of its most remarkable documents was *Nostra Atate,* or "The Declaration on the Relations of the Church to Non-Christian Religions."

Unofficially, many Catholic theologians and missionaries had already adopted some version of Logos theology. In 1962 the Jesuit theologian Karl Rahner was so bold as to speak of "anonymous Christians." The anonymous believers were nearly all of humanity. Anyone with an ear at- tuned to God, even as the world sparkled with diverse beliefs and doubts, was on the path to salvation, Rahner suggested.

Going into the council, Rahner was in hot water for allowing salva- tion to the unbaptized. But as liberal and conservative forces at the coun- cil vied mightily, his ideas rose to the surface and were finally reflected in

Nostra Atate. Because Christ had worked behind all faiths, the document said, the church "rejects nothing of what is true and holy in these religions," although the church is "duty bound to proclaim without fail, Christ who is the way." Catholics could now prudently talk across the interfaith divide, and even "encourage the spiritual and moral truths found among non-Christians," which included "their social life and culture."[22]

The first Gifford lecturer to capitalize on Vatican II was Robert Charles Zaehner, a convert to Catholicism and Oxford's leading professor on world religions. His talks at St. Andrews beginning in 1967 were also the first dedicated entirely to religious pluralism. Holding the chair of Eastern religions and ethics at Oxford since 1952, Zaehner was a former diplomat with a Persian, or Iranian, background. An expert in Zoroastrianism, he also knew Islam by experience, taught courses on Sufi texts, and translated several major sacred writings of Hinduism. His encyclopedia of "living faiths" came out in the late 1950s, a bellwether on the eve of the Second Vatican Council.

Knowing how long it took the Catholic church to make the "volte-face" of *Nostra Atate,* he said that Catholicism's change in mood would have surprised even Lord Gifford. To understand Vatican II, Zaehner reached back to Saint Francis of Sales, the missionary-minded bishop of Geneva in the early seventeenth century. When Francis founded a missionary order, he looked at world religions as both brothers and enemies. He called it "concordant discord." Zaehner said little had changed, except that Christianity had become more open to its rivals. His lectures hoped to show that "there is much in Eastern religion that is still valid," not only to its adherents, but for broadening Christianity.[23]

Zaehner was also caught up in the vision of Pierre Teilhard de Chardin, the Jesuit paleontologist whose work on cosmic and human evolution had been forbidden publication until the 1950s. In Teilhard, he found the centrality of Christ but also the convergence of all the faiths. They were evolving toward a final unity in the Omega Point. As in much Logos theology, this vision set Christians free from having to convert others but made goodwill obligatory. "For me the center of coherence can only be Christ," Zaehner said. Yet he was "not at all convinced that Catholic Christianity need be the true religion for everyone." He had seen enough students of the 1960s to know the trends. Many young people "are far more interested in Zen Buddhism than they are in Christianity," he said. "Good luck to them, if they are at all serious in their exotic aspirations."[24]

Like Gifford, Zaehner saw great promise in the Hindu-Christian relationship, not least because, unlike with orthodox Islam, there were degrees of tolerance on both sides. Kept vague enough, theism could be a significant issue for talking together. Zaehner believed, in fact, that the most significant trend in pantheistic Hinduism was a "movement toward monotheism," belief in a personal God who was distinct from the universe. Whether Gifford could imagine such a day was perhaps beside the point. With ideas like Teilhard's gaining currency, the search for one God had become "curiously topical," Zaehner said.[25]

Zaehner had laid out an inclusive view of truth claims, but a truly pluralistic outlook was still to come. As Zaehner gave his lectures in 1967, a Presbyterian minister and philosophy professor arrived at the University of Birmingham, England, to fill its chair of philosophy for the next fifteen years. His name was John Hick, and by the time he gave the Giffords in Edinburgh twenty years later, he stood for pluralism at its peak. Quite happily, Hick became the lightning rod for a clash over truth claims.

Born in Yorkshire, England, in 1922, Hick went to Edinburgh to study law and there "experienced a strong evangelical conversion." A conscientious objector during the Second World War, he served in a Middle East ambulance corps and afterward switched to philosophy at the Scottish university. With a doctorate in the subject from Oxford, he next went to Westminster College, Cambridge, to train for the ministry, becoming ordained in 1953. He was Presbyterian because that was where "my evangelical friends were." On New Year's Day in 1951, Hick speculated on his accomplishments by 2000, wondering in his diary if he could write "a popular account & defence of Xian belief for the non-Christian." By century's end he was defending all beliefs. "Life would be much less interesting if it held no surprises," he commented in retrospect.[26]

At Cambridge, Hick became aware of Christianity's encounter with other religions through the work of various Cambridge theologians, some more inclusive of world religions and others more exclusive in regard to Christianity. That exclusive stance was taken by Hick's own teacher, Herbert H. Farmer, who said that *only* theism counted. Farmer's Gifford Lectures in 1949, published as *Revelation and Religion,* portrayed Christianity as "the normative concept of religion." He had attended the International Missionary Council in Tambaram in 1938. Under Barthian sway, he returned believing that the great non-Christian religions were utter falsehoods. "If Christianity be right, they must be in a radical and total way wrong," he said.[27]

Farmer typified the sequestered Christian, Hick believed. So far as Hick knew, his teacher lacked "any first-hand encounter with people of those faiths sufficient to take him beyond the letter found in books, to the living spirit, found in people."[28] The touchstone of experience, of actually meeting and talking with non-Christian adherents, would become a new dividing line in the modern debate on pluralism. While some felt that "dialogue" was a slippery slope to compromise, others said religion is not real until it is manifest in real people. Only human contact revealed that world religions were not encyclopedia entries but complex belief systems filled with diversity, local creativity, cultural trappings, and internal disagreements.

Soon after his ordination, Hick began to take academic posts, first as a philosophy professor at Cornell, then teaching at Princeton Theological Seminary, and for a while afterward as a lecturer at Cambridge. By his encounters in parish work and community work, Hick would also take on the religion-as-people view. When he arrived at Birmingham, an industrial city with immigrants and palpable racial tensions, Hick joined the local interfaith council. "As I spent time in the mosques, synagogues, gurudwaras and temples as well as churches something very important dawned on me," he said. "Essentially the same thing was going on in all these different places of worship."[29]

After the 1944 Butler Education Act required all British schools to include religious instruction, each local community designed a curriculum, called the Agreed Syllabus, to carry out the national guidelines. The national syllabi tended to fulfill the religious requirement with Bible study. But a movement led by British scholars of religion such as Ninian Smart—who delivered Giffords in 1979 on "transcendental pluralism"—urged a broader treatment of religious instruction.[30] In Birmingham, the idea of studying more than just Christianity got a cool reception at first. One group lobbied vigorously for exclusive instruction in Church of England doctrine. The city council finally endorsed an inclusion of "several faiths" and "other major ideologies."[31] In this same period, Hick became a leader in "All Faiths for One [Human] Race," which formed in 1970 to protest apartheid during the tour, finally canceled, of an all-white South African cricket team. The group continued as a community relations broker, which included aid to non-Christian immigrants.

Despite his activism, which styled Hick as the kind of "prophetic" cleric so fashionable in the 1960s, he was a theorist at heart. As a student he had held fundamentalist beliefs about the Bible and salvation but gradually gave them up for a broader view. Still, his first original work, *Faith and Knowledge* (1957), ardently defended Christian belief as meaningful knowledge. It was a time during which academic philosophy in Britain and the United States demanded the use of words that were "meaningful," or could be verified by facts. Words such as *God,* these analytical philosophers said, were meaningless. Hick mounted a contrary case, showing how religious propositions, when taken or rejected by people, made profoundly material and factual differences in the world.

Yet from early on, Hick was also a Kantian. He adopted the view that religion was foremost about moral imperatives. Ultimate beliefs, although authentic when held on faith, could never be proved rationally. The result was a kind of egalitarian pluralism that counseled humility in making truth claims, especially to Hick's home territory, Christianity. In pluralism, a person can claim his religious experience to be valid, even unto salvation. But that same person, rationally, may not argue "that our own form of religious experience" is salvation-giving "whilst the others are not."[32] Hick brought his full-blown pluralist "hypothesis" to fruition in the Giffords at Edinburgh in 1986–87, published as *An Interpretation of Religion,* a seminal work of the century.

The trek from his prewar days as a student at Edinburgh back to Edinburgh as a Gifford lecturer in a pluralistic age had its controversies and surprises. Perhaps the biggest, Hick said humorously, was to meet the great-grandson of Lord Gifford and find out he pronounced the family name "Jifford," quite a shock after decades of everyone saying it the other way. "Who is he to presume to know, better than generations of philosophers of religion, how to pronounce his own name?" Hick joked.[33]

More seriously, Hick's religious odyssey had brought him to believe in two core values. The first was universalism in salvation, a belief that God destined the well-being of all souls regardless of religious affiliation. Second was a humanistic approach to interreligious affairs; the focus was on human respect and need, not exclusivist claims of one religion or another. This broader view, meanwhile, was related to how Hick solved the theological problem of evil. Just before arriving in Birmingham, he published his Christian theory on that prickly topic, rejecting the Augustinian view of the Fall, sin, and grace, and reviving a stance taken by the early church father Irenaeus, the bishop of Lyons. Irenaeus said God created man incomplete in a world of good and evil. God also placed man at an "epistemic distance" so he could not entirely know God. Life was a process of "soul-making," so difficult a task that even holy men may have come up short. That led to a plausible argument, Hick said, for the afterlife, where a loving God allows the soul to become complete.

World religions were his next philosophical challenge, and he concluded that "for Christianity the problem of religious plurality hinged on the central doctrine of the incarnation," that God *only* appeared on earth in Christ. In 1977 the historic roots of this doctrine became the quarry of Hick and others, who compiled their essays under the attention-grabbing title *The Myth of God Incarnate,* published by the SCM Press, which knew the value of a good controversy after breaking publishing records with *Honest to God.* The reaction was immediate and intense, producing headlines, news conferences, synod debates, and a counter book, *The Fact of God Incarnate.*

That John Hick was in trouble again, but now in England, was no surprise to conservative Presbyterians in the United States, where he had faced heresy charges twice, first at Princeton Seminary. While there, he was "theologically fairly conservative," but thought the 1640 Westminster

Confession was "completely out of date" on matters such as the virgin birth of Christ. Orthodox critics filed a complaint, and the "Hick case" went all the way up to the high court of the United Presbyterian Church, which cleared him in 1962. Hick said a kind of "revenge" came in 1980, when he moved to Claremont Graduate School in Claremont, California, where he retired.[34] As an ordained minister, he asked to be listed on the roster of the church region, which would make him a cleric in good standing able to preach and vote in the presbyterial governing system. He was strenuously opposed by orthodox believers who remembered his heretical victory of yore, so he withdrew his request, saying that he did not want to divide that body.

Two years into his tenure at Claremont, Hick received the Gifford invitation to speak at Edinburgh. Since being a student there he had "nursed some kind of assumption" that he might one day give the Giffords, and it was naturally satisfying that his belief was not rebuffed.[35] By then Hick's thought was mature, and he would devote the Giffords to a final theory of religious pluralism. In *An Interpretation of Religion: Human Responses to the Transcendent,* he calls it "a religious but not confessional interpretation of religion in its plurality of forms."[36]

As a good Kantian, Hick set the stage by showing how no historical religious or philosophical claim rationally to prove or disprove God was valid. Life was "ambiguous." Various doctrinal claims, dealing with unseen worlds and the afterlife, could not be adjudicated on earth. Even in earthly life, events could be taken as either spiritual or natural, depending on the person. It was thus "rationally appropriate for those who experience their life in relation to the transcendent to trust their own experience," regardless of their tradition. All religious tradition made contact with the ultimate transcendent reality ("the Real"), according to Hick. "The great world traditions constitute different conceptions and perceptions of, and responses to, the Real from within the different cultural ways of being human," he said.[37]

Most important to authentic transcendence were the fruits, not the dogmas, and every religion provided such moral systems and exemplary figures. They all transformed lives, offering a kind of salvation. Adherents changed from self-centeredness to being centered on the higher reality, usually called God. By admitting that God was beyond all claims, how-ever, religions could appreciate, cultivate, and even debate all their ex-ternal differences without lapsing into violence or coercion. For in an ambiguous cosmos where ultimate reality was beyond human grasping, Hick said, "it seems implausible that our final destiny should depend upon our professing beliefs [for] which we have no definitive informa-tion."[38] In a world where people see in a glass darkly, religion was simply to encourage participation in all spiritual and moral dimensions.

Hick enjoyed the return to Edinburgh, with its old friends and famil-iar places. "The lectures themselves were no great event," he said.[39] The audience was about fifty. Their published version, however, became a mainstay, provoking about thirty other books in response, mostly by con-servative Christian theologians and philosophers arguing for an exclu-sivity to salvation. Outside of fundamentalism, Hick said, the "initially disturbing" idea of respectful pluralism would invariably spread. Still, this was the 1980s, the aftermath of the Iranian revolution. It was an era of bold political assertions, from Hindus and Sikhs as well as from evangeli-cals in North America, an era that saw a "wide resurgence of the 'us against them' attitude in the forms of both religious fundamentalism and political nationalism," Hick said.[40] The tribal gods might become bigger than the Real.

Between 1960 and 1990, a whole host of Gifford lecturers would take up the relations between Eastern and Western religion and offer various solutions to pluralism, most of them inclusive. A major task was to recon-cile the personal God of the monotheistic faiths with the impersonal force of much Asian thought. It was a challenge of deciding how to hold one's own beliefs as true, yet offer a degree of recognition to other truth claims in religion.

Edinburgh was particularly keen on the topic. It invited the first Muslim, the Iranian scholar Seyyed Hossein Nasr, to give the Giffords in 1980. An appointment followed to Raimundo Panikkar, a Catholic priest of Indian ancestry, the author of the church-approved book *The Unknown Christ of Hinduism,* and a theologian, who like Gifford, found promise of plurality and unity in the idea of the Trinity.[41] Of them all, Hick was the most controversial. He was the point of attack for those arguing the uniqueness of Christianity or that not all religions can be true.

A decade and a half after Hick delivered his Giffords, one of his most reasonable and worthy rivals had a chance to challenge his liberal claims about pluralism. In 1993 Keith Ward, the regius professor of divinity at Oxford, devoted a considerable segment of his Glasgow Giffords to re-butting Hick's "hard pluralism," as he described the stance that all salva-tions are equal. Ward was a proponent of "soft pluralism." It recognized degrees of truth in other faiths but could not brook so vague a synonym for God as *the Real.* "Hick's attitude to the Real is ambivalent in the ex-treme," Ward said. By speaking of religion that is authentic and choices that are appropriate, Hick had opened the door to true and false claims. Not everything is acceptable even under Hick's concept of the Real. "There is little reason to think that different traditions and beliefs must be equally valid or true," Ward said in his Giffords on "revelation in the world's religions."[42]

Ward is an advocate of comparative theology, looking at the world's truth claims in order rationally to improve all of them, but certainly to expose falsehood as well and to show something better in Christian thought. It is only logical, he went on, that not every report of experi-encing God can be true. "There must be some distinction between true and false, between authentic and unauthentic manifestations of the Real. That entails that we have some true information about the Real, and therefore that some beliefs in religion must be false."[43]

As a Kantian, Hick had put limits on "true information" available to the mind. For example, he said that ultimate reality, the Real, "cannot be said to be one or many, person or thing, conscious or unconscious, pur-posive or non-purposive, substance or process, good or evil, loving or hating." However, the reported experiences of the Real can be "mytho-logically true," Hick argued, and thus an array of conflicting beliefs—belief in Vishnu, the Tao, or Christ, for example—could share equal veracity as practical truths. They are true in their ability to evoke "an ap-propriate dispositional attitude to" the Real.[44]

For a start, Ward rejected this as pragmatism, which he did not con-sider a final arbiter of what is true. He asserted that even in Hick's Real, it is impossible to reconcile two such entirely opposite ideas as a personal God, seen in monotheism, and an impersonal force, as seen in Asian thought. The idea of Allah or Yahweh cannot be a synonym for the Tao. A second point, Ward said, is that moral virtue or heroism could not be the common mark of religious truth, for even antireligious Marxists and existentialists, for example, could show such character. Finally, since people must decide which version of the god-reality is authentic, they could certainly argue that a "blind, purposeless source of energy is less authentic than an idea of it as a person, which is in turn less adequate than the idea of it as 'a being of unlimited perfection.'"[45] Ward was on his way, according to another Gifford lecturer, to a "non-exclusive view of the Incarnation as the climactic revelation of divine love."[46]

On one matter, however, Ward could agree with Hick. Whatever is ultimate, it comes to human minds as symbols, requiring interpretation of similarities and differences in truth claims. In addition, since no one can just sit down one day and figure out the truth, it is rational to begin with belief in one's religion by birth, whether it be Christian, Hindu, or Jewish. Despite that commitment, however, Ward said people must continue a rational exploration of belief. They should accept that conversion to another faith can indeed be a rational step, but also accept that God "has not been silent in the other religions of the world." Such a view "delivers one from a myopia which confines God to one small sector of human history."[47]

By now, Ward had waded into the swampy debate on whether all human beings use the same kind of rational faculty, enabling them to move toward a single higher truth. One published article on religious pluralism, which Ward challenged, went under the title "Whose Objectivity? Which Neutrality?" An earlier Gifford lecturer, the philosopher Alasdair MacIntyre, had written an entire book on moral truth claims titled *Whose Justice? Which Rationality?* The debate was this: did human beings even have a common rationality on which they could agree (let alone a common religion)? Whereas MacIntyre found in Aristotle and Aquinas the best basis of universal rationality, Ward anchored it in the testimony of human nature itself. With that tool of rationality, people find reasons to prefer one religion over another, Ward said. They "find a more adequate view of the Real in some traditions, and perhaps in one tradition, than in others."[48] Yet nobody starts from a neutral position since everyone is born into a cultural context. Once located in a culture, however, everybody operates by the norms of reason, by the very nature of facts, perceptions, morals, and human cooperation in every society. "There is therefore a minimal level of rationality present in all societies, which does not vary from one society to another," Ward argued. In addition to being compatible with such reason, truth claims should match experience, should not be delusory, and ideally will offer a unified outlook. Ward called this "rational believing" built on universal rational criteria. "Every person uses them even while denying it, to articulate and render more coherent one's own view of human existence."[49]

Religion and rationality had long been a theme in the Gifford Lectures. World religions too had been a popular motif in the first decade or two, but then disappeared, only to reappear after the 1960s and 1970s. "I think it's a topic that's come back," said one Gifford officer. "It had disappeared within mainstream Christian theology from the 1930s to the 1960s, and since then it's made something of a return." He said the modern-day work on world faiths is more "sensitive," not presuming Christian dominance, as was the case in the Victorian heyday of progress, proselytizing, and empire.[50]

The rise of religious pluralism has its secular parallel as well, and this has become the second great crisis in Western thought: the relativity of knowledge. As the West became secular, pluralism of truth was no longer just a parlor game for theologians. When the sociologist Berger spoke in the 1970s about the "heretical imperative," or requirement that everyone choose their belief system, he was talking about religion. But he could have been speaking as well about all kinds of knowledge, scientific, historical, and political. "In the premodern situation there is a world of religious certainty, occasionally ruptured by heretical deviations," Berger said.[51] By contrast, the modern world is home to religious uncertainty, occasionally jarred by religious affirmation.

The same had become true for Western culture as a whole. In the Victorian age, a consensus on reality was easier to find in the general public, especially regarding the nature of truth and the inevitability of progress. Today, the voice of a single culture is drowned out by congeries of different voices, each of them a different point of view. At the end of the twentieth century, part of the great appeal of religious fundamentalism and religious nationalism, whether seen among Hindus in India, Buddhists in Thailand, Muslims in Indonesia, or evangelical Christians in the United States, was the promise of a unified cultural viewpoint that shut out all other viewpoints. Still, despite such Christian efforts to regain a monopoly in the West, its societies have remained more a garden of flowers than a field with a single crop.

That garden, however, was not to everyone's taste. So in the 1980s philosopher Alasdair MacIntyre used his Gifford Lectures to decry the way in which uncontrolled pluralism had invaded both common society and the more exclusive modern university. The plight, he said, almost made the Gifford project impossible. The public no longer held all the same assumptions as a public lecturer. That was a product of a bygone era, he said, a Victorian belief that was no longer true. "Ought we to cease and desist from delivering lectures of this kind?" he asked.[52]

Not only had the public lost its common assumptions, but so had modern university education, MacIntyre claimed. This had led to a sad dichotomy. On one hand, the modern university still operates on the Victorian illusion of the unity of knowledge. Yet on the other, the modern campus is a hotbed of vying ideologies and worldviews, as far apart as Aristotle is from Friedrich Nietzsche. To illustrate the change, MacIntyre pointed to a publication that had bridged Gifford's era with the present, the *Encyclopaedia Britannica*. At the close of the Victorian age, the *Britannica* had presumed a unified knowledge, agreed to by all mainly by virtue of science and reason. But after 1911 the *Britannica* lost that unifying vision. Like future society, he said, it became a hodgepodge of separate stories and facts. In a flight of hopeful speculation, he urged a return to solid traditions, even at the heart of universities, which could harbor a few rival schools in a forum of "constrained disagreement" and "systematically conducted controversy." Even this, he concluded, would be taken as "a piece of fancy" by the modern educational system.[53]

When it came to modern relativism, MacIntyre was not the only Gifford lecturer to join the battle. The University of Glasgow joined it in full in 2001. To celebrate its 550th year, the university dedicated a unique series of Giffords, featuring five speakers, to the topic of "The Nature and Limits of Human Understanding." What the large and fascinated audiences learned was this: the possibilities of human knowledge were not even exhausted by a group of experts representing psychological, scientific, evolutionary, philosophical, and theological points of view. They all had a different answer to relativism.

A psychologist explained why the brain invariably made errors, and a linguist doing cognitive science said human knowledge developed according to the shape, biology, and computer-like functions of the brain itself. A Darwinian philosopher said that human knowledge developed for its survival value, but its absolute certainty was finally a philosophical, not a biological, question. Then, a philosopher argued that for all science's great achievements, it had a limited kind of knowledge, and that most people handled life quite well with a "commonsense" and subjective way of knowing the world. She was followed by a theologian who naturally argued that "a theistic metaphysic makes most sense of the world as we know it to be."[54]

Taken as a whole, the Glasgow Gifford forum gave the public both reasons for, and doubts about, placing confidence in the modern search for truth. It was not the sort of ambiguity that would have appeared in Gifford's time, when science, reason, and certitude were synonyms. When act one of the Gifford legacy opened, the two great systems of philosophical idealism and scientific materialism were spectacular for their clear and towering profiles, as if two giants astride Western thought. Then science tried to create the proverbial worldview that would gather together all peoples, but at midcentury the great protest of "subjectivism" bridled at such scientific hegemony, which, with the rapid specialization of science, turned in a kaleidoscope of variety, hiding forever the quest of an early era for a "unified" mind in the West. As the twenty-first century approached, many Gifford lecturers, no less than any other thinking persons, were forced to address the new mood of this intellectual pluralism and relativism.

It may be no surprise, meanwhile, that as a single way of thinking began to evaporate, the West became a landscape of negotiations between

conflicting ideas and interest. In a word, the West became a culture of litigation. Courtroom dramas and battles, although around since medieval
times in the West, now became the foremost pattern of culture, filling not
only real courtrooms but also novels and television shows. The mushrooming of litigation in the West, while very real in its own right, has
become a metaphor for the search for truth, even for the "knowledge of
God." The search is not for a revelation from on high, as if Moses had
gone to Mount Sinai, but takes place in a world in which everyone must
decide amid conflicting testimony. It is a world not too different from that
of Adam Gifford so long ago, when, as a young prosecutor, he presented his
best evidence so that a jury could arrive at a decision of guilty, not guilty,
or even "not proven" as in the violent murder of poor Jess McPherson.

Today, a visitor to Glasgow University can walk south, down a hill,
and about half a mile to Sandyford Place, the scene of the McPherson
murder. The Sandyford "mystery" was for all practical purposes solved in
a Glasgow courtroom in 1862, thanks in part to the prosecution by Adam
Gifford. The murder mystery is still revived in the occasional newspaper
story, anthology of British murders, or BBC docudrama. The Victorian
row house, with its high ceilings, ample hallways, and wide staircase, has
long been home to a Glasgow shipping firm, which is happy to provide a
packet of details. A file cabinet houses news clippings, book entries, and
even floor plans of the murder scene. Generations after Jessie McLachlan
was convicted of the crime, the Strathcylde Police Museum in Glasgow
still displays a floorboard stained with her bloody footprint.

Yet for the fastidious mind, a mystery still lingers. In a world of conflicting truth claims and perceived limits on human understanding, the
Sandyford mystery could become a maze of doubt. Who really did kill the
maid? There was no eyewitness. Testimonies conflicted. Had the jury
found out exactly? As the story is told, Old Fleming was probably the real
culprit. The mob certainly thought so, and it ran him out of town. Yet in
turn, McLachlan had engaged in such an elaborate cover-up that she too
looked guilty enough, and that was certainly what the judge and jury
believed. To them it was a murder and robbery by a woman in need of
money. Maybe she did rob and kill the maid. Maybe Old Fleming was too
feeble or drunk to play a part. As Albert Einstein and Werner Heisenberg

said to each other when they faced the debate on the "uncertainty" involved in locating atomic particles, only "the dear Lord" knows for sure.

The controversy surrounding the Sandyford murder trial hinged on
the Crown's use of Old Fleming to prosecute Mrs. McLachlan, a tradition
of immunity that, in such a case, made no sense to many in the public
and indeed a host of sympathetic members of Parliament in the House of
Commons. What might have made more sense at the time was that other
great quirk of Scots Law, the option of a jury declaring "not proven,"
which had the same practical effect as "not guilty." The suspect went free.
As a scholar of Scots Law has said, "Not Proven is appropriate when there
is grave suspicion but evidence not sufficient or convincing for conviction."[55]

Beyond the courtroom, the phrase "not proven" has become useful
for talking about the ambiguities in life—or even the quest for knowledge of God in the Gifford Lectures. As one Scottish thinker would say
in his Giffords, "It is quite impossible, I believe, to refute theism. A verdict
of 'not proven' is easier to obtain, largely because proof is so difficult and
its standards so exacting."[56] This was a verdict made by, in fact, a philosophical freethinker and skeptic whose research for the Giffords nevertheless produced in him an affection for the probability of the existence
of God. If plausibility is the test, he said, then theism is far more plausible
than most other metaphysical conclusions about life.

As the anthropologist James G. Frazer had obviously enjoyed saying in
his great work *The Golden Bough,* the sun was setting, the ship's sails were
furled, and a "long voyage of discovery" was over. At the beginning of the
twenty-first century, the Gifford Lectures had made an equally long and
romantic journey, leaving behind ages that are hard to remember: the age
when philosophical idealism battled materialism; the age when the natural sciences, like young thoroughbreds, set out to conquer the world; the
age of wars and the flight from reason to the God of revelation; and finally—and easier to remember—the contemporary age of technology,
litigation, and religious pluralism.

With all of those ages to remember, it is easy to overlook just how
often Glasgow, that city of commerce and Calvinism, had been the scene
of such cosmic considerations. In the mid-1700s, the age of the Scottish

Enlightenment, the cleric and philosopher Thomas Reid, the father of "common sense" philosophy, came to Glasgow and said theism was more than plausible. Belief in God was the basic intuition of humanity, he said, a kind of innate belief that need not be questioned. By his success at the University of Aberdeen, Reid was invited to be the chief philosopher at the University of Glasgow, where he continued his argument for belief in the God of the Bible and the self-evident truths of science, morality, and human relationships.

In some ways, the Reid of old Glasgow is the polar opposite of the 2001 Gifford forum in that city so many years later. Although Reid never set out to prove God, which he deemed an ultimate mystery, he boiled all other knowledge down to an innate common sense shared by all normal people. People know God exists in the same way they know other people have minds, Reid said. They just know it. Case closed. At the Gifford forum in 2001, there was a sympathetic hearing for this view, but it was just one view among many. For the prevailing theme was the limits of knowledge and, therefore, the final ambiguity of life.

Even the citizens of the twenty-first century, however, do not prefer ambiguity when they can have clarity, and so the quest continues. In the Giffords, the quest traveled through philosophy, anthropology, psychology, physics, sociology, and the science of history into an even more complex future. At such times, Reid's simple premise can be quite refreshing, as it long has been for religious believers in America, and that premise of "self-evident" truth is worth considering in this last act of the Gifford legacy.

ELEVEN

THE MEASURE *of* GOD

Scotland, America, and the Giffords

A CERTAIN MYSTIQUE flowed from the Gifford Lectures, like a river out of Scotland. The mystique was enhanced by many lofty titles, appellations such as *Mystery of Being, Process and Reality,* and *Space, Time and Deity.* For an American view of the Gifford legacy, however, one subtitle of the 1990s seems custom-made. The subject was the eighteenth-century Scottish philosopher Thomas Reid and the subheading, "Mr. Reid Returns to Scotland."

Among the things that most Americans know about the Scots—the surnames, kilts, castles, bagpipes, whiskey, and ballads—neither Reid nor the Scottish Enlightenment is prominent. Yet Reid's ideas had come to America at about the time of its founding and had probably made religious belief there easier to maintain. His "return" to Scotland, according to Yale University philosopher Nicholas Wolterstorff, suggested how much secular Europe had lost Reid's legacy, even as that influence had continued in the United States.

Responding to David Hume's skepticism, Reid produced a "common sense" philosophy, once called the Scottish philosophy, urging people to trust their everyday senses that the world was real and that a Creator had put its physical and moral order in place. Such Reidian notions as "self-evident" truth *may* have found their way into the Declaration of Independence. Now, in the twenty-first century, Reid's forgotten work may give America—a nation that shuns the word *metaphysics*—an appreciative window on the Gifford legacy.

Back in Reid's day, when a skeptical Hume made his case against proofs for God's existence, the Enlightenment's full-throated espousal of atheism had actually died down. Reid did not think piety was in much trouble. "To a thinking man," he said, "there must appear [so much of divine] design in the universe, that all the objections to it cannot even make it dowbtfule."[1] So Reid focused his energy on Hume's skepticism about the mind knowing what is ultimately real. By "common sense" Reid did not mean opinion on the street, although he felt "vulgar" hearsay was often as reliable as a philosopher's speculations. By "common" Reid meant a God-given ability, an innate quality of the mind to accurately judge plain things, large and small, from business dealings and morals to God's existence.

Once the Creator was acknowledged, Reid said, there was no use speculating endlessly. The goal was to *do* something, as Wolterstorff explained in his 1995 Giffords at St. Andrews. "It is evidently the intention of our Maker, that man should be an active, and not merely a speculative being," Reid said. "Our business is to manage these powers, by proposing to ourselves the best ends, planning the most proper system of conduct that is in our power, and executing it with industry and zeal."[2]

At first blush, Reid's philosophy squares well with the American temperament. Common sense ideas gave early American philosophy, nascent as it was, a fertile soil for theistic assumptions and practical applications. Reid's confidence that belief and so much else was obvious was not in the air a century later, however, when Lord Gifford looked out at Scottish society. Gifford thought that being busy was fine, but that more speculation on God was only for the better. They were two believing Scotsmen holding two different views about the inquiring mind.

By the time of Reid's death in 1796, his influence in the United States was undisputed. Just what impact common sense had, however, is at

issue in three vying interpretations. The first says that Reid was the
source of the "self-evident" truth idea in the Declaration of Independence, a founding political assumption that God's existence was beyond
question. A second view sees Reid's common sense bolstering Bible
fundamentalism, a unique intensity of American Protestantism, which
fought for the face-value meaning of texts. The third view moves Reid's
influence north to Harvard in the late 1800s. There, American pragmatism was born, nurtured on Reid's idea that what works is true.

What some have called a Reid revival in more recent years is part of
a larger Scottish backdrop in America, a story that stretches back two
centuries. The American founding has long been credited to the great
English figure John Locke, who argued for liberty and religious tolerance. Although Wolterstorff is an enthusiastic Reid scholar, he agrees that
Locke was the mainspring of colonial ideas about political liberty, reasonable Christianity, and religious tolerance. Yet the Scottish influence has
grown in historical hindsight. When the clash with England began, not
only was Scotland an English-speaking ally, but the colonies were filled
with Scottish immigrants. The Scottish Enlightenment of which Reid
was a member was also boiling over with political theories, scientific
methods, and proposals on education, religion, and morals just waiting to
be tried out.

That story began with the founder of the Scottish Enlightenment,
Francis Hutcheson, the principal moral philosopher at Glasgow University. He was a lighthouse among "moderate" or "new light" thinkers in
Calvinist Scotland, not quoting Scripture at every turn but arguing an
Enlightenment sort of Christianity, using Newton and philosophy to explain the proper roles of natural science, morals, liberty, commerce, and
governance. Along with Locke, Hutcheson spoke of some first principles
about the world that were self-evident, known by the light of reason.

Hutcheson had made an early argument for why a colony may secede
from a sovereign and become a friendly state, and he used such suggestive
words as "consent of the people," echoed roughly in the Declaration. His
ideas were lectured on in American colleges and read by some American
founders, suggesting more significance than once believed, says historian
William Brock. "There seems to be a good ground for raising the stock

of Hutcheson as a source of American political theory, though not perhaps for lowering that of Locke," he said.[3]

David Hume and Thomas Reid came soon after Hutcheson, two men
born on April 26 but a year apart. Hume's writings on political liberty and
religious skepticism were known in the colonies, for he was a great ally of
American independence, but Reid's works, about two decades behind
Hume's outpouring, were less widely circulated. He was known mostly by
secondhand accounts or through common sense disciples who came over
from Scotland. When the American propagandist and patriot Thomas
Paine needed a name for his fateful revolutionary pamphlet, the Scottish
doctor and signer of the Declaration Benjamin Rush, it is said, advised the
title "common sense." Whatever Jefferson's source when he wrote the Declaration in 1776, "It is very probably from Reid that he borrowed the idea
of 'self-evident truths,' " says the historian Arthur Herman.[4]

At that early stage, common sense had filtered into the colonies by
way of Reid's disciples. Chief among them was the Scottish Presbyterian
clergyman John Witherspoon, the sixth president of the College of New
Jersey, now Princeton. Back in Scotland, Witherspoon had led the evangelical "Popular Party," which warred with the moderate "new light"
clergy. The moderates viewed Hutcheson as a guiding light, and on
occasion called Hume an ally against the Calvinists. It was Witherspoon's
losing battle in moderate Edinburgh that persuaded him in 1768 to come
to America, where an evangelical revival had transpired. At Princeton he
educated more than one hundred ministers, thirteen college presidents,
many future professors, twenty U.S. senators, a vice president, and James
Madison, a future president.

Who actually put Scottish fingerprints on the U.S. Constitution—
Hume or Witherspoon—is up for debate. The young Madison, "the
father of the Constitution," got his ideas from somewhere. He read
Hume but he had studied under Witherspoon. In his Federalist Paper
No. 10, Madison said that because "the nature of man" is to form factions, they should be put to good use. When a society encompasses "a
greater variety of parties and interests," it becomes "less probable that a
majority of the whole will have a common motive to invade the rights of
other citizens."[5] Historian Herman argued that Madison borrowed from

the Great Infidel, not the Calvinist cleric: "It was above all to Hume, Witherspoon's avowed nemesis, that Madison found himself drawn."[6]

Others give Witherspoon more weight. They suggest that Madison got his formative ideas at Princeton, turning the Scottish minister's teaching into an American political structure by way of the Constitution.[7] In his Princeton lectures, Witherspoon had espoused the faction model. "Every good form of government must be complex," he said. It "must be so balanced, that when every one draws to his own interest or inclination, there must be a poise upon the whole." To achieve a final balance of conflicting interests, Witherspoon said, the balance must have a center. "There must be always some *nexus imperii,* something to make one of them necessary to the other."[8] He was speaking of a division of the ruling powers into different branches.

The first two Americans to give the Giffords, the Harvard philosophers Josiah Royce and William James, traveled to Scotland around 1900, 124 years after the Declaration. They were products of the success of the U.S. system of government, with its concept of God above and factions below. Royce came from an evangelical environment, lived in California, studied in Europe, and finally became enamored of German idealism and American pragmatism. James was reared in a household of Swedenborgian spiritualism and then went headlong into materialist sciences. He finally decided that individualist religious experience, whatever the "over-belief" about God, was what made human nature so alive.

Both Royce and James were products of the American design, a freedom of thinking with something arching overhead, which has generally been the idea of God. The Declaration and the Constitution created this intellectual hothouse. God was above and freedom of conscience, and of the mind, was down below, a kind of metaphysical *nexus imperii* that kept a nation together. This "one and the many" is an age-old philosophical problem, of course, worked out in America with the best ideas of the time. In a way, the Gifford bequest is a kindred spirit, with things of God and the Infinite as the common purpose. The ways of arriving there, however, were as diverse as could possibly be imagined. Although Americans have never made much hay over having "intellectuals," it was such a country that produced a William James and a Josiah Royce, something new in the world. But it was the Gifford Lectures that gave them a high place to expound.

At about the time of Royce and James, Reid's influence may have been reborn in America, but now in the form of pragmatism. This is the thesis preferred by Wolterstorff, the Gifford lecturer. William James is widely considered the founder of pragmatism, but as a man who always gave credit, he himself pointed to his colleague Charles Peirce, the brilliant but eccentric philosopher, for its first articulation. "This is the principle of Peirce, the principle of pragmatism," James testified to his Edinburgh audience in 1902.[9] The son of a Harvard mathematician, Peirce was a classmate of James. Peirce was also the man who invented the name "Metaphysical Club" for the informal meetings he and his speculative peers held on occasion at Peirce's or James's homes. "Charles Peirce knew Reid quite well," Wolterstorff said. "So my own view is that American pragmatism is the successor of Scots common sense philosophy."[10]

In the late nineteenth century, Peirce was enamored of the new mathematical universe, a world not of machinelike determinism but of probability and statistics and what Peirce called "the law of errors"—the impossibility of precision and the need to average and compensate. He spent his life asking how the mind can know what is exactly real. In such a world, he decided, what better way to determine truth than to test it by experience. Peirce first proposed to define truth as the practical worth of an idea.

More than anyone else at the time, Peirce had digested Reid's available writings. Reid had been asking the same kinds of questions in the eighteenth century, using mathematics and physics as his tools, but backing them up with the Bible, with its God of design and reason but also of the deepest mystery. Whereas William James would touch on Reid, or use the words *common sense,* Peirce extensively showed his indebtedness to the Scottish thinker. Reid's ideas were "reborn once more in the pragmaticism" of Peirce, who called it "critical common-sensism."[11]

Peirce did not take Reid's philosophy intact, however. As a whole package, Reid's philosophy had God above, so to speak, and down on earth the truth of a common sense in human life was verified by both reason and application. God was the Author of both, and so what was reasonable was also fruitful and successful in the world, providing a kind of proof that God had put reason and common sense application in harmony with each other. Peirce did not draw any such theological implications, though he was more comfortable than most philosophers of his day in speaking of God. For Peirce, the age of science, Darwinism, probability, and much else had undercut the certainty that reason could always be proved in practice, or vice versa. Hence, Peirce took only one part of Reid's common sense—the part about application verifying the truth—and from this he produced the theory of pragmatism. In a word, pragmatism said that experience tested and proved what is true.

It could be said that pragmatism became *the* American philosophy, for nothing works like success. The high rates of religious belief in the United States may certainly go back to "nature and nature's God" in the Declaration, the separation of powers including church and state, or confidence in a literal Bible. A case has also been made, however, that the pragmatic temperament of American culture has guaranteed that religious ideas would flourish: religious belief works so it is taken as true. A belief in God as the higher authority, for example, has been a brake on tyranny. The value of the individual before God has buttressed individualism and populism. Belief in a chosen nation has helped in all kinds of American adventures. Finally, the idea that every person has equal access to the mind of God has been a very strong engine for democratic opinion. Modern studies have shown, meanwhile, that the world's people become *more* religious-minded when they immigrate to the United States.

The part of Reid that Peirce had left behind—that God would not deceive us—was also abandoned by modern philosophy. The guarantee of God was seen as just so much "dogmatism."[12] In the 1970s, when Reid was "coming into modest vogue" for his more scientific ideas of perception and common language, the Harvard philosopher Hillary Putnam explained the enduring dogmatism problem. "Reid's followers, even more than Reid, tended to 'solve' philosophical problems the easy way, by taking to be innate whatever they needed for their purposes—the existence of God, the immortality of the soul, free will, or whatever."[13] Still, when Putnam gave his Giffords in 1990 at St. Andrews, he was far more amenable to religion playing a role in philosophy, for he had rediscovered the value of his own Jewish heritage.

Although Reid was a kind of scientific genius of his age, having taught math and physics for years at Aberdeen, he was also a clergyman. He served a parish for about two decades before moving to Glasgow to teach full-time. Yet Reid was part of the Enlightenment project, which wanted to reconcile science and religion. He did not simply quote the Bible. Reid made his case about perception and psychology by presenting it in hundreds of pages of careful analysis. He might have been surprised, in fact, that even though he was not a Bible scholar, his scientific logic and his psychology would have a profound effect on American approaches to the Bible a century later.

Bible controversies had sputtered in America from an early time. Jefferson produced his own pick-and-choose Bible to keep the morals of Jesus while expunging supernatural events and ancient cruelties. Witherspoon, in a best-selling sermon that might be hard to preach with such certainty today, argued that the God of the Bible presided over the American Revolution. Off and on at Princeton, common sense would be commandeered to make evangelical claims for a literal truth in the Bible. Certainty about the Bible met its most monumental test during the Civil War, when both sides quoted Scripture. "Both read the same Bible and pray to the same God," said Abraham Lincoln, "and each invokes His aid against the other."

The ranks of professionals in biblical studies were beginning to grow in America, represented by the founding of the Society of Biblical Literature in 1880, when a furious controversy in Scotland reverberated across the Atlantic. At a church college in Aberdeen, the Old Testament scholar William Robertson Smith was submitted to a five-year investigation for using historical criticism on the Bible. Although he was found not guilty of heresy, he was dismissed in 1881. He had tried to strike an evangelical balance, using the new science on texts while also affirming the supernatural, miracles, and the divine inspiration of Bible writers.

The German scholarship used by Smith had already reached the United States, but his case brought the debate into the open there, in particular among Presbyterians. As an orthodox Presbyterian wrote the editor of *Presbyterian Review,* "discussion [of the Smith case] is demanded and would make the *Review* famous."[14] So for the next two years, orthodox and modernist Bible scholars escalated the debate, with the vying *Review* articles beginning in April 1881. Both groups said that science was on their side. The conservatives drew upon common sense science, arguing that the Bible gave plain facts and God would not deceive human understanding. The liberals adopted the new science, with its notions of historical development, hypothesis, and testing for error and uncertainty.

The leading conservative figure was Benjamin B. Warfield, a theologian at Princeton and its future president, while the liberal banner was carried by Charles A. Briggs, who in a decade would fill the chair of biblical theology at Union Theological Seminary in New York City. Briggs would also come to face heresy charges after his inaugural lecture at Union, "The Authority of Holy Scripture," which located authority also in scholarship, church tradition, and experience. The Presbyterian denomination suspended him from ministry, an opening shot in the widening American war over biblical interpretation. While Presbyterians, Anglicans, and others in Britain worked out a liberal-conservative coexistence, churches in the United States broke apart. The breach was started by the conservative Bible scholars of the 1880s and 1890s, but it was consummated by their successors, the founders of American fundamentalism.

Bible fundamentalism in essence opposed projects such as the Gifford Lectures. During their first century, Anglicans, Methodists, Lutherans, Presbyterians, Baptists, Congregationalists, Unitarians, and Catholics in Europe all contributed to the metaphysical search for God. But few if any American evangelicals were interested. Only modernist Protestants were willing to mingle science, philosophy, and biblical tradition. Since the 1970s, however, some orthodox Protestants have employed metaphysical philosophy in hopes of reviving a Christian natural theology, arguing that creeds and Bible texts alone will not sustain belief in a scientific age.

Some of the credit goes to Thomas Reid. A generation of theistic believers has returned to his common sense, or more broadly, the argument that the external world—describable by science, philosophy, and metaphysics—is a reliable testimony of an intelligent Creator. In this respect,

North America would make an unexpected contribution to the Gifford legacy. It was with some daring, therefore, that Wolterstorff went to give the St. Andrews Giffords with the long-forgotten Thomas Reid under his arm. The Europeans might think it was American fundamentalism. Still, the exchange of ideas between Americans and Europeans is not as one-sided as it used to be.

When James arrived there, for example, he spoke of Edinburgh as soil "sacred" to the American imagination. He confessed to "trepidation" at addressing well-educated Europeans. "It seems the natural thing for us [Americans] to listen whilst the Europeans talk," he said. "The contrary habit, of talking whilst the Europeans listen, we have not yet acquired; and in him who first makes the adventure it begets a certain sense of apology being due for so presumptuous an act."[15]

The religious vibrancy of the United States, the maturing of its intellectual traditions, plus a shift in world power, have changed the American contribution to the Giffords. Of the nearly eighty appointments before the Second World War, eight hailed from American universities (and two were émigrés). Since the war, the number of lecturers from the United States has been around forty. The infusion has brought the more traditional questions of natural theology back into the foreground. When Aberdeen University hosted the first-ever Gifford conference on natural theology, welcoming academic papers, the North American response was dramatic. "We had an unbelievably high number of submissions," said Gordon Graham, the regius professor of philosophy. "So it was hugely successful."[16]

A feature of the "new" natural theology is a revival of realism, the idea that the external world reliably reveals a cosmos created by God. It is a tradition going back to Aristotle and Thomas Aquinas that runs through English natural theology, Reid, and the Scottish Enlightenment. A growing interest in God-accepting realism has been central to the story of science and religion over a century; it has revealed itself across the four acts of the Gifford legacy—from the battle of idealism with materialism, to the new sciences, the revolt against reason, and finally the new age of pluralism and technology.

Thomas Reid has always represented a kind of God-believing realism of the past, and a new interest in his work is therefore telling. Yet in many ways, he has been written out of history. The traditional story is that the realism of medieval times—God created an external world that reflected his nature—was undercut by the materialism of science and the skepticism of people such as Hume. The reaction to skepticism came in the form of German idealism, which then set the new trajectory of Western philosophy. What was left out, however, was Reid's answer to skepticism and materialism. If the victors write the history, Wolterstorff said, then philosophy books have portrayed all thought as jumping from Hume to Hegel, with no mention of Reid, who had apparently disappeared for a century.

The new interest in a realism that includes God has dragged Reid out of obscurity and revived the study of many of his arguments. Realism, which includes the Reid revival, is gaining popularity as an antidote to the thing that makes rational debate about God impossible, the nihilism, deconstruction, and subjectivity that pervades modern thought. As a result, Reid has returned to Scotland in more than a few Gifford Lectures. Two such American appointees, in fact, hailed from the same school, the University of Notre Dame, one being a Catholic thinker and the other a Protestant philosopher.

For years, Ralph McInerny was the resident expert on Thomas Aquinas at Notre Dame. He was also widely known for his Father Brown mystery novels, many of which were adapted for television. McInerny headed the Jacques Maritain Center, named for the French neo-Thomist who missed his Giffords because of diplomatic duties after the war. McInerny arrived at Glasgow for the 1999–2000 lectures, dedicated to showing the en-during merits of natural theology. His allies in this were Aristotle and Aquinas, of course. But on two occasions he brought in a Calvinist such as Thomas Reid to make the same kinds of points.

When people hear Reid's common sense assertions, "They know he is right," McInerny told his audience. Ordinary people like to hear that ordinary folk have a grasp on reality. Thomas Aquinas also gave much credit to ordinary people's ability to understand a universe that had an ultimate cause and an apparent purpose. With Aquinas, "the amazing assumption is that everybody already knows all sorts of things," McInerny said. "Thomas Aquinas emphasized that it is from truths known to all that philosophical thinking too takes its origin."[17]

McInerny had only one "Thomistic regret" with the other Thomas, Thomas Reid, that is, and that was the Scotsman's failure to make God's existence a point of rational argument. According to Wolterstorff, Reid said God was the ultimate dark mystery, and indeed God had made things that way. In McInerny's words, Reid had thus failed to "make the *very* first principle stand out from" other principles of common sense. Aristotle's first principle was that men would disagree on first principles, and Thomas Aquinas similarly argued that because some propositions are more valid than others, rational dispute may follow. "Once that has been established, natural theology becomes a possibility," McInerny said. "A disagreement between the theist and atheist is possible, since one of them is right and the other is wrong."[18]

The other Notre Dame professor who brought Reid back to Scotland was Alvin Plantinga, a noted philosopher in the Reformed, or Calvinist, tradition. In 1982 he had moved down to Notre Dame from the Protestant enclave of Calvin College to the north. Under the likes of McInerny, Plantinga, and others, Notre Dame earned a reputation as a training ground for young theistic philosophers, both Catholic and Protestant.

After tasting secular philosophy at Harvard, and then studying its Christian forms at Calvin, the young Plantinga aired his first comprehensive argument for God in 1967 in a book titled *God and Other Minds*. He expanded on Reid's idea that recognizing God was no different from recognizing a human mind. Both were "properly basic" as a human belief requiring no justification. "Since belief in other minds is clearly rational, the same goes for belief in God," Plantinga summarized.[19] But as he knew well, philosophers were now attacking rationality itself.

In Scotland today, Plantinga and Richard Swinburne, who once held a chair in Christian philosophy at Oxford, have been credited with making rational theism a vibrant new field. Plantinga would become the only American to deliver the Giffords twice, first at Aberdeen in 1987, and then at St. Andrews in the next century. He told his Aberdeen audience, which came out faithfully at the dinner hour, that speaking on Thomas Reid "only a few miles from his birthplace, and the scene of his early work" was a pleasure indeed.[20] A North Sea oil worker and his wife came to hear his American accent, but Plantinga's topic was not easy in any dialect. He tried to show that even though Western thinkers had stirred up much skepticism, "properly basic" beliefs were still a good philosophical argument.

In all his work, Plantinga was mindful of the arguments made by John Calvin, the French lawyer and religious reformer who, like Reid, had argued that God was an ordinary belief. Calvin did not use the plain words *common sense,* but rather spoke in ornate Latin of a *sensus divinitatis,* a "sense of deity inscribed in the hearts of all." On a list of normal human beliefs, Plantinga said, "certain beliefs about God are also properly basic." Other proper beliefs are that external objects are real, memory is true, reason is effective, and people's testimonies can for the most part be trusted. "Reformed thinkers such as John Calvin have held that God has implanted in us a tendency . . . toward accepting belief in God under certain widely realized conditions," he said.[21]

Over the years, Plantinga has become well known for giving more backbone to theism by his evolutionary "argument against metaphysical naturalism." Naturalism, which rejects all supernatural gods, spirits, or minds in the universe, had been given much of its own modern spine by Darwinism, which claimed that unguided evolution formed all things, including the mind. There seemed to be no longer a role, Darwin suggested, for a Creator in a universe whose "fixed laws" produced no more design or purpose than "the course which the wind blows."[22]

Although Reid never heard of Darwin, common sense arguments remain pertinent to the Darwinian challenge, Plantinga argues. If the mind developed solely for survival value—getting food, fending off predators, and reproducing—as Darwin suggested, where did art, philosophy, mathematics, and science come from? Even Darwin wondered how a mind arising from apes could be trustworthy. For Plantinga, himself a philosopher who is dubious about the scientific claims of Darwinian evolution, strict naturalism cannot guarantee a reliable mind. Reid had a plausible solution. A person could presume that God created the mind to know the world reasonably well. Plantinga's Gifford Lectures suggested that this had been exactly the historical case with science. For confidence about knowledge, modern science should not rely on the Darwinian rise of ape minds, but "prefer theism to metaphysical naturalism" as a view of reality.[23] And if there is theism, Plantinga would conclude, then something besides materialist evolution could explain the origins and designs of life.

The Gifford Lectures pose a final question: amid this sampling of Scottish influence on the United States—from Hutcheson to Hume, Reid, Witherspoon, and Robertson Smith—where might Adam Gifford rank?

Even in Scotland, the Gifford name in regard to the lectures is not exactly common knowledge. Americans in general may know even less; the clergy as well as some philosophers and scientists in the United States, as in Scotland, are perhaps the largest tier of cognoscenti—those who would recognize the Gifford moniker. In terms of carrying out the Gifford bequest, the lectures have been remarkably regular in the best and worst of times. The 1980s was the most productive decade, with about thirty series delivered at the four schools, followed by the 1990s (with twenty-five). Yet even in the 1930s, the decade of the Depression, twenty sets of lectures were presented, the next best tally.

With such modern distractions as television and paperbacks, some might have argued that the days of voluntary lecture attendance were over. The evidence has been mixed, though the best and worst cases can always be cited. The grand-opening Gifford lecture, delivered by Friedrich Max Müller inside Glasgow University's massive Bute Hall in November 1888, drew 1,400 listeners (according to family memoirs). On the other extreme, one thinks of how almost nobody showed up to hear Alfred North Whitehead complete his difficult talks. When the liberal Catholic theologian Hans Küng canceled his 1974–75 appointment at St. Andrews on "short notice," perhaps because of his growing conflict with the Vatican, his replacement was Dutch historian of science and religion Reijer Hooykaas, who would suffer "dwindling attendance" at his eight lectures in November of 1976. "The Lecturer himself is depressed, and I fear that the series may be cut short unless numbers increase," according to a memo.[24]

Yet there is no proof that ordinary Gifford attendance has not grown with modern times, in spite of the distractions, organizers say. "We don't know the attendance for Giffords in the past, whether in the nineteen-twenties or the fifties," said New College historian Stewart J. Brown.[25] In many cases, a growing attendance has been reported, apparently boosted by wider educational interests, modern techniques of publicity, and the shortening of the series to ten and now six lectures. The audience tends to be retirees, some students, and professionals downtown who can make the 5:15 P.M. events—a time nearly carved in stone by tradition at all four schools. As each era's loyal following of the Giffords passes with time, another is expected to emerge. "I would hope that twenty years from now there would still be a good public constituency," says Brown. "It will just be a slightly different aging group; including myself."

In this sense, the Gifford Lectures seem to be a perpetual motion machine, as long as ordinary people wonder, professional thinkers are ambitious to talk and publish, and the bequests at the four universities keep up a minimum of funds, as they have, despite a century that included economic depressions and two world wars.

In Edinburgh, organizers of the Giffords are playing on the strengths of the city and the university. In more recent years, for example, the Gifford Lectures have moved to the famed Playfair Library at the Old College, a large hall that is one of the most beautiful examples of interior architecture in Scotland. While the eye is pleased by the Playfair's high, ornate ceiling, the ear can hear the lectures by way of a modern sound system—a small but important revolution in the Gifford legacy. More consequential even than good sound, the new cultural role that Edinburgh is playing may increase the sway of the Gifford Lectures.

That cultural shift began after the Second World War. The war had left many of Europe's most beautiful cities in ruins, including Munich and Salzburg, which had long been hosts to an international arts festival. Now that municipal role was wide open, and Edinburgh decided to compete for the honors. In 1947 Edinburgh launched its own arts festival, which today is the world's largest—and Scotland's biggest tourism event. Gifford organizers cannot help but wonder how to dovetail their event with the cultural explosion each year, a sort of dream come true in light of Gifford's wish "that the lectures shall be public and popular, that is, open not only to students of the Universities, but to the whole community."

Although nothing would have pleased Gifford more than to see the lectures ascend to festival-like popularity, they are inevitably filled with difficult material for the general public. Yet hope springs eternal, even after a century of the Gifford legacy. What may become more popular is the simpler import of the lectures, and that is the name of Gifford himself, his story, and the entire drama of the Western mind trying to understand God. On one hand, Gifford's name stands out as one of many Scottish contributions to transatlantic goodwill. He dared the world to find God by way of science, reason, and experience—by way of natural theology. To the extent that natural theology is having a revival in the West—among both the experts and the learned reading public—Gifford has also left an important legacy.

Before and after Gifford, the story of natural theology in the West seemed to be one of rise and fall, thrust and parry on a grand scale. First came the philosophical proofs of God so beloved of medieval theology. These were eventually joined by the so-called scientific ones of British natural theology. Then science destroyed all such allusions to the "God hypothesis," saying they were no longer necessary. Slowly but surely, thanks often to the Gifford Lectures, the many and sundry reasons why science did not necessarily demand atheism were brought to the fore. As science became a more complex and more questioned pursuit, a reversal seemed to take place. The once-orphaned child called philosophical theology was adopted and given a new home. It grew strong and built up arguments for making sense of the universe with a God hypothesis. Metaphysics died but was reborn, no doubt to face ever-new scientific and skeptical challenges.

For natural-theology boosters who herald a new age, a time when reason and scientific evidence can lead a confused Western world to some basic conclusions about the existence of God, Adam Gifford would certainly make a fitting patron saint.

The previous herculean attempt to merge Christian philosophy with the science of the ages was the Renaissance, when medieval culture wanted to build bridges between faith and reason. The best Renaissance minds of the fifteenth and sixteenth centuries turned to the Greek legacy in philosophy and science. In the decades before the Reformation and the Enlightenment rocked the so-called medieval synthesis, the church was obviously excited about uniting the heavenly and the earthly.

Beginning in 1509, the painter Raphael Sanzio took up this theme for Pope Julius II, who wanted the Papal Palace library, the Stanza della Segnatura, at the Vatican covered with beautiful murals. Raphael labored on the wet plaster for three years, and the result was *The School of Athens.* Art historians call it the culminating revolution in art materials, realistic drawing, and Christian homage to Greece. Raphael probably drew inspiration from Dante's fourth book of the *Inferno,* which described how the ancient philosophers were spared hell but were still found in a dim antechamber. They had lacked the final truth of Christianity, according to Dante, but they had at least lived by noble philosophical principles.

Raphael dispersed the darkness and portrayed the ancients amid the brightness of architecture, statues, and dramatic open spaces. Gathered in a great hall, which seems to recede to infinity, the figures of thinking men, stooped in conversations, languid on the marble steps, pointing at important documents, fill the entire panorama. At the center Plato, the Idealist, points upward with his right hand, as if to the transcendent things, and Aristotle, the Realist, reaches out toward the things of the earth. Around them bustle other great names. Heraclitus, Zeno, Pythagoras, Anaximander, and Socrates are gathered here and there, while Epicurus, Parmenides, Diogenes, Ptolemy, Euclid, Zoroaster, and Averroës have also made the scene in colorful Renaissance-style togas.

When Lord Gifford created his bequest, he also would create a sort of school of Athens. It is not too strained an analogy. Edinburgh, surrounded by the three other ancient universities, was after all the "Athens of the North" for a time. Over the 117 years of the Gifford Lectures, its participants would hardly have made Dante's list of the truly pious either, even if they had the benefit of Christianity all around them. What Gifford had done, very much like Raphael, was suspend doctrinal judgment in appreciation of sincere thinking itself. Like Raphael, Gifford brought earnest inquirers out of a purgatorial dimness and into the open light of speculative yet reverent, contentious yet fascinating, talk about God. They took the measure of God, but without the hubris that might entail. As Gifford had said, it could only be for the good.

In his master composition, Raphael crowded 40 figures into the *School of Athens,* whereas a mural of the Gifford legacy would, if taken literally, need to encompass 220 souls, spread with equal ease and drama across an idyllic landscape. Raphael put Aristotle and Plato at center stage, but a Gifford Lectures mural would have to presume that two millenia of philosophical debate and science had interevened since their heyday in Ancient Greece. The mural could certainly be a panorama of buildings, landscapes, open spaces, and nooks from any romantic age it chose, much in the Renaissance manner of composition. A few Scottish castles would be apropos, and perhaps the towers of some old Scottish university quadrangles.

The people in the landscape, however, would not begin to appear until 1888, as if opening a four-act drama of God and science for the next century. It began with the clash of idealist philosophy and materialism. Then came the dramatic rise of the natural sciences—anthropology, psychology, physics, sociology, and the science of history. The reaction at midcentury came in the "subjectivism" of revelation and existentialism, and the century finished out in a flurry of diverse ideas that were new soil for taking the measure of God.

In such a painting of the Scottish "school of Athens," the foreground might include a large contingent of idealists, the Caird brothers, Bernard Bosanquet, Andrew (Seth) Pringle-Pattison, and Royce, with his large head and red hair. The anthropologists would be crowding in nearby, with such men as Friedrich Max Müller, Edward B. Tylor, James G. Frazer, and Andrew Lang prominent among them, added to by later comers in the field, such as Mary Douglas. All the Gifford philosophers were not idealists, of course, and those who came later included John Dewey, Frederick Copleston, Gabriel Marcel, Iris Murdoch, and Alvin Plantinga, surrounded by so many others standing here and talking there.

Somewhere on a flight of marble steps, surrounded by discussants, would be the famous psychologist William James, his eyes sparking blue and his beard grown long and gray. Many of his coterie were friends: James Ward of Cambridge, Conwy Lloyd Morgan, and, from France, Emile Boutroux and Henri Bergson. The vitalists and emergentists stand nearby, the likes of Samuel Alexander, the Haldanes, and Whitehead.

Across the way, a tall, bald figure stands out. He is Reinhold Niebuhr. He is conversing with an old rival, the stocky and confident Karl Barth, and joining them are other Protestant theologians of their common era, including Paul Tillich and Rudolf Bultmann. All of them pay a kind of deference to the old man himself, Albert Schweitzer, who stands large, white-haired, and somewhat hunched in their midst. The theologians number as many as the philosophers. Among them are two archbishops, William Temple and Nathan Söderblom, and also the thinkers and writers Emil Brunner, Charles Raven, John Baillie, Eric Mascall, Richard Swinburne, John Macquarrie, John Hick, Raimundo Panikkar, Keith Ward, and Stanley Hauerwas.

Not too far away is a billowing tree, and in its shade have gathered the historians. These scions of history are legion, many of them experts in the natural theologies of ancient Greece and Rome and finally of early and medieval Christianity. The tall figure is Arnold Toynbee, but standing there also are Etienne Gilson, Robert Zaehner, Christopher Dawson, Owen Chadwick, and Herbert Butterfield.

At a great building nearby, the portico has been occupied by lecturers in the hard sciences, with Niels Bohr and Arthur Eddington as the two stately older figures. Nearby are Werner Heisenberg and Carl Sagan, biologists such as Alister Hardy, and two neuroscientists with Nobel Prizes, Charles Sherrington and John Eccles. Others include Michael Polanyi, Carl Friedrich von Weizsäcker, Stanley Jaki, Freeman Dyson, Ian Barbour, John Barrow, and John Polkinghorne. Cambridge's George G. Stokes, the first scientist to give the Giffords, wears his long Victorian coat and collar, a balding man with great sideburns. Not so remarkably, many of these people hold company here and there with the freethinkers, skeptics, and atheists. Gifford deemed them just as earnest and sincere, and among that company stand materialist philosopher Alfred J. Ayer and atheist scholar Anthony Flew—who ended up a theist, more or less.

In a corner somewhere is Alasdair MacIntyre, the probing critic who said the Gifford Lectures were a "magnificent array of fundamental and unresolved disagreements, a kind of museum of intellectual conflict."[26] He said they resolved nothing. The Bible fundamentalists, in turn, would warn their young people away from reading such metaphysical confusion. Yet, in some significant number, many Gifford appointees were reared in strict religious homes, where piety and diligence were rewarded and mental laziness disowned. Whatever the confusion, everyone in the mural played a role in holding up the Gifford theme, as if it were the sky above, the *nexus imperii,* a kind of divine canopy that sheltered the whole enterprise.

Gifford himself is hard to find in the mural. But finally there he is, off in the background. He is skating on a frozen lake, his long dark locks flowing in the breeze. He has given away all his money, but he is clearly happy. He is back with his adoring wife and joined by their son. His limbs are agile and free, and he moves swiftly over the ice, swirling at times, then on again through the cold, the air, and whole atmosphere of his Scotland, as if indeed he had more of God. If a recollection of great paintings in history is the way to end this story of the Gifford Lectures, it is not a "school of Athens" mural in which Adam Gifford might be most at home, but a Scottish painting of a minister gliding across the winter ice.

When the world visits Edinburgh, one of its most popular destinations is the National Gallery of Scotland, where hangs the famous painting *Reverend Walker Skating on Duddingston Loch.* In 1997, when Scottish history changed, the painting still had its resonating power. That year, Scotland voted to restore its own parliament, which had been dissolved in the Union of 1707, the great political event that preceded the export of the Scottish Enlightenment. Although dissolved for 290 years, the parliament had left its traces in other ways. The old Parliament House was taken over by the courts, and became the workplace of Lord Gifford when he served the bench. At Parliament House, David Hume had landed a job as legal librarian, allowing him to write his histories and thus become rich and famous. In 1997, in fact, the city erected a large bronze statue of the seated Hume on the Old Town street that passes Parliament House.

That year also, plans were begun to build a new parliament building, a modern structure that would be located at the bottom of Old Town, only a block from Old Calton Cemetery, where Hume lies in his mausoleum and Lord Gifford rests in the earth under a pinkish marble stela. Until the new building's completion, the members of the Scottish parliament repaired to the great General Assembly Hall of the Presbyterian Church, which stood right behind the gothic towers of New College, scene to so many Gifford Lectures in the twentieth century. When the design of the new parliament house was approved, it had one particularly modernesque feature—and that was an attempt to abstract the themes of the *Reverend Walker Skating on Duddingston Loch* onto the building exterior. This was done in a most abstract way, a series of identical bent shapes that captured the feeling of Reverend Walker, who leans into the wind and glides on the ice—happy enough in a wintry world with a stern God above.

Scotland has had many favorite sons, but at the start of the twenty-first century, the great infidel Hume, the common sense Reid, the skating Reverend Walker, and Lord Gifford with his lectures all retain their particularly unique appeals. Not long after Lord Gifford died, an acquaintance wondered about the old judge's impact on the world. In his own life, Gifford had enjoyed the risks of "skating over thin ice, round awkward corners, and dangerous holes," the acquaintance wrote in 1889. Yet what would become of his lectures? "The name Adam Gifford is not likely to sink into oblivion in a century or two," he ventured. "It will probably survive a few reformations or at least revolutions in religion."[27]

Metaphysical Scotland is not what it used to be. But every Western nation has changed, becoming more secular, probably less enchanted than in times of yore. Yet Scotland keeps its endearing metaphysical images for the world, images of good-natured skeptics, dark-clothed ice skaters, and inquirers after God, all of whom live high on the globe, in a land of northern lights and short winter days. The winter afternoons of Scotland still cast very long shadows, much as the Gifford legacy has stretched over a century, and may go on still more.

APPENDIX

The Gifford Lectures, 1888 to 2005

This list dates Gifford Lectures by calendar year (1890) or academic sessions (1890–92). Published lectures are italicized. The academic position (which here omits names of chairs and other honors) is for the date of the lecture. The list is chronological and then alphabetical: Aberdeen, Edinburgh, Glasgow, and St. Andrews. (For further research on the Gifford lecturers see www.giffordlectures.org.)

Frederich Max Müller, 1888, 1890–92 at Glasgow. *Natural Religion, Physical Religion, Anthropological Religion,* and *Theosophy: Psychological Religion.* Professor of Comparative Philology, Oxford University. (1823–1900.)

James Hutchison Stirling, 1889 and 1890 at Edinburgh. *Philosophy and Theology.* Author and Philosopher. (1820–1909.)

Edward Burnett Tylor, 1889–91 at Aberdeen. Unpublished. Keeper of the University Museum and Reader in Anthropology, Oxford University. (1832–1917.)

Andrew Lang, 1889 and 1890 at St. Andrews. *The Making of Religion.* Fellow of Merton College, Oxford. (1844–1912.)

Edward Caird, 1890–92 at St. Andrews. *The Evolution of Religion,* 2 vols. Professor of Moral Philosophy, University of Glasgow. (1835–1908.)

Andrew Martin Fairbairn, 1891–93 at Aberdeen. *The Philosophy of the Christian Religion.* Principal, Mansfield College, Oxford. (1838–1912.)

George Gabriel Stokes, 1891–93 at Edinburgh. *Natural Theology,* Professor of Mathematics, Cambridge University. (1819–1903.)

Otto Pfleiderer, 1892–94 at Edinburgh. *Philosophy and Development of Religion,* 2 vols. Professor of Theology, University of Berlin. (1839–1908.)

John Caird, 1892–93 and 1895–96 at Glasgow. *The Fundamental Ideas of Christianity.* Principal, University of Glasgow. (1820–98.)

William Wallace, 1894 and 1895 at Glasgow. *Lectures and Essays on Natural Theology and Ethics.* Professor of Moral Philosophy, Oxford University. (1844–97.)

Alexander Campbell Fraser, 1894–96 at Edinburgh. *Philosophy of Theism,* 2 vols. Professor of Logic and Metaphysics, University of Edinburgh. (1819–1914.)

Lewis Campbell, 1894–96 at St. Andrews. *Religion in Greek Literature.* Professor Emeritus of Greek, University of St. Andrews. (1830–1908.)

James Ward, 1896–98 at Aberdeen. *Naturalism and Agnosticism,* 2 vols. Professor of Mental Philosophy and Logic, Cambridge University. (1843–1925.)

Cornelius Petrus Tiele, 1896–98 at Edinburgh. *Elements of the Science of Religion,* 2 vols. Professor of the History and Philosophy of Religion, University of Leiden. (1830–1902.)

Alexander Balmain Bruce, 1896–98 at Glasgow. *The Providential Order of the World* and *The Moral Order of the World.* Professor of Exegesis, Free Church College, Glasgow. (1831–99.)

Josiah Royce, 1899 and 1900 at Aberdeen. *The World and the Individual,* 2 vols. Professor of History of Philosophy, Harvard University. (1855–1916.)

Rudolfo Amadeo Lanciani, 1899–1901 at St. Andrews. *New Tales of Old Rome.* Professor of Ancient Topography, University of Rome. (1846–1929.)

Archibald Henry Sayce, 1900–1902 at Aberdeen. *The Religions of Ancient Egypt and Babylonia.* Professor of Assyriology, Oxford University. (1845–1933.)

Edward Caird, 1900–1902 at Glasgow. *The Evolution of Theology in the Greek Philosophers,* 2 vols. Master of Balliol College, Oxford. (1835–1908.)

William James, 1901 and 1902 at Edinburgh. *The Varieties of Religious Experience: A Study of Human Nature.* Professor of Philosophy, Harvard University. (1842–1910.)

Richard Burdon Haldane, 1902–4 at St. Andrews. *The Pathway to Reality,* 2 vols. Member of Parliament. (1856–1928.)

Henry Melvill Gwatkin, 1903–5 at Edinburgh. *The Knowledge of God and Its Historical Development,* 2 vols. Professor of Ecclesiastical History, Cambridge University. (1844–1916.)

Emile Boutroux, 1903–5 at Glasgow. *Science and Religion in Contemporary Philosophy.* Professor of Philosophy, Sorbonne University, Paris. (1845–1921.)

James Adam, 1904–6 at Aberdeen. *The Religious Teachers of Greece.* Fellow and Senior Tutor, Emmanuel College, Cambridge. (1860–1907.)

Simon Somerville Laurie, 1905–6 at Edinburgh. *On God and Man.* Professor of Education, University of Edinburgh. (1829–1909.)

Hans Driesch, 1907–8 at Aberdeen. *The Science and Philosophy of the Organism,* 2 vols. Professor of Biology, University of Heidelberg. (1867–1941.)

Andrew Cecil Bradley, 1907–8 at Glasgow. *Ideals of Religion.* Professor of Poetry, Oxford University. (1851–1935.)

James Ward, 1907–9 at St. Andrews. *The Realm of Ends, or Pluralism and Theism.* Professor of Mental Philosophy and Logic, Cambridge University. (1843–1925.)

William Ridgeway, 1909–11 at Aberdeen. *The Evolution of Religions of Ancient Greece and Rome.* Reader in Classics, Cambridge University. (1853–1926.)

William Warde Fowler, 1909–10 at Edinburgh. *The Religious Experience of the Roman People from the Earliest Times to the Age of Augustus.* Tutor and Fellow of Lincoln College, Oxford. (1847–1921.)

John Watson, 1910–12 at Glasgow. *The Interpretation of Religious Experience,* 2 vols. Professor of Moral Philosophy, Queen's University, Kingston, Canada. (1847–1939.)

Bernard Bosanquet, 1911 and 1912 at Edinburgh. *The Principle of Individuality and Value* and *The Value and Destiny of the Individual.* Professor of Moral Philosophy, University of St. Andrews. (1848–1923.)

James George Frazer, 1911 and 1912 at St. Andrews. *The Belief in Immortality and the Worship of the Dead, Vol. I.* Professor of Social Anthropology, University of Liverpool. (1854–1941.)

Andrew (Seth) Pringle-Pattison, 1912 and 1913 at Aberdeen. *The Idea of God in the Light of Recent Philosophy.* Professor of Logic and Metaphysics, University of Edinburgh. (1856–1931.)

William Ritchie Sorley, 1914–15 at Aberdeen. *Moral Values and the Idea of God.* Professor of Moral Philosophy, Cambridge University. (1855–1935.)

Henri Bergson, 1914 at Edinburgh. "The Problem of Personality." Professor of Philosophy, College of France, Paris. (1859–1941.)

Arthur James Balfour, 1914 at Glasgow. *Theism and Humanism.* Member of Parliament. (1848–1930.)

John Arthur Thomson, 1914–16 at St. Andrews. *The System of Animate Nature,* 2 vols. Professor of Natural History, University of Aberdeen. (1861–1933.)

William Mitchell Ramsay, 1915–16 at Edinburgh. *Asianic Elements in Greek Civilization.* Professor of Humanity, University of Aberdeen. (1851–1939.)

Clement C. J. Webb, 1917–19 at Aberdeen. *God and Personality* and *Divine Personality and Human Life.* Fellow of Magdalen College, Oxford University. (1865–1954.)

Samuel Alexander, 1917 and 1918 at Glasgow. *Space, Time and Deity,* 2 vols. Professor of Philosophy, University of Manchester. (1859–1938.)

William Ralph Inge, 1917–19 at St. Andrews. *The Philosophy of Plotinus,* 2 vols. Dean of St. Paul's, London, and former Professor of Divinity, Oxford University. (1860–1954.)

Frederick George Stout, 1919–21 at Edinburgh. *Mind and Matter* and *God and Nature.* Professor of Logic and Metaphysics, University of St. Andrews. (1860–1944.)

Henry Jones, 1919–21 at Glasgow. *A Faith That Enquires.* Professor of Moral Philosophy, University of Glasgow. (1853–1922.)

Lewis Richard Farnell, 1919–20 at St. Andrews. *Greek Hero Cults and Ideas of Immortality.* Rector of Exeter College and Vice-Chancellor of Oxford University. (1856–1934.)

Ernest William Hobson, 1920–22 at Aberdeen. *The Domain of Natural Science.* Professor of Pure Mathematics, Cambridge University. (1850–1933.)

Andrew (Seth) Pringle-Pattison, 1921–23 at Edinburgh. *The Idea of Imortality.* Professor of Logic and Metaphysics, University of Edinburgh. (1856–1931.)

Conwy Lloyd Morgan, 1921–23 at St. Andrews. *Emergent Evolution* and *Life, Mind, and Spirit.* Professor Emeritus of Zoology and Geology, University of Bristol. (1852–1936.)

Arthur James Balfour, 1922–23 at Glasgow. *Theism and Thought.* Member of Parliament. (1848–1930.)

William P. Paterson, 1923–25 at Glasgow. *The Nature of Religion.* Professor of Divinity, University of Edinburgh. (1860–1939.)

William Mitchell, 1924–26 at Aberdeen. *The Place of Minds in the World* and "The Power of Minds in the World." Professor of Philosophy, University of Adelaide, Australia. (1861–1962.)

James George Frazer, 1924 and 1925 at Edinburgh. *The Worship of Nature.* Professor of Anthropology, Cambridge University. (1854–1941.)

Lewis Richard Farnell, 1924–25 at St. Andrews. *The Attributes of God.* Rector of Exeter College and Vice-Chancellor of Oxford University. (1856–1934.)

Alfred Edward Taylor, 1926–28 at St. Andrews. *The Faith of a Moralist,* 2 vols. Professor of Moral Philosophy, University of Edinburgh. (1869–1945.)

Ernest William Barnes, 1927–29 at Aberdeen. *Scientific Theory and Religion: The World Described by Science and Its Spiritual Interpretation.* Bishop of Birmingham. (1874–1953.)

Arthur Stanley Eddington, 1927 at Edinburgh. *The Nature of the Physical World.* Professor of Astronomy, Cambridge University. (1882–1944.)

John Scott Haldane, 1927 and 1928 at Glasgow. *The Sciences and Philosophy.* Fellow of New College, Oxford. (1860–1936.)

Alfred North Whitehead, 1928 at Edinburgh. *Process and Reality: An Essay in Cosmology.* Professor of Philosophy, Harvard University. (1861–1947.)

John Dewey, 1928 and 1929 at Edinburgh. *The Quest for Certainty: A Study of the Relation of Knowledge and Action.* Professor of Philosophy, Columbia University. (1859–1952.)

John Alexander Smith, 1929–31 at Glasgow. *The Heritage of Idealism.* Professor of Mental Philosophy, Oxford University. (1863–1939.)

Charles Gore, 1929 and 1930 at St. Andrews. *The Philosophy of the Good Life.* Former Bishop of Oxford. (1853–1932.)

Robert Ranulph Marett, 1931–32 at St. Andrews. *Faith, Hope and Charity in Primitive Religion* and *Sacraments of Simple Folk.* Rector of Exeter College, Oxford. (1866–1943.)

Etienne Gilson, 1931 and 1932 at Aberdeen. *The Spirit of Mediaeval Theology,* (2 vols. in the original French). Professor, College of France, Paris. (1884–1978.)

Nathan Söderblom, 1931 at Edinburgh. *The Living God: Basal Forms of Personal Religion.* Archbishop of Upsala. (1866–1931.)

Edwyn Robert Bevan, 1932–34 at Edinburgh. *Symbolism and Belief* and *Holy Images.* Lecturer in Hellenistic History and Literature, King's College, London. (1870–1943.)

William Temple, 1932–34 at Glasgow. *Nature, Man and God.* Archbishop of York. (1881–1944.)

Albert Schweitzer, 1934 and 1935 at Edinburgh. "The Problem of Natural Theology and Natural Ethics." Missionary Doctor. (1875–1965.)

William David Ross, 1935–36 at Aberdeen. *Foundations of Ethics.* Provost of Oriel College, Oxford. (1877–1971.)

William Macneile Dixon, 1935–37 at Glasgow. *The Human Situation.* Professor Emeritus of English Language and Literature, University of Glasgow. (1866–1946.)

Herbert Hensley Henson, 1935 and 1936 at St. Andrews. *Christian Morality: Natural, Developing, Final.* Bishop of Durham. (1863–1947.)

Werner Jaeger, 1936–37 at St. Andrews. *The Theology of the Early Greek Philosophers.* Professor of Greek and Ancient Philosophy, University of Chicago. (1899–1961.)

Karl Barth, 1937 and 1938 at Aberdeen. *The Knowledge of God and the Service of God According to the Reformation: Recalling the Scottish Confession of 1560.* Professor of Theology, University of Basel. (1886–1969.)

Charles Scott Sherrington, 1937 and 1938 at Edinburgh. *Man on His Nature.* Professor Emeritus of Physiology, Oxford University. (1857–1952.)

William George De Burgh, 1937–38 at St. Andrews. *From Morality to Religion.* Professor Emeritus of Philosophy, University of Reading. (1866–1943.)

William Ernest Hocking, 1938 and 1939 at Glasgow. "Fact and Destiny." Professor of Philosophy, Harvard University. (1873–1966.)

John Laird, 1938–40 at Glasgow. *Theism and Cosmology* and *Mind and Deity.* Professor of Moral Philosophy, Aberdeen. (1887–1946.)

Joseph Bidez, 1938–39 at St. Andrews. *Eos, ou, Platon et l'Orient.* Professor of Classical Philology and History of Philosophy, University of Ghent, Belgium. (1867–1945.)

Arthur Darby Nock, 1939–40 at Aberdeen. "Hellenistic Religion: The Two Phases." Professor of History of Religion, Harvard University. (1902–63.)

Reinhold Niebuhr, 1939 at Edinburgh. *The Nature and Destiny of Man: A Christian Interpretation,* 2 vols. Professor of Theology and Ethics, Union Theological Seminary, New York City. (1892–1971.)

Richard Kroner, 1939–40 at St. Andrews. *The Primacy of Faith.* Former Professor of Philosophy, University of Prague. (1872–1942.)

Oscar Kraus, 1940–41 at Edinburgh. "New Meditations on Mind, God, and Creation." Former Professor of Philosophy, University of Prague. (1872–1942.)

Lectures for 1941–46 were canceled.

Arthur Darby Nock, 1946–47 at Aberdeen. "Hellenistic Religion: The Two Phases." Professor of History of Religion, Harvard University. (1902–63.)

Ralph Barton Perry, 1946–48 at Glasgow. *Realms of Value: A Critique of Human Civilization.* Professor Emeritus of Philosophy, Harvard University. (1876–1957.)

Emil Brunner, 1946–48 at St. Andrews. *Christianity and Civilization: Foundations* and *Christianity and Civilization, Specific Problems.* Professor of Systematic and Practical Theology, University of Zurich. (1899–1966.)

Christopher Dawson, 1947–49 at Edinburgh. *Religion and Culture* and *Religion and the Rise of Western Civilization.* Writer and Historian. (1889–1970.)

John Wisdom, 1948 and 1950 at Aberdeen. "The Mystery of the Transcendental" and "The Discovery of the Transcendental." Lecturer in Moral Science, Cambridge University. (1904–93).

Alexander Murray Macbeath, 1948–49 at St. Andrews. *Experiments in Living: A Study of the Nature and Foundation of Ethics and Morals in Light of Recent Work in Social Anthropology.* Professor of Logic and Metaphysics, Queen's University, Belfast. (1888–1964.)

Gabriel Marcel, 1949 and 1950 at Aberdeen. *The Mystery of Being: Reflection and Mystery* and *The Mystery of Being: Faith and Reality.* Writer and Philosopher. (1889–1973.)

Niels Henrik David Bohr, 1949 at Edinburgh. "Causality and Complementarity: Epistemological Lessons of Studies in Atomic Physics." Physicist, Copenhagen, Denmark. (1885–1962.)

Herbert H. Farmer, 1949–50 at Glasgow. *Revelation and Religion: Studies in the Theological Interpretation of Religions.* Professor of Systematic Theology, Cambridge University. (1892–1981.)

Herbert James Paton, 1949–50 at St. Andrews. *The Modern Predicament: A Study in the Philosophy of Religion.* Professor Emeritus of Moral Philosophy, Oxford University. (1887–1969.)

Charles Earle Raven, 1950–52 at Edinburgh. *Natural Religion and Christian Theology,* 2 vols. Professor Emeritus of Divinity, Cambridge University. (1885–1964.)

Michael Polanyi, 1951–52 at Aberdeen. *Personal Knowledge: Towards a Post-Critical Philosophy.* Professor of Social Studies, University of Manchester. (1891–1976.)

Brand Blanshard, 1951–53 at St. Andrews. *Reason and Goodness* and *Reason and Belief.* Professor of Philosophy, Yale University. (1892–1966.)

Arnold Joseph Toynbee, 1952–53 at Edinburgh. *An Historian's Approach to Religion.* Research Professor of International History, University of London. (1889–1975.)

John Macmurray, 1952–54 at Glasgow. *The Self as Agent* and *Persons in Relation.* Professor of Moral Philosophy, University of Edinburgh. (1891–1976.)

Paul Tillich, 1953 and 1954 at Aberdeen. *Systematic Theology,* vol. 2: *Existence and the Christ* and *Systematic Theology,* vol. 3: *Life and the Spirit.* Professor of Philosophical Theology, Union Theological Seminary, New York City. (1886–1965.)

Charles Arthur Campbell, 1953–55 at St. Andrews. *On Selfhood and Godhood.* Professor of Logic and Rhetoric, University of Glasgow. (1897–1974.)

Rudolf Bultmann, 1955 at Edinburgh. *History and Eschatology*. Professor of Theology, University of Marburg. (1884–1976.)

Leonard Hodgson, 1955–57 at Glasgow. *For Faith and Freedom,* 2 vols. Professor of Divinity, Oxford University. (1889–1969.)

Werner Carl Heisenberg, 1955 at St. Andrews. *Physics and Philosophy: The Revolution in Modern Science.* Director, Max Planck Institute, Göttingen. (1901–76.)

Herbert Arthur Hodges, 1956–57 at Aberdeen. "The Logic of Religious Thinking." Professor of Philosophy, University of Reading. (1905–76.)

Austin Marsden Farrer, 1956–57 at Edinburgh. *The Freedom of the Will.* Fellow and Chaplain, Trinity College, Oxford. (1904–68.)

Vigo Auguste Demant, 1956–58 at St. Andrews. "The Religious Climate" and "The Moral Career of Christendom." Professor of Moral and Pastoral Theology, Oxford University. (1893–1983.)

Wolfgang Kohler, 1957–59 at Edinburgh. "The Psychology of Values" and "Psychology and Physics." Professor of Psychology, Swarthmore College, Pennsylvania. (1887–1967.)

Georg Henrik von Wright, 1958–60 at St. Andrews. *Norm and Action: A Logical Inquiry* and *The Varieties of Goodness.* Professor of Philosophy, University of Helsinki, Finland. (1916–2003.)

Roderick Diarmid McLennan, 1959–60 at Edinburgh. "The Unity of Moral Experience." Minister, Church of Scotland. (1898–1977.)

Carl Fredrich von Weizsäcker, 1959–61 at Glasgow. *The Relevance of Science: Creation and Cosmogony.* Professor of Philosophy, University of Hamburg. (1912–.)

Henry Habberley Price, 1960 at Aberdeen. *Belief.* Professor of Logic, Oxford University. (1899–1984.)

Steven Runciman, 1960–62 at St. Andrews. *The Great Church in Captivity: A Study of the Patriarchate of Constantinople from the Eve of the Turkish Conquest to the Greek War of Independence.* Fellow of Trinity College, Cambridge. (1903–2000.)

David Daube, 1962–64 at Edinburgh. "The Deed and the Doer in the Bible" and "Law and Wisdom in the Bible." Professor of Civil Law, Oxford University. (1909–94.)

Charles William Hendel, 1962–63 at Glasgow. "Politics: The Trial of a Pelagian Faith" and "The Limit of Human Power." Professor Emeritus of Moral Philosophy, Yale University. (1890–1982.)

Henry Chadwick, 1962–64 at St. Andrews. "Authority in the Early Church." Professor of Divinity, Oxford University. (1920–.)

Alister Clavering Hardy, 1963 and 1965 at Aberdeen. *The Living Stream* and *The Divine Flame.* Professor Emeritus of Zoology, Oxford University. (1896–1985.)

Donald MacKenzie MacKinnon, 1964–66 at Edinburgh. "The Problem of Metaphysics." Professor of Divinity, Cambridge University. (1913–94.)

John Niemeyer Findlay, 1964–66 at St. Andrews. *The Discipline of the Cave* and *The Transcendence of the Cave*. Professor of Philosophy, King's College, University of London. (1903–87.)

Raymond Aron, 1965 and 1967 at Aberdeen. "On Historical Consciousness in Thought and Action." Professor, Institut d'études politiques and Sorbonne University, Paris. (1905–84.)

Herbert Butterfield, 1965–67 at Glasgow. "Human Beliefs and the Development of Historical Writing" and "Historical Writing and Christian Beliefs." Professor of Modern History, Cambridge University. (1900–1979.)

Malcolm Knox, 1966 and 1968 at Aberdeen. *Action* and *A Layman's Quest*. Principal of the University of St. Andrews. (1900–1980.)

Hywel David Lewis, 1966–68 at Edinburgh. *The Elusive Self* and *Freedom and Alienation*. Professor of History and Philosophy of Religion, University of London. (1910–92.)

Robert Charles Zaehner, 1967–69 at St. Andrews. *Concordant Discord: The Interdependence of Faiths*. Professor of Eastern Religions and Ethics, Oxford University. (1913–74.)

Winston H. F. Barnes, 1968–70 at Edinburgh. "Knowledge and Faith." Vice-Chancellor, University of Liverpool. (1909–84.)

William Homan Thorpe, 1969–71 at St. Andrews. *Animal Nature and Human Nature*. Professor Emeritus of Animal Ethology, Cambridge University. (1902–86.)

Arend Theodoor van Leeuwen, 1970–72 at Aberdeen. *Critique of Heaven* and *Critique of Earth*. Associate Professor of Christian Ethics, Catholic University of Nijmegen, Holland. (1918–93.)

Eric Lionel Mascall, 1970–71 at Edinburgh. *The Openness of Being: Natural Theology Today*. Professor of Historical Theology, University of London. (1905–93.)

Richard William Southern, 1970–72 at Glasgow. "The Rise and Fall of the Medieval System of Religious Thought." President, St. John's College, Oxford. (1912–2001.)

Anthony J. P. Kenny, 1971–73 at Edinburgh. *The Nature of Mind* and *The Development of Mind*. Fellow, Balliol College, Oxford. (1931–.) Included panel discussion with H. C. Longuet-Higgins, J. R. Lucas, and C. H. Waddington.

Hannah Arendt, 1972–74 at Aberdeen. *The Life of the Mind*, 2 vols. Professor of Philosophy, New School for Social Research, New York City. (1906–75.)

Alfred J. Ayer, 1972–73 at St. Andrews. *The Central Questions of Philosophy*. Professor of Logic, Oxford University. (1910–89.)

Owen Chadwick, 1973–74 at Edinburgh. *The Secularization of the European Mind in the Nineteenth Century*. Professor of Modern History, Cambridge University. (1916–.)

Stanley L. Jaki, 1974–76 at Edinburgh. *The Road of Science and the Ways to God*. Professor, Seton Hall University, South Orange, New Jersey. (1924–.)

Basil G. Mitchell, 1974–75 at Glasgow. *Morality, Religion and Secular: The Dilemma of the Traditional Conscience*. Professor of the Philosophy of the Christian Religion, Oxford University. (1917–.)

John Zachary Young, 1975–77 at Aberdeen. *Programs of the Brain.* Professor Emeritus of Anatomy and Embryology, University College, London. (1907–97.)

Reijer Hooykaas, 1975–77 at St. Andrews. "Fact, Faith and Fiction in the Development of Science." Professor of History of Science, University of Utrecht, Holland. (1906–94.)

Jean Pierre Jossua, 1976–77 at Edinburgh. *Pierre Bayle ou l'obsession du mal.* Professor of Dogmatics, Faculté de théologie, Saulchoir, France. (1930–.)

John C. Eccles, 1977–79 at Edinburgh. *The Human Mystery* and *The Human Psyche.* Professor Emeritus of Neurobiology, State University of New York, Buffalo. (1903–97.)

David Stafford-Clark, 1977–78 at St. Andrews. "Myth, Magic and Denial: The Treacherous Allies." Department of Psychiatry, Guy's Hospital, London. (1916–99.)

Frederick Charles Copleston, 1979–81 at Aberdeen. *Religion and the One: Philosophies of East and West.* Professor Emeritus of the History of Philosophy, University of London. (1907–94.)

Ninian Roderick Smart, 1979–80 at Edinburgh. *Beyond Ideology: Religion and the Future of Western Civilization.* Professor of Religious Studies, University of California, Santa Barbara. (1927–2001.)

Sydney Brenner, 1979–80 at Glasgow. "The New Biology." Director, MRC Laboratory of Molecular Biology, Oxford University. (1927–.)

Seyyed Hossein Nasr, 1980–81 at Edinburgh. *Knowledge and the Sacred.* Professor of Religion, Temple University, Philadelphia. (1933–.)

Gregory Vlastos, 1980–81 at St. Andrews. *Socrates, Ionist and Moral Philosopher.* Professor Emeritus of Philosophy, Princeton University. (1907–91.)

Åke Hultkrantz, 1981–83 at Aberdeen. "The Veils of Religion: Religion in Its Ecological Forms." Professor of Comparative Religion, University of Stockholm. (1920–.)

Richard G. Swinburne, 1982–84 at Aberdeen. *The Evolution of the Soul.* Professor of Philosophy, University of Keele, England. (1934–.)

Iris Murdoch, 1982 at Edinburgh. *Metaphysics as a Guide to Morals.* Novelist and Fellow of St. Ann's College, Oxford. (1919–99.)

David Daiches, 1982–83 at Edinburgh. *God and the Poets.* Director, Institute of Advanced Studies in the Humanities, University of Edinburgh. (1912–.)

Stephen R. L. Clark, 1982 at Glasgow. *From Athens to Jerusalem: The Love of Wisdom and the Love of God.* Lecturer in Philosophy, University of Glasgow. (1945–.)

Christina J. Larner, 1982 at Glasgow. *The Thinking Peasant: Popular and Educated Belief in Pre-Industrial Culture.* Reader in Psychology, University of Glasgow. (1933–83.)

Donald Geoffrey Charlton, 1982–83 at St. Andrews. *New Images of the Natural in France: A Study in European Cultural History, 1750–1800.* Professor of French, University of Warwick. (1925–.)

Freeman J. Dyson, 1983–85 at Aberdeen. *Infinite in All Directions.* Professor, Institute for Advanced Study, Princeton, New Jersey. (1923–.)

Michael Arbib, 1983–84 at Edinburgh. "The Construction of Reality." Professor of Computer Science, University of Massachusetts, Amherst. (1940–.) Given jointly with Mary Brenda Hesse, Professor of Philosophy of Science, Cambridge University. (1924–.)

Anthony J. Sanford, 1983 at Glasgow. *Models, Mind and Man.* Professor of Psychology, University of Glasgow. (1944–.)

Philip Drew, 1983 at Glasgow. "The Literature of Natural Man." Professor of English Literature, University of Glasgow. (1925–.)

John Macquarrie, 1983 at St. Andrews. *In Search of Deity: An Essay in Dialectical Theism.* Professor of Divinity, Oxford University. (1919–.)

Allan Douglas Galloway, 1984 at Glasgow. "God's Other Nature: The Humanity of God." Professor Emeritus of Divinity, University of Glasgow. (1920–.)

Jürgen Moltmann, 1984–85 at Edinburgh. *God in Creation: A New Theology of Creation and the Spirit of God.* Professor of Systematic Theology, University of Tübingen. (1926–.)

Adolf Grunbaum, 1984–85 at St. Andrews. "Psychoanalytic Theory and Science." Professor of Philosophy, University of Pittsburgh. (1923–.)

Paul Ricoeur, 1985–86 at Edinburgh. "On Selfhood: The Question of Personal Identity" and "Oneself as Another." Philosopher, University of Paris. (1913–.)

Carl Sagan, 1985 at Glasgow. "The Search for Who We Are." Professor of Astronomy and Space Sciences and Director of the Laboratory for Planetary Studies, Cornell University. (1934–96.)

John Hick, 1986–87 at Edinburgh. *An Interpretation of Religion: Human Responses to the Transcendent.* Professor of Religion, Claremont College, California. (1922–.)

Donald M. MacKay, 1986 at Glasgow. *Behind the Eye.* Professor Emeritus, Department of Communication and Neuroscience, University of Keele, England. (1922–87.)

Anthony Flew, 1986 at St. Andrews. *The Logic of Mortality.* British Philosopher and Author. (1923–.)

Alvin Plantinga, 1987 and 1988 at Aberdeen. *Warrant: The Current Debate* and *Warrant and Proper Function.* Professor of Philosophy, University of Notre Dame. (1932–.)

Alasdair MacIntyre, 1987–88 at Edinburgh. *Three Rival Versions of Moral Enquiry: Encyclopedia, Genealogy and Tradition.* Professor of Philosophy, University of Notre Dame. (1929–.)

Raimundo Panikkar, 1988–89 at Edinburgh. "Trinity and Atheism: Housing of the Divine in the Contemporary World." Professor Emeritus of Religious Studies, University of California, Santa Barbara. (1918–.)

The 1988 Gifford Centenary at the University of Glasgow. *Humanity, Environment, God: Glasgow Centenary Gifford Lectures.* Lectures by John David Barrow, Richard Dawkins, Don Cupitt, Anthony J. P. Kenny, John Morris Roberts, and John S. Habgood.

Ian Barbour, 1989 and 1990 at Aberdeen. *Religion in an Age of Science* and *Ethics in an Age of Technology.* Professor Emeritus of Science, Technology, and Society, Carleton College, Minnesota. (1923–.)

Mary Douglas, 1989–90 at Edinburgh. "Claims on God." Professor of Humanities, Northwestern University, Evanston, Illinois (1921–.)

Mary Midgley, 1989–90 at Edinburgh. *Science as Salvation: A Modern Myth and Its Meaning.* Senior Lecturer, University of Newcastle-upon-Tyne. (1919–.)

Walter Burkert, 1989 at St. Andrews. *Creation of the Sacred: Tracks of Biology in Early Religions.* Professor of Classical Philology, University of Zurich. (1931–.)

James Barr, 1990–91 at Edinburgh. *Biblical Faith and Natural Theology.* Professor of the Hebrew Bible, Vanderbilt University. (1924–.)

George Steiner, 1990 at Glasgow. *Grammars of Creation.* Professor of English and Comparative Literature, University of Geneva, Switzerland. (1929–.)

Hilary Putnam, 1990–91 at St. Andrews. *Renewing Philosophy.* Professor of Philosophy, Harvard University. (1926–.)

Annemarie Schimmel, 1991–92 at Edinburgh. *Islam: A Phenomenological Study.* Professor of Indo-Muslim Culture, Harvard University. (1922–2003.)

Jaroslav Pelikan, 1992 and 1993 at Aberdeen. *Christianity and Classical Culture.* Professor of History, Yale University. (1923–.)

Martha Craven Nussbaum, 1992–93 at Edinburgh. *Upheavals of Thought: A Theory of the Emotions.* Professor of Philosophy, Classics, and Comparative Literature, Brown University. (1947–.)

Mary Warnock, 1992 at Glasgow. *Imagination and Time.* Mistress of Girton College, Cambridge University. (1924–.)

Arthur Robert Peacocke, 1992–93 at St. Andrews. "Nature, God and Humanity: Towards a Christian Theology for a Scientific Age." Warden Emeritus of the Society of Ordained Scientists, Exter College, Oxford. (1924–.)

Roger Penrose, 1992–93 at St. Andrews. "The Question of Physical Reality." Professor of Mathematics, Oxford University. (1931–.)

John Polkinghorne, 1993–94 at Edinburgh. *The Faith of a Physicist: Reflections of a Bottom-Up Thinker.* President, Queen's College, Cambridge University. (1930–.)

Keith Ward, 1993–94 at Glasgow. *Religion and Revelation: A Theology of Revelation in the World's Religions.* Professor of Divinity, Oxford University. (1938–.)

The 1994–95 Giffords celebrating the 550th anniversary of the University of Aberdeen. "Natural Theology in Scotland Over 500 Years." Lectures by Alexander Broadie, John W. Rogerson, John H. Burns, Michael Alexander Steward, and Peter Jones.

Nicholas Wolterstorff, 1995 at St. Andrews. *Thomas Reid and the Story of Epistemology.* Professor of Philosophical Theology, Yale University. (1932–.)

John Hedley Brooke and Geoffrey N. Cantor, 1995 at Glasgow. *Reconstructing Nature: The Engagement of Science and Religion Historically Considered.* Brooke, Professor of Science and Religion, Oxford University (1944–); Cantor, Leeds University (1943–).

Gerald Allen Cohen, 1996 at Edinburgh. "The Production of Equality: From History,

Through Politics to Morals." Professor of Social and Political Theory, Oxford University. (1941–.)

Michael Dummett, 1996–97 at St. Andrews. "Thought and Reality." Professor Emeritus of Logic, Oxford University. (1925–.)

Richard R. Sorabji, 1997 at Edinburgh and 1997 at St. Andrews. *Emotion and Peace of Mind: From Stoic Agitation to Christian Temptation.* Professor of Ancient Philosophy, Kings College, University of London. (1934–.)

Russell Stannard, 1997–98 at Aberdeen. *The God Experiment.* Former Professor of Physics, the Open University, England. (1931–.)

Holmes Rolston III, 1997–98 at Edinburgh. *Genes, Genesis and God: Values and Their Origins in Natural and Human History.* Professor of Philosophy, Colorado State University. (1932–.)

Robert James (Sam) Berry, 1997–98 at Glasgow. "God, Genes, Greens and Everything." Professor of Genetics, University College, London. (1934–.)

Charles Taylor, 1998–99 at Edinburgh. "Living in a Secular Age." Professor Emeritus of Philosophy, McGill University, Montreal. (1931–.)

David Tracy, 1999–2000 at Edinburgh. "This Side of God." Professor of Catholic Studies, Theology, and Philosophy of Religion, University of Chicago. (1939–.)

Ralph McInerny, 1999–2000 at Glasgow. *Characters in Search of Their Author.* Professor of Philosophy, Jacques Maritain Center, University of Notre Dame. (1929–.)

Marilyn McCord Adams, 1999 at St. Andrews. "The Coherence of Christology." Professor of Historical Theology, Yale Divinity School. (1943–.)

Robert Merrihew Adams, 1999 at St. Andrews. "God and Being." Professor of Moral Philosophy and Metaphysics, Yale University. (1937–.)

John Stapylton Habgood, 2000 at Aberdeen. "The Concept of Nature." Former Archbishop of York. (1927–.)

Onora O'Neill, 2000–2001 at Edinburgh. *Autonomy and Trust in Bioethics.* Principal of Newnham College, Cambridge University. (1941–.)

Mohammed Arkoun, 2001 at Edinburgh. *The Unthought in Contemporary Islamic Thought.* Professor Emeritus of Islamic Studies, Sorbonne University, Paris. (1928–.)

The 2001 Giffords celebrating the 550th anniversary of the University of Glasgow. *The Nature and Limits of Human Understanding.* Lectures by Phil Johnson-Laird, George Lakoff, Michael Ruse, Lynne Baker Rudder, and Brian Hebblethwaite.

Stanley Hauerwas, 2001 at St. Andrews. *With the Grain of the Universe: The Church's Witness and Natural Theology.* Professor of Theological Ethics, Duke Divinity School. (1940–.)

Eleonore Stump, 2003 at Aberdeen. "Wandering in Darkness: Narrative and the Problem of Suffering." Professor of Philosophy, St. Louis University. (1947–.)

Michael Ignatieff, 2003 at Edinburgh. *The Lesser Evil: Political Ethics in an Age of Terror.*

Professor of Human Rights Policy, Kennedy School of Government, Harvard. (1947–.)

Simon Blackburn, 2003 at Glasgow. "Reason's Empire." Professor of Philosophy, Cambridge University. (1944–.)

Peter van Inwagen, 2003 at St. Andrews. "The Problem of Evil." Professor of Philosophy, University of Notre Dame. (1942–.)

J. Wentzel van Huyssteen, 2004 at Edinburgh. "Alone in the World? Science and Theology on Human Uniqueness." Professor of Theology and Science, Princeton Theological Seminary. (1942–.)

John Haldane, 2005 at Aberdeen. "Mind, Soul, and Deity." Professor of Philosophy, University of St. Andrews. (1954–.)

The 2005 Gifford Lectures at Edinburgh. Lectures by Margaret Anstee, "Peacebuilding in a Shrinking World," Stephen Toulmin, "Orientalism and Occidentalism," and Noam Chomsky, "Illegal but Legitimate: A Dubious Doctrine."

Jean Bethke Elshtain, 2005–6 at Edinburgh. "Sovereignties." Professor of Social and Political Ethics, University of Chicago Divinity School. (1941–.)

Other Gifford Appointments

Robert Flint, 1907–8 at Edinburgh. Resigned appointment due to health. Professor of Philosophy, University of Edinburgh.

Rudolph Otto, 1933 at Glasgow. Declined invitation due to health. Professor of Theology, Marburg, Germany.

Friedrich von Hügel, 1924–26 at Edinburgh. Resigned appointment due to health. His lecture materials were posthumously published as *The Reality of God* (1931).

James Jeans was appointed in 1931 by Aberdeen but resigned, probably due to family illnesses. English Mathematician, Physicist, and Astronomer.

Jacques Maritain, 1940–42 at Aberdeen. War intervened. Resigned appointment for diplomatic duties after the war.

John Baillie, 1961 at Edinburgh. Died in 1961. Lectures published as *The Sense of the Presence of God*. Professor of Divinity and Principal of New College, Edinburgh.

Hans Küng, 1974–75 at St. Andrews. Resigned the appointment.

David Braine, 1983 at Aberdeen. Gifford Research Fellow. "The Reality of Time and the Existence of God: The Project of Proving God's Existence."

C. J. Arthur, 1985 at St. Andrews. Gifford Research Fellow. *In the Hall of Mirrors: Some Problems of Commitment in a Religiously Plural World*.

Edward Said, 2004–5 at Edinburgh. Died in 2003. Professor of Literature, Columbia University.

NOTES

Introduction

1. For Gifford's will, see Stanley L. Jaki, *Lord Gifford and His Lectures: A Centenary Retrospect* (Edinburgh: Scottish Academic Press, 1986), 66–76.

2. Robert Fraser, "Introduction," in James George Frazer, *The Golden Bough: A Study in Magic and Religion,* new abridgement, ed. Robert Fraser (London: Oxford Univ. Press, 1994), xliii.

3. Owen Chadwick, *The Secularization of the European Mind in the Nineteenth Century* (Cambridge: Cambridge Univ. Press, 1975), 161.

4. The German theologian Otto Pfleiderer is quoted in Jaki, *Lord Gifford and His Lectures,* 10.

5. Robert Charles Zaehner, *Concordant Discord: The Interdependence of Faiths* (Oxford: Clarendon Press, 1970), 3.

6. Quoted in Michael de la Bedoyere, *The Life of Baron von Hügel* (New York: Scribner, 1951), 343.

7. John Hick, *John Hick: An Autobiography* (Oxford: Oneworld, 2002), 262.

8. Chadwick, *Secularization,* 1.

9. Interview with Alvin Plantinga, April 2004.

10. Victor Lowe, "The Gifford Lectures," *Southern Journal of Philosophy* 7 (Winter 1969–70): 329.

11. Rudolph Metz, *A Hundred Years of British Philosophizing* (New York: Macmillan, 1938), 779.

12. Neil Spurway, "Preface," in *Humanity, Environment and God: Glasgow Centenary Gifford Lectures,* ed. Neil Spurway (Oxford: Blackwell, 1993), vii.

13. Barth quoted in Wilhelm Pauck, *From Luther to Tillich: The Reformers and Their Heirs* (San Francisco: Harper & Row, 1984), 147; and in Karl Barth, *The Knowledge of God and the Service of God According to the Teachings of the Reformation: Recalling the Scottish Confession of 1560,* trans. L.L.M. Haire and Ian Henderson (New York: AMS Press, 1979), 6.

14. Lewis said that Arthur Balfour's *Theism and Humanism,* the 1914 Giffords at Glasgow, was one of ten books that influenced his views. See Michael W. Perry, "Second Edition Foreword," in Arthur J. Balfour, *Theism and Humanism,* ed. Michael W. Perry (Seattle: Inkling Books, 2000), 8.

15. See Iris Murdoch, *Metaphysics as a Guide to Morals* (New York: Viking Peguin, 1993); Bernard E. Jones, ed., *Earnest Enquirers After Truth: A Gifford Anthology; Excerpts from Gifford Lectures 1888–1968* (London: Allen & Unwin, 1970), 14.

16. Alasdair MacIntyre, *Three Rival Versions of Moral Enquiry: Encyclopedia, Genealogy, and Tradition* (Notre Dame, IN: Univ. of Notre Dame Press, 1990), 10.

Chapter One

1. This account relies primarily on William N. Roughead, *Classic Crimes* (Westport, CT: Hyperion Press, 1975), 171–206. Other sources used are two booklets in the Glasgow Univ. Library archives: *The Mysterious Murder at Sandyford, Glasgow: Complete Report* (Glasgow, ca. 1862); *The Sandyford Murder Case. Revised and Corrected Report of the Trial of Mrs. M'Lachlan for the Murder of Jessie M'Pherson* (Glasgow: Sept. 1862). I have also quoted from two chapters in crime anthologies handed out by the current occupant of 17 Sandyford Place with no citation.

2. J. Campbell Smith, review of *Lectures Delivered on Various Occasions by Adam Gifford,* by Alice Raleigh and Herbert James Gifford (eds.), *The Judicial Review: A Journal of Legal and Political Science* 1 (1889): 401.

3. Smith, review of *Lectures Delivered,* 397.

4. John Gifford, "Recollections of a Brother, and of His Homes," in Stanley L. Jaki, *Lord Gifford and His Lectures: A Centenary Retrospect* (Edinburgh: Scottish Academic Press, 1986), 78–99.

5. James Hutchison Stirling, *Philosophy and Theology, Being the First Edinburgh University Gifford Lectures* (Edinburgh: T. & T. Clark, 1890), 4.

6. "Illness of Lord Gifford," *Times* (London), Jan. 5, 1881, 9.

7. In London, Charles Volsey of the Theistic Church gave at least two sermons (and promised more) on the Gifford bequest in 1887; William James, *The Varieties of Religious Experience: A Study in Human Nature* (New York: Simon & Schuster, 1997), 164.

8. Arthur Herman, *How the Scots Invented the Modern World* (New York: Random House, 2001), 189–90, 191–92.

9. Alice Raleigh and Herbert James Gifford, eds., *Lectures Delivered on Various Occasions by Adam Gifford* (Edinburgh, June 1889), 3. This privately published booklet is available at the Edinburgh Public Library.

10. Ralph L. Rush, *The Life of Ralph Waldo Emerson* (New York: Scribner, 1949), 337–39.

11. On the four cases, see Peter Hinchliff, *God and History: Aspects of British Theology, 1875–1914* (Oxford: Oxford Univ. Press, 1992), 180–84.

12. Alexander Balmain Bruce, *The Training of the Twelve*, 4th ed. (New York: A. C. Armstrong & Son, 1894), 6.

13. Raleigh and Gifford, *Lectures,* 152, 155.

14. Smith, review of *Lectures Delivered,* 398.

15. Raleigh and Gifford, *Lectures,* 30–31.

16. "Extract from the Last Will and Testament of the Late Rev. John Bampton," in Alan Richardson, *History Sacred and Profane* (Philadelphia: Westminster Press, 1964), 15.

17. Thomas H. Huxley, "Evolution and Ethics," in *Evolution and Ethics and Other Essays* (New York: AMS Press, 1970), 81.

18. Raleigh and Gifford, *Lectures,* 4.

19. The 80,000 pounds sterling was divided thus: 25,000 for Edinburgh, 20,000 each for Glasgow and Aberdeen, and 15,000 for St. Andrews.

20. Quoted in Amelia Hutchison Stirling, *James Hutchison Stirling: His Life and Work* (London: T. Fisher Unwin, 1912), 311.

Chapter Two

1. James Hutchison Stirling, *Philosophy and Theology, Being the First Edinburgh University Gifford Lectures* (Edinburgh: T. & T. Clark, 1890), 15.

2. Hume quoted in Ernest Campbell Mossner, *The Life of David Hume,* 2d ed. (New York: Oxford Univ. Press, 1980), 161, 250.

3. Hume quoted in Andrew Seth Pringle-Pattison, *Scottish Philosophy; a Comparison of the Scottish and German Answers to Hume,* 2d ed. (New York: B. Franklin, 1971), 11.

4. Hume quoted in Daniel S. Robinson, ed., *The Story of Scottish Philosophy: A Compendium of Selections from the Writings of Nine Pre-eminent Scottish Philosophers* (New York: Exposition Press, 1961), 131–32; Reid quoted in Robinson, *Story of Scottish Philosophy,* 133.

5. G. A. Johnston, "Introduction," in *Selections from the Scottish Philosophy of Common Sense,* ed. G. A. Johnston (Chicago: Open Court Publishing, 1915), 7.

6. Reid quoted in Nicholas Wolterstorff, *Thomas Reid and the Story of Epistemology* (Cambridge: Cambridge Univ. Press, 2001), 248.

7. David Hume, *Dialogues Concerning Natural Religion* (New York: Penguin, 1990), 138.

8. Hamilton quoted in Robinson, *Story of Scottish Philosophy,* 223.

9. Kant quoted from the introduction to his *Prolegomena to Any Future Metaphysics* in Wolterstorff, *Thomas Reid,* 215.

10. Heine quoted in Edward Caird, *The Critical Philosophy of Immanuel Kant* (Glasgow: James Maclehose & Sons, 1889), 63.

11. Pringle-Pattison, *Scottish Philosophy,* 154.

12. Kant quoted in Gerald R. Cragg, *The Church and the Age of Reason, 1648–1789* (New York: Penguin, 1970), 252.

13. James McCosh, *The Scottish Philosophy: Biographical, Expository, Critical, from Hutcheson to Hamilton* (Hildesheim, Germany: Georg Olms, 1966), 9, 454.

14. Pringle-Pattison, *Scottish Philosophy,* 1–2.

15. Letter of the Glasgow University Academic Senate, Jan. 14, 1890, in university archives. (Author's copy.)

16. Traveler quoted in Elisabeth Fraser, *An Illustrated History of Scotland* (Norwich, England: Jarrold, 1997), 125; William R. Sorely, *A History of English Philosophy* (New York: Putnam, 1921), 291.

17. Memo by Principal Stewart, "Gifford Lectureship: Additional Suggestions," St. Andrews University, Dec. 10, 1905, in university archives. (Author's copy.)

18. Quoted in Amelia Hutchison Stirling, *James Hutchison Stirling: His Life and Work* (London: T. Fisher Unwin, 1912), 315.

19. James Hutchison Stirling, "Preface to New Edition," *The Secret of Hegel: Being the Hegelian System in Origin, Principle, Form, and Matter* (Edinburgh: Oliver & Boyd, 1898), xviii.

20. Stirling, *Secret,* 720–21.

21. Stirling, *Secret,* xxii.

22. Roger Scruton quoted in Scruton et al., *German Philosophers: Kant, Hegel, Schopenhauer, Nietzsche* (Oxford: Oxford Univ. Press, 2001), 163.

23. Quoted in Stirling, *Secret,* xxvii.

24. Eugene Thomas Long, "The Gifford Lectures and the Scottish Personal Idealists," *Review of Metaphysics* 49 (1995): 365.

25. Stirling, *Philosophy and Theology,* 31.

26. Werner Jaeger, *The Theology of the Early Greek Philosophers* (Oxford: Clarendon Press, 1947), 184.

27. Stirling, *Philosophy and Theology,* 33.

28. Kant quoted in Stirling, *Philosophy and Theology,* 322.

29. Charles Darwin, *The Autobiography of Charles Darwin, 1809–1882,* ed. Nora Barlow (New York: Norton, 1963), 52, 56, 57.

30. Darwin, *Autobiography,* 87; Stirling, *Philosophy and Theology,* 325.

31. Translator quoted in Stirling, *Philosophy and Theology,* 342.

32. Darwin quoted in Stephen Jay Gould, *The Flamingo's Smile: Reflections in Natural History* (New York: Norton, 1985), 132.

33. Weismann quoted in Ernst Mayr, *One Long Argument: Charles Darwin and the Genesis of Modern Evolutionary Thought* (Cambridge: Harvard Univ. Press, 1991), 126.

34. Stirling, *Philosophy and Theology,* 359, 375.

35. Stirling, *Philosophy and Theology,* 376, 353.

36. Pringle-Pattison, *Scottish Philosophy,* 195, 216.

37. Frederick Copleston, *A History of Philosophy: Bentham to Russell,* vol. 8 (Mahway, NJ: Paulist Press, 1966), 182.

38. John Caird, *Introduction to the Philosophy of Religion* (Glasgow: James Maclehose & Sons, 1901), 165, 176.

39. Edward Caird, *A Critical Account of the Philosophy of Kant* (Glasgow: James Maclehose, 1877), 667.

40. Edward Caird, *The Evolution of Religion,* vol. 1 (Glasgow: James Maclehose & Sons, 1894), 68, 175.

41. John Macquarrie, *Twentieth-Century Religious Thought,* 3d ed. (New York: Scribner, 1981), 25, 27n4.

42. Henry Jones, *A Faith That Inquires* (New York: Macmillan, 1922), 278.

43. Hamilton Ellis, *British Railway History* (London: Allen & Unwin, 1959), 28.

44. Josiah Royce, *The World and the Individual,* vol. 1 (New York: Macmillan, 1900), 335; Royce quoted (suffering) in John Clendenning, *The Life and Thought of Josiah Royce,* rev. and expanded ed. (Nashville: Vanderbilt Univ. Press, 1999), 229; Andrew Seth Pringle-Pattison, *The Idea of God in Light of Recent Philosophy* (New York: Oxford Univ. Press, 1920), iii.

45. Royce, *The World and the Individual,* vol. 1, 324, 468; Josiah Royce, *The World and the Individual,* vol. 2 (New York: Macmillan, 1901), 417, 172.

46. Royce, *The World and the Individual,* vol. 1, 295, 339; vol. 2, 378.

47. Casey Nelson Blake, *Beloved Community* (Chapel Hill: Univ. of North Carolina Press, 1990), 117.

48. Josiah Royce, *The Philosophy of Loyalty* (New York: Macmillan, 1908), 351.

49. Royce, *The World and the Individual,* vol. 1, 474.

50. Hobhouse quoted in Henry Jones, "Idealism and Politics," *Contemporary Review* 92 (Nov. 1907): 619, 620.

51. Jones, "Idealism and Politics," 618, 617.

52. Leonard T. Hobhouse, *The Metaphysical Theory of the State: A Criticism* (Westport, CT: Greenwood Press, 1984). (Originally published in 1918.)

53. Anthony M. Quinton, "Absolute Idealism," *Proceedings of the British Academy* 75 (1971): 303.

54. Bernard Bosanquet, "Preface," *The Principle of Individuality and Value* (London: Macmillan, 1912), v.

55. Arthur James Balfour, *Theism and Humanism: Being the Gifford Lectures Delivered at the University of Glasgow, 1914* (London: Hodder & Stoughton, 1915), 24.

Chapter 3

1. Mary Douglas, "Introduction," in James G. Frazer, *The Illustrated Golden Bough,* ed. Mary Douglas (Garden City, NY: Doubleday, 1978), 10.

2. Andrew Lang, *The Making of Religion* (London: Longmans, Green, 1898), 44.

3. Frazer quoted in Robert Fraser, "Introduction," in James George Frazer, *The Golden Bough: A Study in Magic and Religion,* new abridgement, ed. Robert Fraser (London: Oxford Univ. Press, 1994), xxiv.

4. F. Max Müller, *Natural Religion* (London: Longmans, Green, 1889), 21.

5. F. Max Müller, *Theosophy: Psychology of Religion* (London: Longmans, Green, 1893), 52–53.

6. F. Max Müller, *Physical Religion* (London: Longmans, Green, 1891), 366. For an account of Müller's Giffords and the public controversy they stirred in 1891, see *The Life and Letters of the Right Honourable Friedrich Max Muller,* vol. 2., ed. Mrs. Müller (London: Longmans, Green, 1902), 246–48, 274–77.

7. Mircea Eliade, *Myths, Dreams and Mysteries: The Encounter Between Contemporary Faiths and Archaic Realities* (New York: Harper Torchbooks, 1967), 23.

8. R. R. Marett, *Tylor* (New York: Wiley, 1936), 69.

9. Tylor's Gifford Lectures outline cited in Barbara W. Freire-Marreco, "A Bibliography," in *Anthropological Essays Presented to Edward Burnett Tylor,* ed. H. Balfour et al. (Oxford: Clarendon Press, 1907), 396.

10. Marett, *Tylor,* 14.

11. Lang, *Making of Religion,* 200, 209, 175.

12. Wilhelm Schmidt, *The Origin and Growth of Religion: Facts and Theories,* trans. H. J. Rose (New York: Dial Press, 1931), 83; Julien Ries, *The Origins of Religion* (Grand Rapids, MI: Eerdmans, 1994), 25.

13. Lang, *Making of Religion,* 2.

14. Lang, *Making of Religion,* 181, 327.

15. Lang, *Making of Religion,* 334, 200, 257.

16. Lang, *Making of Religion,* 321, 328.

17. Robert Ackerman, *J. G. Frazer: His Life and Work* (Cambridge: Cambridge Univ. Press, 1987), 63.

18. Abram Kardiner and Edward Preble, *They Studied Man* (Cleveland: World, 1961), 89; Bronislaw Malinowski, *A Scientific Theory of Culture and Other Essays* (Chapel Hill: Univ. of North Carolina Press, 1944), 186.

19. Smith quoted in Ackerman, *J. G. Frazer,* 63.

20. Frazer, *Illustrated Golden Bough,* 252.

21. Douglas, "Introduction," 10.

22. Servius quoted in Fraser, "Introduction," xvii.

23. Frazer quoted in Sabine MacCormack, "Editorial Notes," *Illustrated Golden Bough,* 251.

24. Douglas, "Introduction," 11.

25. Frazer quoted in MacCormack, "Editorial Notes," 251.

26. Frazer quoted in Ackerman, *J. G. Frazer,* 234.

27. Frazer quoted in Ackerman, *J. G. Frazer,* 157.

28. Frazer quoted in Ackerman, *J. G. Frazer,* 157.

29. Frazer quoted in Ackerman, *J. G. Frazer,* 15.

30. Malinowski, *Scientific Theory of Culture,* 182.

31. Fraser, "Introduction," x.

32. Fraser, "Introduction," xli.

33. Ackerman, *J. G. Frazer,* 171.

34. Lang quoted in Ackerman, *J. G. Frazer,* 171.

35. Andrew Lang, *Magic and Religion* (London: Longmans, Green, 1901), 214, 103.

36. Lang, *Magic and Religion,* 216, 202.

37. Frazer quoted in Ackerman, *J. G. Frazer,* 175.

38. Frazer quoted in Ackerman, *J. G. Frazer,* 172, 173.

39. Daniel L. Pals, *Seven Theories of Religion* (New York: Oxford Univ. Press, 1996), 9.

40. Douglas, "Introduction," 14.

41. Schmidt, *Origin and Growth of Religion,* 88.

42. Frazer quoted in Joseph Campbell, *The Masks of God: Primitive Mythology* (New York: Penguin, 1976), 353.

43. Robert A. Segal, *Joseph Campbell: An Introduction* (New York: Garland, 1987), ix.

44. Joseph Campbell, *The Hero with a Thousand Faces* (New York: Pantheon, 1949), 382.

45. James quoted in Ackerman, *J. G. Frazer,* 175.

46. James quoted in Ackerman, *J. G. Frazer,* 175.

47. William James, *The Varieties of Religious Experience: A Study in Human Nature* (New York: Simon & Schuster, 1997), 388.

Chapter 4

1. William James, *The Varieties of Religious Experience: A Study in Human Nature* (New York: Simon & Schuster, 1997), 262.

2. Quoted in Linda Simon, *Genuine Reality: A Life of William James* (New York: Harcourt Brace, 1999), 309.

3. E-mail interview with John Snarey, March 2004; John Snarey, "Editorial, The Gifford Lectures," *Journal of Moral Education* 32 (Dec. 2003): 323–24, 327nn2,4,5.

4. "The Edinburgh Gifford Lectures," *Scotsman,* May 17, 1901, 8.

5. William James, "Preface," *The Will to Believe and Other Essays in Popular Philosophy* (Cambridge: Harvard Univ. Press, 1979), 5, 6.

6. Quoted in "Laboratories of Psychology," in *Oxford Companion to the Mind,* ed. Richard L. Gregory (Oxford: Oxford Univ. Press, 1998), 416.

7. James quoted in Amedeo Giorgi, "The Implications of James's Plea for Psychology as a Natural Science," in *Reflections on the Principles of Psychology: William James After a*

Century, ed. Michael G. Johnson and Tracy B. Henley (Hillsdale, NJ: Lawrence Erlbaum Associates, 1990), 69.

8. James, *Varieties,* 25, 29, 30.

9. James, *Varieties,* 35.

10. James, *Varieties,* 123.

11. James, *Varieties,* 139–40.

12. James, *Varieties,* 54.

13. James, *Varieties,* 393, 54.

14. James, *Varieties,* 401.

15. James, *Varieties,* 404, 405; Simon, *Genuine Reality,* 296–97, 376.

16. James quoted in Daniel N. Robinson, *Toward a Science of Human Nature: Essays on the Psychologies of Mill, Hegel, Wundt, and James* (New York: Columbia Univ. Press, 1982), 176.

17. Robinson, *Toward a Science,* 173.

18. James, *Will to Believe,* 13, 31.

19. James quoted in Simon, *Genuine Reality,* 276.

20. Alice James quoted in Simon, *Genuine Reality,* 276.

21. James quoted in Simon, *Genuine Reality,* 276.

22. Royce quoted in John Clendenning, *The Life and Thought of Josiah Royce,* rev. and expanded ed. (Nashville: Vanderbilt Univ. Press, 1999), 235.

23. James quoted in Simon, *Genuine Reality,* 290.

24. Alice James quoted in Simon, *Genuine Reality,* 291.

25. James quoted in Simon, *Genuine Reality,* 296.

26. James quoted in Simon, *Genuine Reality,* 298.

27. James quoted in Simon, *Genuine Reality,* 309, 312, 313.

28. Royce quoted in Clendenning, *Life and Thought,* 246.

29. Royce quoted in Clendenning, *Life and Thought,* 247, 248.

30. William James, "Monistic Idealism," in *The Writings of William James: A Comprehensive Edition,* ed. John J. McDermott (Chicago: Univ. of Chicago Press, 1977), 506.

31. C. Lloyd Morgan, *Emergent Evolution* (New York: Henry Holt, 1927), 36; C. Lloyd Morgan, "Mind in Evolution," in *Creation by Evolution,* ed. Frances B. Mason (New York: Macmillan, 1928), 345.

32. James Ward, "Psychology," *Encyclopaedia Britannica,* vol. 20 (Edinburgh: Adam & Charles Black, 1886), 85.

33. William James, *The Principles of Psychology,* vol. 1 (New York: Dover, 1950), 181, 182, 239.

34. James, *Principles of Psychology,* vol. 1, 278, 454.

35. James quoted in Simon, *Genuine Reality,* 378, 385.

36. James, *Varieties,* 198.

37. William James, "Does Consciousness Exist?" in *Writings of William James,* 169, 170, 171, 178, 179.

38. Peter J. Bowler, *Reconciling Science and Religion: The Debate in Early Twentieth-Century Britain* (Chicago: Univ. of Chicago Press, 2001), 44, 241.

39. Quoted in "Henri Bergson," *Britannica Guide to Nobel Prizes,* at http://www.britannica.com/nobel/micro/64_69.html.

40. Perry Miller, *The American Transcendentalists: Their Prose and Poetry* (New York: Anchor, 1957), 218; Frederic Ives Carpenter, *Emerson Handbook* (New York: Hendricks House, 1953), 243.

41. William James, "Bergson and His Critique of Intellectualism," in *Writings of William James,* 561, 566, 569.

42. Henri Bergson, *Creative Evolution* (Mineola, NY: Dover, 1998), 248.

43. Bergson, *Creative Evolution,* xiv.

44. S. Alexander, "Preface to the New Impression," *Space, Time and Deity,* vol. 1 (London: Macmillan, 1920; new impression 1927), xxiii; S. Alexander, *Space, Time and Deity,* vol. 2 (London: Macmillan, 1920), 429; Alexander quoted in John Laird, "Memoir," in *Samuel Alexander, Philosophical and Literary Pieces,* ed. John Laird (Freeport, NY: Books for Libraries Press, 1969), 64 (first published in 1940).

45. Alexander quoted in Laird, "Memoir," 60.

46. Alexander quoted in Laird, "Memoir," 96.

47. Inge quoted in Bowler, *Reconciling Science and Religion,* 275; and in Laird, "Memoir," 65.

48. Whitehead quoted in William A. Barker et al., "Alfred North Whitehead," *Dictionary of Scientific Biography,* vol. 14, ed. Charles Coulston Gillispie (New York: Scribner, 1976), 306.

49. Arthur S. Eddington, *The Nature of the Physical World* (Cambridge: Cambridge Univ. Press, 1929), 249.

50. Whitehead quoted in Victor Lowe, *Alfred North Whitehead: The Man and His Work,* vol. 1, ed. J. B. Schneewing (Baltimore: Johns Hopkins Univ. Press, 1990), 24; Russell and Mrs. Whitehead quoted in Victor Lowe, *Alfred North Whitehead: The Man and His Work,* vol. 2, ed. J. B. Schneewing (Baltimore: Johns Hopkins Univ. Press, 1990), 188.

51. Alfred North Whitehead, *Science and the Modern World* (New York: Macmillan, 1953), viii, 143; Alfred North Whitehead, "Preface," in *Process and Reality: Corrected Edition,* ed. David Ray Griffin and Donald W. Sherburne (New York: Free Press, 1978), vii.

52. Whitehead, "Preface," xii, xiii.

53. Whitehead quoted in Victor Lowe, "The Gifford Lectures," in *Southern Journal of Philosophy* 7 (Winter 1969–70): 330, 332. On Whitehead's Giffords, see also Lowe, *Alfred North Whitehead,* vol. 2, 215–52.

54. "Gifford Lecture: Process and Reality," *Glasgow Herald,* June 2, 1927, 11; Whitehead quoted in Lowe, "Gifford Lectures," 337.

55. Whitehead, *Process and Reality,* 18, 23.

56. Whitehead, *Process and Reality,* 351.

57. Whitehead quoted in Lowe, "Gifford Lectures," 336, 337.

58. John Macquarrie, *In Search of Deity: An Essay in Dialectical Theism* (New York: Crossroad, 1985), 145, 146.

59. John Polkinghorne, *The Faith of a Physicist* (Minneapolis: Fortress Press, 1996), 65, 68.

Chapter 5

1. Arthur S. Eddington, *The Nature of the Physical World* (Cambridge: Cambridge Univ. Press, 1929), vii.

2. Einstein quoted in "News and Views," *Nature,* March 26, 1927, 467.

3. Niels Blaedel, *Harmony and Unity: The Life of Niels Bohr* (Madison, WI: Science Tech, 1988), 169.

4. Eddington, *Nature of the Physical World,* 207, 350.

5. William James, "Some Metaphysical Problems Pragmatically Considered," in *Pragmatism* (New York: Dover, 1995), 39.

6. Eddington, *Nature of the Physical World,* 2, 1.

7. G. G. Stokes, *Natural Theology* (London: Adam & Charles Black, 1891), 98–100.

8. Peter J. Bowler, *Reconciling Science and Religion: The Debate in Early Twentieth-Century Britain* (Chicago: Univ. of Chicago Press, 2001), 90.

9. Quoted in A. Vibert Douglas, *The Life of Arthur Stanley Eddington* (London: Thomas Nelson & Sons, 1956), 40.

10. Eddington, *Nature of the Physical World,* 341.

11. Eddington, *Nature of the Physical World,* 281, 275, 282.

12. Eddington, *Nature of the Physical World,* 1.

13. Eddington, *Nature of the Physical World,* 1.

14. Niels Bohr, *Essays 1958–1962 on Atomic Physics and Human Knowledge,* vol. 3, *The Philosophical Writings of Niels Bohr* (Woodbridge, CT: Ox Bow Press, 1987), 31; Francis Crick, *What Mad Pursuit: A Personal View of Scientific Discovery* (New York: Basic, 1988), 58.

15. Rutherford quoted in Lawrence Badas, "Ernest Rutherford," in *Dictionary of Scientific Biography,* vol. 12, ed. Charles Coulston Gillispie (New York: Scribner, 1975), 31.

16. Blaedel, *Harmony and Unity,* 34.

17. Quoted in Blaedel, *Harmony and Unity,* 60.

18. Quoted in Blaedel, *Harmony and Unity,* 110.

19. Heisenberg quoted in Blaedel, *Harmony and Unity,* 111.

20. Pauli quoted in Arthur I. Miller, "On the Origins of the Copenhagen Interpretation," in *Niels Bohr: Physics and the World, Proceedings of the Niels Bohr Centennial Symposium,* ed. Herman Feshbach, Tetsuo Matsui, and Alexandra Oleson (Chur, Switzerland: Harwood Academic, 1988), 35.

21. Eddington, *Nature of the Physical World,* 210, 199.

22. Bohr quoted in Blaedel, *Harmony and Unity,* 117.

23. Bohr quoted in Blaedel, *Harmony and Unity,* 109.

24. Rutherford quoted in Blaedel, *Harmony and Unity,* 119.

25. Bohr, *Essays 1958–1962,* 2.

26. Bohr paraphrased in Aage Peterson, "The Philosophy of Niels Bohr," in *Niels Bohr: A Centenary Volume,* ed. A. P. French and P. J. Kennedy (Cambridge: Harvard Univ. Press, 1985), 301, 302.

27. Bohr, *Essays 1958–1962,* 12.

28. Blaedel, *Harmony and Unity,* 268–69.

29. A. Pais, "Niels Bohr and the Development of Physics," in *Niels Bohr: Physics and the World,* 18.

30. Werner Heisenberg, *Physics and Philosophy: The Revolution in Modern Science* (Amherst, NY: Prometheus, 1999), 181. (Originally published in 1958.)

31. Heisenberg, *Physics and Philosophy,* 200–201.

32. For the Jeans appointment, see the *Glasgow Herald,* Feb. 25, 1931, 8; quoted from James Jeans, *The Mysterious Universe* (New York: Macmillan, 1930), 144.

33. Eddington, *Nature of the Physical World,* 332, 338.

34. Eddington, *Nature of the Physical World,* 294, 302.

35. Eddington, *Nature of the Physical World,* 339, 333.

36. Arthur S. Eddington, *The Expanding Universe* (Cambridge: Cambridge Univ. Press, 1946), 86–87, 51.

37. Eddington, *Expanding Universe,* 59.

38. Eddington, *Expanding Universe,* 61.

39. Einstein quoted in Gunter Stent, "Does God Play Dice?" *The Sciences* 19 (March 1979): 23.

40. Heisenberg quoted in Gerald Holton, "Werner Heisenberg and Albert Einstein," paper presented at the symposium "Creating Copenhagen," March 27, 2000, Graduate Center of the City University of New York.

41. Quoted in Werner Heisenberg, *Beyond Physics: Encounters and Conversations,* trans. Arnold J. Pomerans (New York: Harper & Row, 1971), 82.

42. Rabbi quoted in Stent, "Does God Play Dice?" 22.

43. Stent, "Does God Play Dice?" 21–22, 23.

44. Bohr quoted in Blaedel, *Harmony and Unity,* 26, 28.

45. Bohr, *Essays 1958–1962,* 49; Douglas Murdoch, *Niels Bohr's Philosophy of Physics* (New York: Cambridge Univ. Press, 1987), 243–44.

46. Bohr quoted in Max Jammer, *Einstein and Religion* (Princeton: Princeton Univ. Press, 1999), 29.

47. Pais, "Niels Bohr and the Development of Physics," 20.

48. Heisenberg, *Beyond Physics,* 87, 91.

49. Heisenberg, *Beyond Physics,* 215.

50. Quoted in "Discussion with Professor Heisenberg," in *The Nature of Scientific Discovery: A Symposium Commemorating the 500th Anniversary of the Birth of Nicholas Copernicus,* ed. Owen Gingerich (Washington, DC: Smithsonian Institution Press, 1975), 559.

51. Thomas F. Torrance, *Space, Time and Incarnation* (Oxford: Oxford Univ. Press, 1969), 56, 53, 67.

52. Keith Ward, *Religion and Revelation: A Theology of Revelation in the World's Religions* (Oxford: Clarendon Press, 1994), 321; for other examples, see Jammer, *Einstein and Religion,* 229–30.

53. Eddington, *Nature of the Physical World,* 353.

54. Heisenberg, *Physics and Philosophy,* 159, 160.

55. John Barrow, "Inner Space and Outer Space: The Quest for Ultimate Explanation," in *Humanity, Environment and God: Glasgow Centenary Gifford Lectures,* ed. Neil Spurway (Oxford: Blackwell, 1993), 50, 82.

56. Werner Heisenberg, "Tradition in Science," in *Nature of Scientific Discovery,* 224.

Chapter 6

1. John Baillie, "Gifford Lecturer: Professor R. Niebuhr," *Scotsman,* April 22, 1939, 17; Niebuhr quoted in Richard W. Fox, *Reinhold Niebuhr: A Biography* (New York: Pantheon, 1985), 188.

2. Reinhold Niebuhr, "Leaves from the Notebook of a War-Bound American," *Christian Century,* Dec. 27, 1939, 1607.

3. Reinhold Niebuhr, *The Nature and Destiny of Man: Human Nature,* vol. 1 (Louisville, KY: John Knox Press, 1996), 49.

4. "The Mass Man: Dr. Niebuhr's Gifford Lecture," *Scotsman,* April 25, 1939, 14.

5. Reinhold Niebuhr, "Intellectual Autobiography," in *Reinhold Niebuhr: His Religious, Social, and Political Thought,* vol. 2, ed. Robert W. Bretall and Charles W. Kegley (New York: Macmillan, 1956), 4.

6. Niebuhr, *Nature and Destiny,* vol. 1, 255.

7. Niebuhr, "Intellectual Autobiography," 9.

8. Hans Gerth and Saul Landau, "The Relevance of History to the Sociological Ethos," in *Sociology on Trial,* ed. Maurice Stein and Arthur Vidich (Englewood Cliffs, NJ: Prentice-Hall, 1963), 26.

9. Niebuhr, *Nature and Destiny,* vol. 1, 108–9.

10. Niebuhr, *Nature and Destiny,* vol. 1, 93.

11. John Stuart Mill, *System of Logic: Rationcinative and Inductive* (London: Longmans, Green, 1896), 573.

12. Emile Durkheim, *Suicide: A Study in Sociology,* trans. John A. Spaulding and George Simpson (New York: Free Press, 1951), 299; Emile Durkheim, *The Rules of Sociological Method* (New York: Free Press, 1964), 8, 9, 10.

13. Max Weber, *The Theory of Social and Economic Organization,* trans. A. M. Henderson and Talcott Parsons (New York: Oxford Univ. Press, 1947), 88.

14. Charles Howard Hopkins, *The Rise of the Social Gospel in American Protestantism, 1865–1915* (New Haven: Yale Univ. Press, 1940), 267; Strong quoted on 257.

15. Niebuhr, "Intellectual Autobiography," 20.

16. Adolf von Harnack, *What Is Christianity?* trans. Thomas Bailey Saunders (New York: Harper & Bros., 1957), 51.

17. Niebuhr quoted in Fox, *Reinhold Niebuhr,* 92.

18. Fox, *Reinhold Niebuhr,* 90.

19. Niebuhr, "Intellectual Autobiography," 5.

20. Fox, *Reinhold Niebuhr,* 217.

21. Niebuhr, "Intellectual Autobiography," 5.

22. John Dewey, *The Quest for Certainty: A Study of the Relation of Knowledge and Action* (New York: Putnam, 1929), 306, 304.

23. Joseph Fletcher, *William Temple: Twentieth Century Christian* (New York: Seabury Press, 1963), 319n140.

24. William Temple, *Nature, Man and God* (London: Macmillan, 1956), 306.

25. Gary Bullert, *The Politics of John Dewey* (Buffalo, NY: Prometheus, 1983), 182.

26. Niebuhr, "Intellectual Autobiography," 8.

27. H. Richard Niebuhr quoted in Fox, *Reinhold Niebuhr,* 212.

28. Reinhold Niebuhr, *Moral Man and Immoral Society: A Study in Ethics and Politics* (New York: Scribner, 1960), xv, 35.

29. Niebuhr, *Moral Man and Immoral Society,* xxv, xix.

30. Coffin quoted in Fox, *Reinhold Niebuhr,* 159.

31. Niebuhr, "Intellectual Autobiography," 8.

32. Fox, *Reinhold Niebuhr,* 204.

33. Niebuhr quoted in Fox, *Reinhold Niebuhr,* 166.

34. Niebuhr quoted in Fox, *Reinhold Niebuhr,* 178.

35. Quoted in Fox, *Reinhold Niebuhr,* 189.

36. Niebuhr, "Intellectual Autobiography," 10.

37. Calhoun quoted in Fox, *Reinhold Niebuhr,* 203.

38. Niebuhr, "Intellectual Autobiography," 15, 16.

39. Reinhold Niebuhr, *The Nature and Destiny of Man: Human Destiny,* vol. 2 (Louisville, KY: John Knox Press, 1996), 85.

40. Fox, *Reinhold Niebuhr,* 213.

41. Gerth and Landau, "Relevance of History," 29.

42. David Martin, *Reflections on Sociology and Theology* (Oxford: Oxford Univ. Press, 1997), 95, 97–98, 102.

43. Calhoun quoted in Fox, *Reinhold Niebuhr,* 204.

44. Niebuhr, *Nature and Destiny,* vol. 1, 170–71.

45. Stanley Hauerwas, *With the Grain of the Universe: The Church's Witness and Natural Theology* (Grand Rapids, MI: Brazos Press, 2001), 38, 139.

46. Niebuhr, "Preface to the 1964 Edition," in *Nature and Destiny,* vol. 2, xiii–xiv.

47. Paul Tillich, *A History of Christian Thought,* ed. Carl E. Braaten (New York: Simon & Schuster, 1967), 521.

48. Albert Schweitzer, *Out of My Life and Thought: An Autobiography* (Baltimore: Johns Hopkins Univ. Press, 1998), 54.

Chapter 7

1. "Modern Thought Problems: Dr. Schweitzer on Danger of Skepticism," *Glasgow Herald,* Nov. 6, 1934, 11.

2. Herbert Butterfield, "God in History," in *Herbert Butterfield: Writings on Christianity and History,* ed. C. T. McIntire (New York: Oxford Univ. Press, 1979), 3, 4.

3. Herodotus quoted in Edward Hallett Carr, *What Is History?* (New York: Knopf, 1962), 113.

4. Ranke quoted in Carr, *What Is History?* 5.

5. Albert Schweitzer, *Out of My Life and Thought: An Autobiography* (Baltimore: Johns Hopkins Univ. Press, 1998), 44.

6. Albert Schweitzer, *The Quest of the Historical Jesus: A Critical Study of Its Progress from Reimarus to Wrede* (New York: Macmillan, 1968), 208.

7. Schweitzer, *Quest,* 399.

8. Schweitzer, *Out of My Life,* 115.

9. Schweitzer, *Out of My Life,* 125, 128.

10. Schweitzer, *Out of My Life,* 54.

11. H. Stuart Hughes, "Preface to the Present Edition," in Oswald Spengler, *The Decline of the West,* abr. Helmet Werner, trans. Charles Francis Atkinson (New York: Oxford Univ. Press, 1991), xv.

12. Schweitzer, *Out of My Life,* 155.

13. "Man's Progress in Thought: Dr. Schweitzer's Lecture," *Glasgow Herald,* Nov. 8, 1934, 11.

14. The following quotes are from ten reports in the *Glasgow Herald,* Nov. 6–27, 1934, and twelve reports in the *Glasgow Herald,* Nov. 5–30, 1935.

15. "Modern Thought and Religion: The Conception of a Personal God," *Glasgow Herald,* Nov. 30, 1935, 7.

16. There is some dispute about whether Toynbee read, and even copied, Spengler's idea of cycles in history. Toynbee biographer William H. McNeill believes he did both. See McNeill, "Toynbee's Life and Thought: Some Unresolved Questions," in *Toynbee: Reappraisals,* ed. C. T. McIntire and Marvin Perry (Toronto: Univ. of Toronto Press, 1989), 34.

17. McNeill, "Toynbee's Life and Thought," 65.

18. McNeill, "Toynbee's Life and Thought," 40.

19. Quoted in "Historian's Approach to Religion," *Glasgow Herald,* Oct. 25, 1952, 9.

20. Arnold Toynbee, *An Historian's Approach to Religion* (London: Oxford Univ. Press, 1956), 17.

21. Toynbee, *Historian's Approach,* 292, 277.

22. Toynbee, *Historian's Approach,* 298, 274, 275, 265.

23. Toynbee, *Historian's Approach,* 280–81, 282, 284.

24. Toynbee, *Historian's Approach,* 299.

25. Christian B. Peper, "Toynbee: An Historian's Conscience," in *Toynbee: Reappraisals,* 81.

26. C. T. McIntire, "Preface," in *Toynbee: Reappraisals*, vii.

27. Harbison quoted in C. T. McIntire, ed., *God, History, and Historians: An Anthology of Modern Christian Views of History* (New York: Oxford Univ. Press, 1977), 4.

28. Roger Lincoln Shinn, *Christianity and the Problem of History* (New York: Scribner, 1953), 166.

29. Acton quoted in Carr, *What Is History?*, 147.

30. Butterfield quoted in McIntire, *Herbert Butterfield*, xxxi; Owen Chadwick, *Freedom and the Historian: An Inaugural Lecture* (Cambridge: Cambridge Univ. Press, 1969), 37.

31. Herbert Butterfield, "The Establishment of a Christian Interpretation of World History," in *Herbert Butterfield*, 111.

32. Augustine quoted in Pardon E. Tillinghast, *Approaches to History: Selections in the Philosophy of History from the Greeks to Hegel* (Englewood Cliffs, NJ: Prentice-Hall, 1963), 68, 69.

33. Butterfield, "Establishment of a Christian Interpretation," 184.

34. Butterfield, "God in History," 13, 16.

35. Herbert Butterfield, "The Modern Historian and New Testament History," in *Herbert Butterfield*, 96.

36. Butterfield, "Modern Historian," 97, 102, 104.

37. Robin G. Collingwood, *The Idea of History*, rev. ed. (New York: Oxford Univ. Press, 1993), 213.

Chapter 8

1. Owen Chadwick, *The Secularization of the European Mind in the Nineteenth Century* (Cambridge: Cambridge Univ. Press, 1975), 226.

2. See Grant Wacker, "The Demise of Biblical Civilization," in *The Bible in America: Essays in Cultural History*, ed. Nathan O. Hatch and Mark A. Noll (New York: Oxford Univ. Press, 1982), 121–38.

3. Karl Barth, *The Knowledge of God and the Service of God According to the Teachings of the Reformation: Recalling the Scottish Confession of 1560*, trans. L.L.M. Haire and Ian Henderson (New York: Scribner, 1939), 6.

4. Buddhist quoted in Ernest Benz, "Ideas for a Theology of the History of Religion," in *The Theology of the Christian Mission*, ed. Gerald H. Anderson (New York: McGraw-Hill, 1961), 136.

5. Frederick C. Copleston, *Existentialism and Modern Man* (Oxford: Blackfriars, 1948), 7.

6. Barth, *Knowledge of God*, 5.

7. Interview with Gordon Graham, Regius Professor of Philosophy at Aberdeen University, Nov. 2003.

8. Letter from St. George Mivart to the Academic Senatus, St. Andrews University, March 8, 1888, in university library archives. (Author's copy.)

9. Michael de la Bedoyere, *The Life of Baron von Hügel* (New York: Scribner, 1951), xii.

10. Etienne Gilson, *The Spirit of Mediaeval Philosophy,* trans. A. H. C. Downs (Notre Dame, IN: Univ. of Notre Dame Press, 1991), 70.

11. Gilson, *Spirit of Mediaeval Philosophy,* 244.

12. James Barr, *Biblical Faith and Natural Theology: The Gifford Lectures for 1991 Delivered in the University of Edinburgh* (Oxford: Clarendon Press, 1993), 116, 116n20.

13. Barth quoted in Wilhelm Pauck, *From Luther to Tillich: The Reformers and Their Heirs* (San Francisco: Harper & Row, 1984), 146.

14. Barth quoted in Stanley Hauerwas, *With the Grain of the Universe: The Church's Witness and Natural Theology* (Grand Rapids, MI: Brazos Press, 2001), 154.

15. Christopher Dawson, *Religion and Culture* (Cleveland: Meridian, 1958), 3.

16. The excerpts from letters exchanged between Barth and the Aberdeen senate were provided in their original German by the Karl Barth Archive in Switzerland. The translation used here was done by Werner Seubert.

17. Barth, *Knowledge of God,* 6–7.

18. Eberhard Busch, *Karl Barth: His Life from Letters and Autobiographical Texts* (London: SMC Press, 1976), 280, 281.

19. Busch, *Karl Barth,* 281.

20. Busch, *Karl Barth,* 281.

21. Barth quoted in John Macquarrie, *Twentieth-Century Religious Thought,* 3d ed. (New York: Scribner, 1981), 323; and in "Witness to an Ancient Truth," *Time,* April 20, 1962, 62.

22. Barth quoted in "Witness to an Ancient Truth," 59.

23. Interview with Professor John Webster at the University of Aberdeen, Nov. 2003.

24. Blanshard and Torrance quoted in Hauerwas, *With the Grain,* 141, 142.

25. Charles E. Raven, *Natural Religion and Christian Theology,* vol. 1 (Cambridge: Cambridge Univ. Press, 1953), 201–2.

26. Barr, *Biblical Faith and Natural Theology,* 7n5.

27. Barr, *Biblical Faith and Natural Theology,* 112, 11, 111.

28. Barr, *Biblical Faith and Natural Theology,* 123, 102, 131.

29. Barr, *Biblical Faith and Natural Theology,* 132, 117.

30. Jean Bethke Elshtain, "Christian Contrarian," *Time,* Sept. 17, 2001, 76; interview with Webster.

31. Hauerwas, *With the Grain,* 9–10, 39.

32. Hauerwas, *With the Grain,* 26, 142–43, 145–46.

33. Hauerwas, *With the Grain,* 206–7, 216–17.

34. Dilthey quoted in Steven Galt Crowell, "Twentieth-Century Continental Philosophy, the Early Decades: Positivism, Neo-Kantianism, Dilthey," in *The Columbia History of Western Philosophy,* ed. Richard H. Popkin (New York: Columbia Univ. Press, 1999), 673.

35. Copleston, *Existentialism and Modern Man,* 17.

36. Thomas R. Flynn, "Existentialism and Beyond, Jean-Paul Sartre, Simone de Beauvoir, Albert Camus, and Maurice Merleau-Ponty," in *The Columbia History of Western Philosophy*, 698.

37. Bultmann quoted in Macquarrie, *Twentieth-Century Religious Thought*, 362.

38. Rudolf Bultmann, *History and Eschatology: The Presence of Eternity* (New York: Harper & Row, 1957), 155.

39. Macquarrie, *Twentieth-Century Religious Thought*, 353.

40. John Herman Randall, Jr., "The Philosophical Legacy of Paul Tillich," in *The Intellectual Legacy of Paul Tillich*, ed. James R. Lyons (Detroit: Wayne State Univ. Press, 1969), 33.

41. Tillich quoted in P. Hamilton, *The Living God and the Modern World* (London: Hodder & Stoughton, 1967), 151.

42. Gilson, *Spirit of Mediaeval Philosophy*, 51.

43. Copleston, *Existentialism and Modern Man*, 20.

44. Emil Brunner, *Christianity and Civilization: Specific Problems*, Part 2 (New York: Scribner, 1949), 132, 142, 134, 139.

45. Eileen Barker, "The Post-War Generation and Establishment Religion in England," in *The Post-War Generation and Establishment Religion: Cross-Cultural Perspectives*, ed. Wade Clark Roof, Jackson W. Carroll, and David A. Roozen (Boulder, CO: Westview Press, 1995), 5. She cites the 1990 European Values Survey.

46. David L. Edwards, ed., *The 'Honest to God' Debate: Some Reactions to the Book 'Honest to God'* (London: SCM Press, 1963), 7.

47. Barker, "Post-War Generation," 12.

48. Bernard E. Jones, ed., *Earnest Enquirers After Truth: A Gifford Anthology* (London: Allen & Unwin, 1970), 15–16. See also Bernard E. Jones, "The Concept of Natural Theology in the Gifford Lectures" (Ph.D. diss., Univ. of Leeds, May 1966). Jones was professor of theology at Wesley College, Bristol, England.

49. Jones, *Earnest Enquirers*, 20.

50. Chadwick, *Secularization of the European Mind*, 171–72.

51. John Gifford, "Recollections of a Brother, and of His Homes," in Stanley L. Jaki, *Lord Gifford and His Lectures: A Centenary Retrospect* (Edinburgh: Scottish Academic Press, 1986), 99.

52. A. L. Brown and Michael Moss, *The University of Glasgow: 1451–2001* (Edinburgh: Edinburgh Univ. Press, 2001), 59.

53. Freeman J. Dyson, *Infinite in All Directions* (New York: Harper & Row, 1988), viii.

Chapter 9

1. C. F. von Weizsäcker, *The Relevance of Science: Creation and Cosmogony* (New York: Harper & Row, 1965), 12, 19, 21.

2. Freeman J. Dyson, *Disturbing the Universe* (New York: Harper & Row, 1979), 256.

3. Charles E. Raven, *Natural Religion and Christian Theology,* vol. 1 (Cambridge: Cambridge Univ. Press, 1953), 18, 14, 197.

4. Raven, *Natural Religion,* 14, 202.

5. Raven, *Natural Religion,* 14.

6. Weizsäcker, *Relevance of Science,* 23.

7. James D. Watson, *The Double Helix: A Personal Account of the Discovery of the Structure of DNA* (New York: Mentor/Penguin, 1969), 18.

8. Francis Crick, "Nucleic Acids," *Scientific American,* Sept. 1957, 200; Francis Crick, *What Mad Pursuit: A Personal View of Scientific Discovery* (New York: Basic, 1988), 138.

9. Raven, *Natural Religion,* 15.

10. Raven, *Natural Religion,* 15.

11. Polanyi quoted in John Polkinghorne, *The Faith of a Physicist: Reflections of a Bottom-Up Thinker* (Princeton: Princeton Univ. Press, 1994), 27.

12. Michael Polanyi, "Life Transcending Physics and Chemistry," *Chemical and Engineering News* 75 (1967): 54–66; Michael Polanyi, "Life's Irreducible Structure," in *Knowing and Being: Essays by Michael Polanyi,* ed. Marjorie Grene (Chicago: Univ. of Chicago Press, 1969), 231.

13. Sherrington quoted in John C. Eccles and William C. Gibson, *Sherrington: His Life and Thought* (Berlin and New York: Springer International, 1979), 183.

14. Sherrington quoted in Eccles and Gibson, *Sherrington,* 155.

15. Charles Sherrington, *Man on His Nature,* 2d ed. (Cambridge: Cambridge Univ. Press, 1951), 251.

16. Charles Sherrington, *The Integrative Action of the Nervous System* (Cambridge: Cambridge Univ. Press, 1948), xxiv.

17. Crick, *What Mad Pursuit,* 151; Francis Crick, *The Astonishing Hypothesis: The Scientific Search for the Soul* (New York: Simon & Schuster, 1994), 3.

18. Crick, *Astonishing Hypothesis,* 6.

19. Crick, *What Mad Pursuit,* 18.

20. John C. Eccles, *The Human Psyche* (New York: Routledge, 1992), 21.

21. Eccles, *Human Psyche,* 49, 25.

22. Quoted in Richard Ostling, "One of the World's Leading Atheists Now Believes in God, More or Less," *Associated Press,* Dec. 9, 2004.

23. Richard Swinburne, *The Evolution of the Soul,* rev. ed. (Oxford: Clarendon Press, 1997), 9, 10, 310.

24. Jack Morrell, *Science at Oxford, 1914–1939: Transforming an Arts University* (Oxford: Oxford Univ. Press, 1997), 271.

25. "Preface," in Alister Hardy, *The Open Sea: Its Natural History* (Boston: Houghton Mifflin, 1956), ix.

26. Alister Hardy, *The Spiritual Nature of Man: A Study of Contemporary Religious Experience* (Oxford: Clarendon Press, 1979), 3; Alister Hardy, *The Biology of God: A Scientist's Study of Man the Religious Animal* (New York: Taplinger, 1976), 193, 233.

27. Hardy, *Biology of God,* 23, 211.

28. Alister Hardy, *The Living Stream: A Restatement of Evolutionary Theory in Its Relation to the Spirit of Man* (London: Collins, 1965), 284.

29. Alister Hardy, *The Divine Flame: An Essay Towards a Natural History of Religion* (London: Collins, 1966), 213; Hardy, *Biology of God,* 212.

30. Hardy, *Biology of God,* 185; see also, Hardy, *Spiritual Nature of Man,* 16–19.

31. Hardy, *Biology of God,* 17, 23.

32. Hardy, *Biology of God,* 20.

33. Sherrington, *Man on His Nature,* 318.

34. Hardy, *Spiritual Nature of Man,* 11.

35. Freeman J. Dyson, *Infinite in All Directions* (New York: Harper & Row, 1988), 5, 119.

36. Ian Barbour, *Religion in an Age of Science: The Gifford Lectures, 1989–91,* vol. 1 (San Francisco: Harper San Francisco, 1990), 232.

37. Barbour, *Religion,* 233.

38. Barbour, *Religion,* 234.

39. James Gleick, *Chaos: Making a New Science* (New York: Viking Penguin, 1987), 11, 30.

40. Gleick, *Chaos,* 4.

41. Gleick, *Chaos,* 8.

42. Gleick, *Chaos,* 55.

43. Polkinghorne, *Faith of a Physicist,* 1.

44. Polkinghorne, *Faith of a Physicist,* 25.

45. Polkinghorne, *Faith of a Physicist,* 25, 26, 77.

46. Pius XII, "Modern Science and the Existence of God," *Catholic Mind* 50 (March 1952): 182–92.

47. Einstein quoted in Robert Jastrow, *God and the Astronomers,* 2d ed. (New York: Norton, 1992), 32.

48. Polkinghorne, *Faith of a Physicist,* 76.

49. Keith E. Yandel, "Protestant Theology and Natural Science in the Twentieth Century," in *God and Nature: Historical Essays on the Encounter Between Christianity and Science,* ed. David C. Lindberg and Ronald L. Numbers (Berkeley and Los Angeles: Univ. of California Press, 1986), 461.

50. Stanley L. Jaki, *The Road of Science and the Ways to God* (Chicago: Univ. of Chicago Press, 1978), 327.

51. Jaki, *Road of Science,* 327.

52. Eric L. Mascall, *The Openness of Being: Natural Theology Today* (London: Darton, Longman, & Todd, 1971), 159, 172.

53. Mascall, *Openness of Being,* 3.

Chapter 10

1. J. Campbell Smith, review of *Lectures Delivered on Various Occasions by Adam Gifford*, by Alice Raleigh and Herbert James Gifford (eds.), *The Judicial Review: A Journal of Legal and Political Science* 1 (1889): 405.

2. Quoted in Charles Clayton Morrison, "The World Missionary Conference, 1910," *Christian Century*, July 4–11, 1984, 660. (Reprinted from the July 7, 1910, issue.)

3. Morrison, "World Missionary Conference," 662.

4. Morrison, "World Missionary Conference," 663.

5. Quoted in Gerald H. Anderson, ed., *The Theology of the Christian Mission* (New York: McGraw-Hill, 1961), 7.

6. William Ernest Hocking, and Laymen's Foreign Missions Inquiry, Commission of Appraisal, *Re-Thinking Missions: A Laymen's Inquiry After One Hundred Years* (New York: Harper & Bros., 1932), 32.

7. Wilfred Cantwell Smith, "Christianity's Third Great Challenge," *Christian Century*, April 27, 1960, 506.

8. Quoted in Wilfred Cantwell Smith, *Religious Diversity: Essay by Wilfred Cantwell Smith*, ed. Willard G. Oxtoby (New York: Harper & Row, 1976), 7.

9. Peter L. Berger, *The Heretical Imperative: Contemporary Possibilities of Religious Affirmation* (Garden City, NY: Anchor Press, 1979), xi.

10. Adam Gifford, "The Ten Avatars of Vishnu," in Stanley L. Jaki, *Lord Gifford and His Lectures: A Centenary Retrospect* (Edinburgh: Scottish Academic Press, 1986), 128.

11. Gifford, "Ten Avatars of Vishnu," 128, 129.

12. Gifford, "Ten Avatars of Vishnu," 130, 131, 132.

13. Acts 14:16–17; 17:23, 27.

14. Ernest Benz, "Ideas for a Theology of the History of Religion," in *Theology of the Christian Mission*, 136.

15. Paul D. Devanandan, "The Resurgence of Non-Christian Religions," in *Theology of the Christian Mission*, 155.

16. "William Ernest Hocking," in *The Oxford Companion to Philosophy* (Oxford: Oxford Univ. Press, 1995), 370–71.

17. Hocking, *Re-Thinking Missions*, 47.

18. William Ernest Hocking, *Living Religions and a World Faith* (New York: Macmillan, 1940), 8.

19. Hocking, *Re-Thinking Missions*, 3, 329.

20. Hocking, *Re-Thinking Missions*, 32.

21. John Baillie, *The Sense of the Presence of God* (New York: Scribner, 1962), 168, 188.

22. Austin P. Flannery, ed., *Documents of Vatican II* (Grand Rapids, MI: Eerdmans, 1975), 739, 740.

23. Robert Charles Zaehner, *Concordant Discord: The Interdependence of Faiths* (Oxford: Clarendon Press, 1970), 20.

24. Zaehner, *Concordant Discord,* 16, 17.

25. Zaehner, *Concordant Discord,* 10, 2.

26. John Hick, *John Hick: An Autobiography* (Oxford: Oneworld, 2002), 323, 78, 86–87.

27. Farmer quoted in Hick, *John Hick,* 84–85.

28. Hick, *John Hick,* 85–86.

29. Hick, *John Hick,* 160.

30. See Ninian Smart, *Beyond Ideology: Religion and the Future of Western Civilization* (San Francisco: Harper & Row, 1981), 14.

31. Quoted in Hick, *John Hick,* 166, 170.

32. John Hick, *An Interpretation of Religion: Human Responses to the Transcendent* (New Haven: Yale Univ. Press, 1989), 235.

33. Hick, *John Hick,* 262.

34. Hick, *John Hick,* 125–26, 129.

35. Hick, *John Hick,* 262.

36. Hick, *Interpretation of Religion,* 1.

37. Hick, *Interpretation of Religion,* 13, 376.

38. Hick, *Interpretation of Religion,* 15, 369–70.

39. Hick, *John Hick,* 262.

40. Hick, *Interpretation of Religion,* 377.

41. For his views on the Trinity, see Raimundo Panikkar, *A Dwelling Place for Wisdom,* trans. Annemarie S. Kidder (Louisville, KY: Westminster John Knox Press, 1993), 147.

42. Keith Ward, *Religion and Revelation* (Oxford: Oxford Univ. Press, 1994), 311, 310.

43. Ward, *Religion and Revelation,* 315.

44. Hick, *Interpretation of Religion,* 350, 348.

45. Ward, *Religion and Revelation,* 316, 311.

46. Brian Hebblethwaite, "The Nature and Limits of Theological Understanding," in *The Nature and Limits of Human Understanding,* ed. Anthony J. Sanford (London: T. & T. Clark, 2003), 247.

47. Ward, *Religion and Revelation,* 324.

48. Ward, *Religion and Revelation,* 318.

49. Ward, *Religion and Revelation,* 319, 322.

50. Interview with Professor David Fergusson, Edinburgh, Nov. 2003.

51. Berger, *Heretical Imperative,* 28.

52. Alasdair MacIntyre, *Three Rival Versions of Moral Enquiry: Encyclopedia, Genealogy, and Tradition* (Notre Dame, IN: Univ. of Notre Dame Press, 1990), 235–36.

53. MacIntyre, *Three Rival Versions,* 231, 232.

54. Lynne Rudder Baker, "Third-Person Understanding," in *Nature and Limits of Human Understanding,* 206; Hebblethwaite, "Nature and Limits of Metaphysical Understanding," in *Nature and Limits of Human Understanding,* 215.

55. David M. Walker, *The Oxford Companion to Law* (Oxford: Clarendon Press, 1980), 1116.

56. John Laird, *Mind and Deity* (London: Allen & Unwin, 1941), 7, 319–20.

Chapter 11

1. Reid quoted in Dale Tuggy, "Reid's Philosophy of Religion," in *The Cambridge Companion to Thomas Reid,* ed. Terrence Cuneo and Rene van Woundeberg (Cambridge: Cambridge Univ. Press, 2004), 307n10.

2. Reid quoted in Nicholas Wolterstorff, *Thomas Reid and the Story of Epistemology* (Cambridge: Cambridge Univ. Press, 2001), 253.

3. William R. Brock, *Scotus Americanus: A Survey of the Sources for Links Between Scotland and America in the Eighteenth Century* (Edinburgh: Edinburgh Univ. Press, 1982), 92.

4. Arthur Herman, *How the Scots Invented the Modern World* (New York: Three Rivers Press, 2001), 263.

5. Alexander Hamilton, James Madison, and John Jay, *The Federalist Papers,* ed. Clinton Rossiter (New York: New American Library, 1961), 79, 83.

6. Herman, *How the Scots Invented,* 258.

7. James H. Smylie, "Madison and Witherspoon: Theological Roots of American Political Thought," *Princeton University Library Chronicle* 22 (Spring 1961): 124.

8. Witherspoon quoted in Garrett Ward Sheldon, *The Political Philosophy of James Madison* (Baltimore: Johns Hopkins Univ. Press, 2001), 17.

9. William James, *The Varieties of Religious Experience: A Study in Human Nature* (New York: Simon & Schuster, 1997), 347.

10. E-mail interview with Nicholas Wolterstorff, April 2004; see also "Note," in William James, *Pragmatism* (New York: Dover, 1995), iii. Peirce is described as having "initially formulated" pragmatism. James says "radical empiricism" is his own main doctrine.

11. Benjamin W. Redekop, "Reid's Influence in Britain, Germany, France, and America," in *Cambridge Companion to Thomas Reid,* 333.

12. Norman Daniels, *Thomas Reid's Inquiry: The Geometry of Visibles and the Case for Realism* (New York: Burt Franklin, 1974), 18.

13. Hillary Putnam, "Foreword," in Daniels, *Thomas Reid's Inquiry,* i, vii.

14. Quoted in Warner M. Bailey, "William Robertson Smith and American Biblical Studies," in *Journal of Presbyterian History* 51 (Fall 1973): 292.

15. James, *Varieties,* 21.

16. Interview with Gordon Graham, Aberdeen, Nov. 2003.

17. Ralph McInerny, *Characters in Search of Their Author: The Gifford Lectures, Glasgow, 1999–2000* (Notre Dame, IN: Notre Dame Univ. Press, 2001), 55, 118.

18. McInerny, *Characters,* 55, 56.

19. Alvin Plantinga, "A Christian Life Partly Lived," in *Philosophers Who Believe: The Spiritual Journeys of 11 Leading Thinkers,* ed. Kelly James Clark (Downers Grove, IL: InterVarsity Press, 1993), 74.

20. Alvin Plantinga, *Warrant: The Current Debate* (New York: Oxford Univ. Press, 1993), viii.

21. Plantinga, *Warrant,* 86; Plantinga, "Christian Life Partly Lived," 75–76.

22. Alvin Plantinga, *Warrant and Proper Function* (New York: Oxford Univ. Press, 1993), 216; Charles Darwin, *The Autobiography of Charles Darwin, 1809–1882,* ed. Nora Barlow (New York: Norton, 1963), 87.

23. Plantinga, *Warrant and Proper Function,* 237.

24. Memo from the Senior Assistant Registrar to the Principal of St. Andrews University, Nov. 5, 1976, in university archives. (Author's notes.)

25. Interview with Stewart J. Brown, Edinburgh, Nov. 2003.

26. Alasdair MacIntyre, *Three Rival Versions of Moral Enquiry: Encyclopedia, Genealogy, and Tradition* (Notre Dame, IN: Univ. of Notre Dame Press, 1990), 10.

27. J. Campbell Smith, review of *Lectures Delivered on Various Occasions by Adam Gifford,* by Alice Raleigh and Herbert James Gifford (eds.), *The Judicial Review: A Journal of Legal and Political Science* 1 (1889): 399.

INDEX